Interfacing PIC Microcontrollers

Interfacing PIC Microcontrollers

Embedded Design by Interactive Simulation

Martin Bates

AMSTERDAM • BOSTON • HEIDELBERG • LONDON
NEW YORK • OXFORD • PARIS • SAN DIEGO
SAN FRANCISCO • SINGAPORE • SYDNEY • TOKYO
Newnes is an imprint of Elsevier

ELSEVIER

Newnes

Newnes is an imprint of Elsevier
The Boulevard, Langford Lane, Kidlington, Oxford OX5 1GB, UK
225 Wyman Street, Waltham, MA 02451, USA

First edition 2008
Second edition 2014

Notice
No responsibility is assumed by the publisher for any injury and/or damage to persons
or property as a matter of products liability, negligence or otherwise, or from any
use or operation of any methods, products, instructions or ideas contained in the material
herein. Because of rapid advances in the medical sciences, in particular, independent
verification of diagnoses and drug dosages should be made.

British Library Cataloguing-in-Publication Data
A catalogue record for this book is available from the British Library.

Library of Congress Cataloging-in-Publication Data
A catalog record for this book is availabe from the Library of Congress.

ISBN: 978-0-08-099363-8

For information on all Newnes publications
visit our Web site at store.elsevier.com

Typeset by MPS Limited, Chennai, India
www.adi-mps.com

100#169648

Printed in the UK

14 15 16 17   10 9 8 7 6 8 4 3 2

Contents

PART 2: PIC Interfacing

PART 3: PIC Systems

Preface

The PIC is one of the biggest selling small microcontrollers. When it first became available, it was not only technically innovative, but helped to make the teaching of microelectronics much more interesting. A small controller with flash memory meant that a great variety of student projects could be realised quickly and easily. It helped that the development toolkit was free as well.

It has always been a problem in electronics that you cannot see a circuit working in the same way that a mechanical engineer can see a steam engine pumping up and down. Sure, we can see the screen flickering on a television, or an electric motor spinning, but you cannot see electrons or volts directly. As a result, it has always been that bit more difficult to learn electronics.

Interactive electronic design software is the answer. The Proteus VSM (Virtual System Modelling) software used in this book has been developed by Labcenter Electronics in the UK. It brings circuits to life on the computer screen and makes learning electronics more effective and more fun. It is also a full-scale professional product, and will take the student electronic engineer seamlessly into commercial design work.

This book is intended to support electronic learning wherever it takes place, at college, at work or in the home. Please enjoy!

Martin Bates
March 2013

Introduction

This book is the second edition of the sequel to 'PIC Microcontrollers, an Introduction to Microelectronics', which attempted to provide introduction to the subject via a single type of microcontroller. It explores the basic techniques for connecting the PIC to peripheral devices and the outside world. It shows how to connect simple input and output devices, such as switches, sensors, displays and motors, as well as demonstrating communication methods that allow the PIC to communicate data with other devices, including intelligent sensors. The second edition has been extensively revised, updated and expanded.

A domestic inkjet printer is an example of a product that contains a range of sensors, drives and displays. It typically has a wireless data link to receive the page data, at least two motors to feed the paper and position the print head, and a microcontroller to output the signals to the print cartridge inkjets and generally coordinate the action. Take an old one apart and have a look! Another good example is the digital camera. In fact, most small electronic products contain a microcontroller that provides its core functions. A smoke detector with a PIC microcontroller is shown in Figure I.1.

The PIC$^©$ microcontroller was the first widely available device to use flash memory, which made it ideal for prototyping and experimental work. Flash memory, as used in memory cards and sticks, allows the application program to be replaced quickly and easily with a new version. Cheap flash memory microcontrollers have also transformed the teaching of microelectronics – they are re-usable and the internal architecture is fixed, making them easier to understand. The small instruction set of the PIC is also a major advantage – there are only 35 instructions to learn in the main microcontroller unit (MCU) used in this book.

The free development system MPLAB$^©$ provided by Microchip Inc.$^©$ is another reason for using the PIC range. In addition to the program editor and project management features, it includes a text-based simulator which allows the program to be tested prior to downloading, potentially saving a lot of time debugging in hardware. However, this only tests the program itself, not the circuit in which it is connected.

Figure I.1
Smoke detector with PIC controller.

Proteus© from Labcenter Electronics© allows a PIC to be simulated in circuit. It consists of two main parts, ISIS and ARES. ISIS is the schematic capture and interactive simulation package used to create the circuit schematic and to test the circuit prior to building the real hardware. On-screen buttons and virtual signal sources provide inputs to the circuit. Output (analogue or digital) can be displayed on a signal probe, a virtual instrument or graph. An MCU can be dropped on the screen, the circuit drawn, a program attached and tested immediately on screen.

When the application is working correctly in simulation mode, a PCB can be designed by exporting a netlist (list of components and connections) from ISIS into the ARES layout package. The resulting PCB files can be output to a production system or sent to a specialist manufacturer. The final stage is then to assemble the board and test the hardware. After using Proteus VSM, it should work first time!

This book is built around particular devices and tools, because it allows specific examples to be used. It is assumed that at a later stage, with more experience, the reader will be able to evaluate these against competing products and choose the most appropriate for any given design task. Each topic is illustrated by designs based on the well-established PIC 16F877A, but it will be replaced in the readers' own designs with a more recent device such as the 16F887 chip.

All the circuits are available on the support website www.picmicros.org.uk. All schematics were produced using ISIS – and you can produce them to the same standard in your

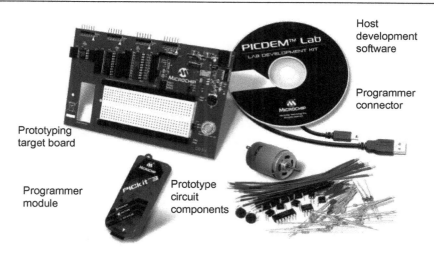

Host
development
software

Programmer
connector

Prototyping
target board

Programmer
module

Prototype
circuit
components

Figure I.2
PICDEM development kit. *Courtesy of Microchip Inc.*

own reports. The microcontroller models can be purchased in packages for institutional or professional use from www.labcenter.com. Currently, a Proteus starter kit including models for the PIC 16F84A, 16F877A and 18F452 can be purchased for only £150.

Microchip provides an extensive range of demonstration and development kits to support their microcontroller product range – see www.microchip.com for all product details. A basic development kit is illustrated in Figure I.2, which consists of a prototyping target board, a selection of hardware components including some small PIC chips, programmer module and development system software that is loaded onto a host PC. A circuit is built on the prototyping area, for example a motor interface, and a suitable PIC chip fitted into one of the sockets on the board. The motor control program can be written in MPLAB on the host computer, debugged in MPSIM, the Microchip simulator, downloaded to the target board and the hardware tested.

However, it is preferable to test the application before constructing the hardware, in case changes are needed. It is much quicker and easier to change the circuit, or the program, on screen, rather than in hardware. ISIS allows the program to be entered, assembled and attached to the on-screen chip for interactive testing within the virtual circuit. A typical simulation screen is shown in Figure I.3.

The book is structured in three parts. Part 1 reviews PIC microcontroller architecture and programming, Part 2 introduces PIC interfacing techniques and Part 3 covers PIC system design and implementation.

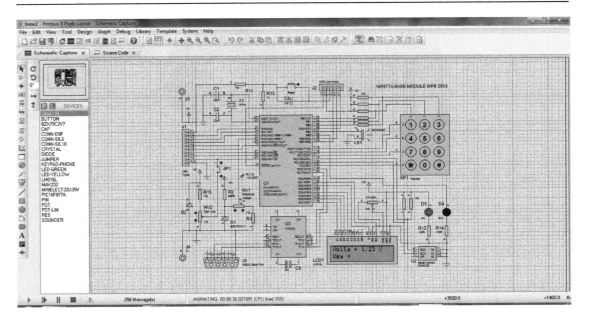

Figure I.3
Base board simulation screenshot.

In Chapter 1, a standard PIC microcontroller, the 16F877A, is described in detail, based mainly on its data sheet. Chapter 2 outlines the programming process, using only MPLAB tools, and Chapter 3 describes the application development process using Proteus VSM.

In Chapter 4, basic input and output devices are introduced, while Chapter 5 describes the techniques for data representation and conversion, Chapter 6 covers all aspects of analogue signal conditioning and, in Chapter 7, power output interfacing is introduced, concentrating on motor drives.

Chapter 8 outlines the operation of the main PIC serial communication ports, and Chapter 9 describes a wide range of sensors for monitoring and measurement applications. Chapter 10 concludes with a consideration of MCU selection and the principles and practice of system design.

The book was originally designed to support project development by students at all levels. It may therefore sometimes state what is obvious to more experienced engineers; hopefully this is not too irritating, and they too will find something of interest within!

Links and Acknowledgements

Support Website

www.picmicros.org.uk
Author's website with related PIC books and application file downloads

Follow the link to demo applications in INTAPPS2.ZIP containing:

- VSM project file **app2.pdsprl**
- PIC source code **app2.asm** (in MCU folder)
- PIC debug file **debug.cof** (in MCU folder)

Labcenter Electronics

www.labcenter.com
Manufacturer and supplier of Proteus VSM electronic design system

- Download demo version of VSM
- Purchase MCU package licence
- Tutorials and product information

Microchip Technology Inc.

www.microchip.com
Manufacturer of the PIC microcontroller range and MPLAB IDE

- Download data sheets
- Information on development tools
- Download MPLAB development system

Custom Computer Services Inc.

www.ccsinfo.com

Manufacturer and supplier of PIC CCS 'C' Compilers

Please search online by name for product datasheets other than Microchip.
Use of all manufacturers' trademarks and data is gratefully acknowledged.
Thanks in particular to Iain Cliffe at Labcenter Electronics.

PART 1

PIC Microcontroller

CHAPTER 1

PIC Hardware

Summary

- The microcontroller contains a processor, memory and input/output devices
- The program is stored in flash ROM memory in numbered locations
- The P16F877A family uses only 35 instructions
- The P16F877A stores a maximum 8k $\times$ 14 instructions in flash ROM
- The P16F877A has 368 bytes of RAM and 5 ports (33 I/O pins)
- The program is executed in sequence, unless there is a jump instruction
- The program counter tracks the current instruction address
- A configuration word is need to select the clock type and other chip options
- The program source code (.ASM) is assembled into machine code (.HEX)
- Machine code is downloaded to the chip and the application hardware tested

The microcontroller is a complete computer on a chip. When introduced, it was one of the most significant developments in electronics since the invention of the microprocessor itself and is essential in the operation of such devices as mobile phones, DVD players, video cameras, and most self-contained electronic systems. Working sometimes with other chips, but often on its own, the microcontroller unit (MCU) provides the key element in the vast range of small, programmed devices that are now commonplace.

Although small, microcontrollers are complex, and we have to look carefully at the way the hardware and firmware (control program) work together to understand the processes at work. This book will then show how to connect the popular PIC range of microcontrollers to the outside world and put them to work. To keep things simple, we will concentrate on one device, the PIC 16F877A, which has a good range of features that allows most of the essential techniques to be explained. It has a set of serial ports built in that are used to transfer data to and from other devices, as well as analogue inputs, which allow measurement of inputs such as temperature. All microcontrollers work in a similar way, so analysis of the PIC MCU will go a long way to understanding all such devices.

This chip has been around some time and is no longer the best choice for new designs. The 16F887 is a more recent equivalent and should be used as a pin compatible

Interfacing PIC Microcontrollers.
DOI: http://dx.doi.org/10.1016/B978-0-08-099363-8.00001-7

replacement in new designs. The reason for continuing to use the '877A' in this edition of the book is that it is still available as a part of a low-cost simulation package that is suitable for students and hobbyists on a budget and has all the main interfaces that are still used in current chips. In any event, when designing a new application, a chip should always be selected from the available range which most closely matches the design requirements at minimum cost, so prototyping with a chip with surplus capabilities is a useful approach. It can be replaced at a later design stage with a chip that matches more closely with the system requirements.

A big advantage of the PIC is that the programming language is relatively simple, as compared with microprocessors such as the Intel series used in the PC. These have a powerful, but complex, instruction set to support a wide range of multimedia applications. The supporting documentation for the PIC MCU range is also clear and well laid out, and a development system, for writing and testing programs, can also be downloaded free from the Microchip website (www.microchip.com).

1.1 Processor System

The microcontroller contains the same main elements as any computer system, namely:

- *Central Processor Unit (CPU)*
- *Data storage (memory) devices*
- *Input and output ports*

In a PC, these features are generally implemented as separate chips, linked together through bus connections on a complex printed circuit board, under the control of the microprocessor. A bus is a set of lines which carry data in parallel form which are shared by the peripheral devices. This type of system can be tailored to suit a particular application, with the type of CPU, size of memory, and selection of input and output (I/O) devices matched to the system requirements. However, even in real-time applications, it has been common for some time to use a standard board based on the Intel processor system running a generic operating system (often Windows and its derivatives). The basic microprocessor system is conveniently summarised in a block diagram (Figure 1.1) which shows the data flow between the main elements using shared bus lines.

In the microcontroller, all these basic elements are on one chip. This means that the MCU for a particular application must be chosen from the available range to suit its operational requirements. In any given circuit, the microcontroller also tends to have a single, specific function. This type of product is frequently referred to as an embedded application.

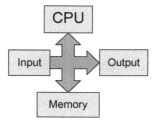

Figure 1.1
Block diagram of a basic microprocessor system.

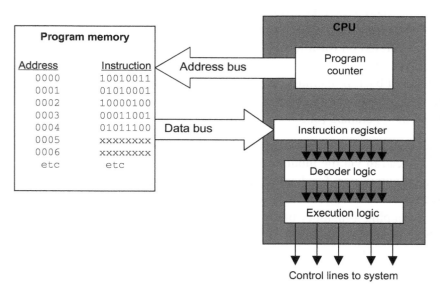

Figure 1.2
Microprocessor program execution.

1.1.1 Processor

In a microprocessor system or a microcontroller, a single processor block is in charge of all input, output, calculations and control. It cannot operate without a program, a list of instructions that is held in memory. The program consists of a sequence of binary codes that are fetched from memory by the CPU and executed in turn (Figure 1.2). The process is driven by a crystal clock circuit producing a fixed frequency that determines the speed of the system.

The instructions are stored in numbered memory locations and copied to an instruction register in the CPU, via a data bus. Here, the instruction controls the selection of the required operation within the control unit of the processor. The program codes are found in memory by the processor by outputting the address number of the instruction on an address bus. The address is generated in the program counter, a register which starts at zero and is incremented or modified during each instruction cycle. The busses are parallel connections which transfer the address or data word in one operation. A set of control lines from the CPU is also needed to assist with this process; these are set up according to the requirements of the current instruction and trigger the data transfer circuits to output and receive the data at the appropriate time. In the conventional microprocessor system, the bus connections consist of parallel tracks on a motherboard but are internal in the microcontroller chip.

Decoding the instruction is a hardware process, using a block of logic gates to set up the control lines of the processor unit, and to fetch the instruction 'operands'. The operands are data to be operated on (or information about where to find it) which follow most instructions. Typically, a calculation or logical operation is carried out on the operands, and a result stored back in memory, or an I/O action set-up. Each complete instruction may be one, two or more bytes long, which includes the operation (instruction) code itself (op-code) and the operand/s (one byte = 8 bits).

For example, compare a word processor and games application. In the word processor, keystrokes are read in via the input keyboard port, stored as character codes in memory and sent to a screen output port for display. In a computer game, input signals from the control pad are processed and used to modify the screen graphics. The graphics are basically generated by mapping a memory block to the screen where the colour of one pixel is controlled by a particular data word. The word processor needs far less memory, and the graphics memory has to be large and fast.

1.1.2 Memory

There are two main types of microprocessor memory, volatile and non-volatile. ROM (Read Only Memory) is non-volatile and retains its data when switched off. In the PC, the main working memory is volatile RAM (Read and Write Memory), normally implemented as plug-in DIMM (Dual In-line Memory Module) modules, which carry a set of dynamic RAM chips. These are used by the CPU to store current working application files and data. RAM originally meant Random Access Memory, referring to the data read-and-write mechanism, but ROM is accessed in exactly the same way, using row and column addresses to identify each storage cell.

In a traditional PC design, a small ROM chip is used to get the system started when it is switched on; it contains the BIOS (Basic Input Output System) program. However, the main operating system (OS), e.g. Windows™, and application program, e.g. Word™, have to be loaded into RAM from hard disk drive (HDD), which takes some time, as you may have noticed! So why not put the OS in ROM, where it would be instantly available? Well, RAM is faster, cheaper and more compact, and the OS can be changed or upgraded on disk. In addition, an OS such as Windows is large compared with the size of RAM generally installed, so elements are only loaded into RAM as needed. Numerous applications can also be stored on disk and loaded only as required.

The ideal memory is non-volatile, read and write, fast, large and cheap. Unfortunately, it does not exist! Therefore, we have a range of memory technologies as shown in Table 1.1, which have different advantages. These are used in combination in the PC to provide the optimum overall performance for a given cost. This also depends on the application being run at any one time. The main trade-off is cost, size and speed of access.

Flash ROM, as used in memory sticks and MP3 players, is closest to the ideal, having the advantages of being non-volatile and rewritable. This is why it is used as program memory in microcontrollers which need to be reprogrammable, such as the PIC 16F877A. The microcontroller uses flash program memory because it is usually dedicated to a particular control task that does not need a large amount of working data storage. The working data registers in the PIC can therefore be implemented as a relatively small block of static RAM (SRAM). As flash ROM technology has improved, conventional one-time programmable ROM has now generally been rendered superfluous but is included for completeness in the comparison.

Table 1.1: Memory and Data Storage Technologies.

	ROM (Read Only Memory)	Flash ROM	RAM (Read and Write Memory)	CD-ROM (Compact Disk-ROM)	DVD-RW (Digital Versatile Disk-Read and Write)	HDD (Hard Disk Drive)
Feature	Chip	Chip	Module	Optical disk	Optical disk	Magnetic disk
Typical size*	128 kb	256 Mb	2 Gb	650 Mb	4.7 Gb	250 Gb
Non-volatile	Yes	Yes	No	Yes	Yes	Yes
Write	Once	Many	Many	Once	Many	Many
Size	Poor	OK	OK	Good	Good	Good
Expandability	OK	OK	Good	None	None	None
Cost per bit	Poor	OK	Good	Good	Good	Good
Speed of access	OK	Good	Good	Poor	Poor	Poor

*1 byte = 8 bits
*1 kb = 1 kilobyte = 1024 bytes
*1 Mb = 1 megabyte = 1024 kb
*1 Gb = 1 gigabyte = 1024 Mb

1.1.3 Input and Output

Without some means of getting information and signals in and out, a data processing or digital control system would not be very useful. Input and output ports generally contain a port data register and a set of control registers that allow data to pass in and out. Serial ports often use standard protocol (method of communication) to format the data.

In a PC, the keyboard, screen and mouse interfaces are the main I/O channels, supported by network, USB, SD Card and disk interfaces. The DIMM memory module has a parallel connector (typically 64 bits), so it attaches directly to the processor busses. This means that access is fast, because complete data blocks can be transferred at one time. USB on the other hand is a serial bus, so data transfer can only occur one bit at a time, which is inherently slower.

Microcontroller ports are generally more basic, especially in the smaller MCUs which cannot accommodate the complex hardware needed for, say, a network port. The basic MCU port consists of a group of 8 bits that can operate as a parallel port, but whose individual pins have alternate functions, often several. The basic parallel operation is straightforward, with an 8-bit data register that holds the I/O data, and an 8-bit data direction register whose individual bits control the data direction, in or out. In the PIC, 0 = output and 1 = input. Alternate functions are selected in the MCU control registers at set-up.

In principle, the parallel port is faster that the serial port but uses more pins to transfer the data. The serial interface must organise the data in groups of bits for transmission. Dedicated registers are used to organise the data stream and control the timing of the data transfer. The serial port is based on a shift register that converts between parallel data on the internal data bus and serial data on the peripheral line. The general principles of parallel and serial data transfer are shown in Figure 1.3. The block arrows represent the 8-bit internal data bus.

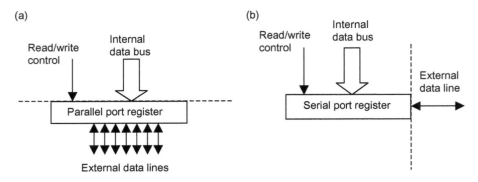

Figure 1.3
(a) Parallel and (b) serial data ports.

In the parallel port operating in output mode, the data byte is loaded from the internal data bus under the control of a read/write pulse from the CPU. The data can then be seen on the output pins by the peripheral line. For testing, a logic probe, logic analyser or just a simple LED indicator can be used. In input mode, data presented at the input pins from a set of switches or other data source is latched into the register when the port is read and is then available on the data bus for collection by the CPU. One of the functions of the port is to separate the internal data bus from the external hardware, and another is to temporarily store the data. The data can then be transferred to memory, or otherwise processed, as determined by the CPU program.

The serial port register also loads data from the internal bus in parallel but then sends it out one bit at a time, operating as a shift register. If an asynchronous serial format is used, such as RS232, start and stop bits are added so that bytes can be separated at the receiving end. An error check bit is also available, to allow the receiver to detect corrupt data. In receive mode, the register waits for a start bit and then shifts in the data at the same speed as it is sent. This means the clock rate for the send and receive port must be the same. The USART (Universal Synchronous/Asynchronous Receive/Transmit) port, which provides RS232, will be described in more detail later.

A USB or network port is a more sophisticated version of the basic serial port and arranges the data bytes in packets of, perhaps, 1k bytes. These are sent in a form which is self-clocking, meaning that there is a transition within each bit (1 or 0), which triggers the bit read into the receiving register. This is synchronous data transmission. An error correction code follows the data, which allows mistakes to be corrected, rather than just detected. This reduces the need for retransmission of incorrectly received data, as required by simple error detection. Addressing information preceding the data allows multiple receivers to be used.

The PIC 16F877A does not have USB or network interfaces built in, so we can avoid detailed consideration of these complex protocols. It does nevertheless have a range of other interfaces that will be discussed in detail and sample programs are provided. If further explanation of the basics of microcontroller operation is required, the reader is invited to refer to the introductory text 'PIC Microcontrollers' by the author.

1.2 PIC Architecture

Microcontrollers contain all the components required for a processor system in one chip: CPU, memory and I/O. A complete system can therefore be built using one MCU chip and a few I/O devices such as a keypad, display and other interfacing circuits. We will now see how this is done in practice in our typical microcontroller.

1.2.1 PIC 16F877A Pin Out

Let us first consider the pins that are seen on the IC package, and then we can discover how they relate to the internal architecture. The chip can be obtained in different packages, such as conventional 40-pin PDIP (Plastic Dual In-Line Package), square surface mount or socket format. The PDIP version, shown in Figure 1.4, is easier for prototyping.

The I/O pins are arranged as 5 ports: A (5), B (8), C (8), D (8) and E (3), giving a total of 32 I/O pins. These can all operate as simple digital I/O pins but mostly have more than one function, and the mode of operation of each is selected by initialising various control registers within the chip. Note, in particular, that Ports A and E become ANALOGUE INPUTS by default (on power-up or reset), so they have to set up for digital I/O if required. The ports which have fewer than 8 bits have corresponding unused bits in their 8-bit data and control registers.

Port B is used for downloading the program to the chip flash ROM, with RB7 receiving the program code in serial form and RB6 receiving a clock which strobes each data bit into the serial register on RB7. MCLR is taken to about 14V to initiate programming mode and supply the programming voltage. RB3 can optionally be used in low-voltage programming mode which does not require the 14V supply. Changes on RB0 and RB4−RB7 can generate an interrupt, which forces a change in the program sequence. Port C gives access to timers and serial ports, while Port D can be used as a slave port, with Port E providing the control pins for this function. All these options will be explained in detail later.

Reset = 0, Run = 1(Vpp)	**MCLR**	1	40	**RB7**	Port B, Bit 7 (Prog. Data, Interrupt)
Port A, Bit 0 (Analogue AN0 in)	**RA0**	2	39	**RB6**	Port B, Bit 6 (Prog. Clock, Interrupt))
Port A, Bit 1 (Analogue AN1in)	**RA1**	3	38	**RB5**	Port B, Bit 5 (Interrupt)
Port A, Bit 2 (Analogue AN2 in)	**RA2**	4	37	**RB4**	Port B, Bit 4 (Interrupt)
Port A, Bit 3 (Analogue AN3 in)	**RA3**	5	36	**RB3**	Port B, Bit 3 (LV Program)
Port A, Bit 4 (Timer 0 I/O)	**RA4**	6	35	**RB2**	Port B, Bit 2
Port A, Bit 5 (Analogue AN4 in)	**RA5**	7	34	**RB1**	Port B, Bit 1
Port E, Bit 0 (AN5, Slave control)	**RE0**	8	33	**RB0**	Port B, Bit 0 (Interrupt)
Port E, Bit 1 (AN6, Slave control)	**RE1**	9	32	**V$_{DD}$**	+5V Power Supply
Port E, Bit 2 (AN7, Slave control)	**RE2**	10	31	**Vss**	0V Power Supply
+5V Power Supply	**V$_{DD}$**	11	30	**RD7**	Port D, Bit 7 (Slave I/O)
0V Power Supply	**Vss**	12	29	**RD6**	Port D, Bit 6 (Slave I/O)
(CR clock) XTAL circuit	**CLKIN**	13	28	**RD5**	Port D, Bit 5 (Slave I/O)
XTAL circuit	**CLKOUT**	14	27	**RD4**	Port D, Bit 4 (Slave I/O)
Port C, Bit 0 (Timer 1 I/O)	**RC0**	15	26	**RC7**	Port C, Bit 7 (USART Receive / Data)
Port C, Bit 1 (Timers 1 & 2)	**RC1**	16	25	**RC6**	Port C, Bit 6 (USART Transmit / Clock)
Port C, Bit 2 (Timer 1 I/O)	**RC2**	17	24	**RC5**	Port C, Bit 5 (I2C Data Out)
Port C, Bit 3 (SPI Clock)	**RC3**	18	23	**RC4**	Port C, Bit 4 (I2C Data In / SPI Data)
Port D, Bit 0 (Slave I/O)	**RD0**	19	22	**RD3**	Port D, Bit 3 (Slave I/O)
Port D, Bit 1 (Slave I/O)	**RD1**	20	21	**RD2**	Port D, Bit 2 (Slave I/O)

Figure 1.4
PIC 16F877A pin out.

The chip has two pairs of power pins, where we usually assume $V_{DD} = +5$ V and $V_{SS} = 0$ V. The chip can actually work down to about 2V supply, for battery and power saving operation. A low-frequency clock circuit using only a capacitor and resistor to set the frequency can be connected to CLKIN, or a crystal oscillator circuit can be connected across CLKIN and CLKOUT. MCLR is the reset input; when cleared to 0, the MCU stops and restarts when MCLR = 1. This input must be tied high, allowing the chip to run if an external reset circuit is not connected, but it is usually a good idea to incorporate a manual reset button in all but the most trivial applications, as it allows the program to be restarted if there is a problem.

1.2.2 PIC 16F877A Architecture

A block diagram of the 16F877A architecture is given in the manufacturer's data sheet (downloadable from www.microchip.com), Figure 1.2. A somewhat simplified version is shown in Figure 1.5, which emphasises the program execution mechanism.

The main program memory is flash ROM, which stores a list of 14-bit instructions. These are fed to the execution unit and used to modify the RAM file registers. These include special control registers, the port registers and a set of general purpose registers which can be used to store data temporarily. A separate working register (W) is used with the Arithmetic Logic Unit (ALU) to hold the current data. Various special peripheral modules provide a range of I/O options.

There are 512 RAM File Register addresses (0−1FFh), which are organised in 4 banks (0−3), each bank containing 128 addresses (Table 1.2). The default (selected on power-up) Bank 0 is numbered from 0 to 7Fh, Bank 1 from 80h to FFh and so on. All banks contain both Special Function Registers (SFRs), which have a dedicated purpose, and the General Purpose Registers (GPRs) that act as data RAM. The SFRs have labels that are specified in the file that is normally combined with the user program. The relevant bank must be selected in the user program to access the SFRs and GPRs (see Chapter 2).

The SFRs are shown in the block diagram as separate from the GPRs, but they are in fact in the same address block. If the SFRs are deducted from the total number of RAM locations (allowing for some registers which are repeated in more than one bank), 368 GPRs are available for user data. The port registers are part of the main RAM block, found at specific SFR addresses, e.g. Port D is at address 08h. If a review of hexadecimal and binary numbering is required, refer to Chapter 5.

1.2.3 The PIC Instruction

The PIC program is written as a source code (a simple text file) on a PC host computer. Any text editor can be used, but an editor is usually provided with the development system.

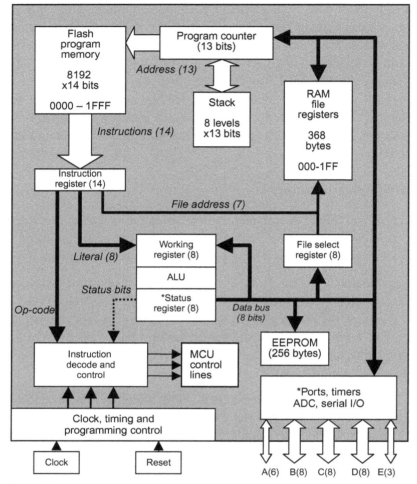

Figure 1.5
16F877A program execution block diagram.

We will assume MPLAB is being used initially. The instructions are selected from the predefined PIC instruction set detailed in Table13.2 in the PIC 16F877A data sheet and summarised in the next chapter in Table 2.4. The demo program listed as Program 2.1 in the same chapter shows the general format. Note that the original source code is preceded on each line by the hexadecimal machine code which is downloaded to program memory. The source code file is saved as PROGNAME.ASM in a suitable folder on the host PC.

The source code is assembled (converted to machine code) by the assembler program MPASM, which creates the list of binary instruction codes. As this is displayed as

Table 1.2: Status Reg Bits.

Bank 0		Bank 1		Bank 2		Bank 3	
00h	Indirect	80h	Indirect	100h	Indirect	180h	Indirect
01h	TMR0	81h	OPTION_REG	101h	TMR0	181h	OPTION_REG
02h	PCL	82h	PCL	102h	PCL	182h	PCL
03h	STATUS	83h	STATUS	103h	STATUS	183h	STATUS
04h	FSR	84h	FSR	104h	FSR	184h	FSR
05h	PORTA	85h	TRISA	105h	—	185h	—
06h	PORTB	86h	TRISB	106h	PORTB	186h	TRISB
07h	PORTC	87h	TRISC	107h	—	187h	—
08h	PORTD	88h	TRISD	108h	—	188h	—
09h	PORTE	89h	TRISE	109h	—	189h	—
0Ah	PCLATH	8Ah	PCLATH	10Ah	PCLATH	18Ah	PCLATH
0Bh	INTCON	8Bh	INTCON	10Bh	INTCON	18Bh	INTCON
0Ch	PIR1	8Ch	PIE1	10Ch	EEDATA	18Ch	EECON1
0Dh	PIR2	8Dh	PIE2	10Dh	EEADR	18Dh	EECON2
0Eh	TMR1L	8Eh	PCON	10Eh	EEDATH	18Eh	[a]
0Fh	TMR1H	8Fh	—	10Fh	EEADRH	18Fh	[a]
10h	T1CON	90h	—	110h		190h	
11h	TMR2	91h	SSPCON2	111h		191h	
12h	T2CON	92h	PR2	112h		192h	
13h	SSPBUF	93h	SSPADD	113h		193h	
14h	SSPCON	94h	SSPSTAT	114h		194h	
15h	CCPR1L	95h	—	115h		195h	
16h	CCPR1H	96h	—	116h		196h	
17h	CCP1CON	97h	—	117h	GPRs	197h	GPRs
18h	RCSTA	98h	TXSTA	118h	16 bytes	198h	16 bytes
19h	TXREG	99h	SPBRG	119h		199h	
1Ah	RCREG	9Ah	—	11Ah		19Ah	
1Bh	CCPR2L	9Bh	—	11Bh		19Bh	
1Ch	CCPR2H	9Ch	CMCON	11Ch		19Ch	
1Dh	CCP2CON	9Dh	CVRCON	11Dh		19Dh	
1Eh	ADRESH	9Eh	ADRESL	11Eh		19Eh	
1Fh	ADCON0	9Fh	ADCON1	11Fh		19Fh	
20h		A0h		120h		1A0h	GPRs
|	GPRs	|	GPRs	|	GPRs	|	75 bytes
|	80 bytes	|	80 bytes	|	80 bytes	1EAh	
|		|		|		1EBh	ICD1
|		|		|		|	|
6Fh		EFh		16Fh		1EFh	ICD4
70h	[b]ICD	F0h	([b]ICD)	170h	([b]ICD)	1F0h	([b]ICD)
71h		F1h		171h		1F1h	
|	GPRs	|	Repeats	|	Repeats	|	Repeats
7Fh	15 bytes	FFh	70h–7Fh	17Fh	70h–7Fh	1FFh	70 h–7Fh

[a]Keep Reserved locations clear.
[b]Reserved ICD locations: 70h (F0, 170h, 1F0h) and 1EBh — 1EFh.

hexadecimal numbers, it is stored by the assembler with the extension PROGNAME.HEX. This is then downloaded from the PC host to the MCU in the application board via a programming module such as the PICkit3 shown in Figure I.1. This is plugged into the USB port of the host and onto a set of 6 pins on the target board that are connected to the programming pins of the MCU. The hex code is then transferred in serial form into the PIC flash program memory.

The '877A' has 8k of program memory, i.e. it can store a maximum of 8192 14-bit instructions, usually starting at address 0. The highest address in this memory is therefore 1FFFh. In real-time applications, the program runs continuously and loops back at the end to repeat the process. Let us look at a typical instruction to see how it works in relation to the internal architecture:

```
Source code: MOVLW 05A
Hex code: 305A (4 hex digits)
Binary code: 0011 0000 0101 1010 (16 bits)
Instruction: 11 00xx kkkk kkkk (14 bits)
```

The instruction MOVLW means Move a Literal (a given number, in this case 5Ah) into the Working register (W). The source code therefore consists of the mnemonic MOVLW and operand 05Ah. This assembles into the hex code 305Ah (see Chapter 5 for number-type conversion) and is stored in binary program memory as 11 0000 0101 1010. Since each hex digit represents four binary bits, the leading two bits are set to zero to fill the first two bits.

In the instruction set, it can be seen that the first 4 bits (11 00) are the instruction code, the next two are unused (xx, appearing as 00 in the binary code) and the last 8 are the literal value (5Ah). The literal is represented as 'kkkk kkkk' since it can have any value from 00000000 to 11111111 (00h—FFh).

The exact format of the PIC instructions depends on the number of bits required for the operand (data to be processed). The number of op-code bits can vary from 3 to 14, depending on the number of bits needed for the operand. This is different from a conventional processor, such as the Pentium, where the op-code and operand are separate and consist of a whole number of bytes. The PIC instruction is more compact, as is the instruction set itself, for greater code efficiency; it is therefore known as an RISC (Reduced Instruction Set Computer) chip.

1.3 Special Function Registers

As we have seen, the file register set is divided into SFRs and GPRs. The SFRs have predetermined functions, as specified in the 16F877A data sheet (Figure 2.3). They occupy locations 00—1F in Bank 0, 80—9F in Bank 1, 100—10F in Bank 2 and 180—18F in Bank 3. The most frequently used are accessible in more than one bank, i.e. it is not necessary to switch banks to read or write these registers. The most frequently used SFRs control some aspect of program execution, but most of the rest are used to set up specific port functions.

1.3.1 Program Counter

The program counter (PCL = 02h) keeps track of program execution by storing the address of the current instruction. The '877A' has 8k of program memory, so each location needs a 13-bit address. PCL contains the low 8 bits of the program counter, while the upper bits (PC < 8−12 >) are accessed via PCLATH. PCL is incremented during each instruction, and the contents replaced during a GOTO, CALL (program address) or RETURN (stack). PCL can be modified directly, for example to implement a data table, but the high bits are only writable indirectly via PCLATH. This causes some complications when making long jumps, or writing the PCL directly, because the program memory is effectively divided into 256 byte blocks and 2k pages (see Chapter 2).

1.3.2 Status Register

The status register (STATUS = 03h) records the result of certain operations and contains the file register bank selection bits. The individual bit functions are summarised in Table 1.3.

1.3.2.1 Zero Flag (Z)

This is set (to 1) when the result of a register operation is zero and cleared (to 0) when it is not zero. The full instruction set must be consulted to confirm which operations affect the Z flag, but it is essentially most register operations. Bit test and skip instructions use this flag for conditional branching, and there are also instructions that perform decrement or increment and skip if zero in one operation as this is such a common requirement. An example of the use of the zero flag is to check if two numbers are the same by subtracting and applying bit test and skip to the Z bit.

Table 1.3: 16F877A File Register Set.

Bit	Label	Active	Name	Function
0	C	1 or 0	Carry bit	Records carry or rotate out of file register MSB or LSB
1	DC	1	Digit Carry	Records carry out of bit 3 in arithmetic operation
2	Z	1	Zero bit	Records a zero result in any file register
3	PD	0	Power Down	Cleared when SLEEP instruction has been executed
4	TO	0	Time Out	Set when watchdog timer has expired
5	RP0	1 or 0	Bank Select 0	Low bank select bit used in direct addressing
6	PR1	1 or 0	Bank Select 1	High bank select bit used in direct addressing
7	IRP	1 or 0	Bank Select	Bank select bit used in indirect addressing

1.3.2.2 Carry Flag (C)

This flag is only affected by add, subtract and rotate instructions. If the result of an add operation generates a carry out, that is, when two 8-bit numbers give a 9-bit sum, this flag is set. The carry bit must then be included in subsequent calculations to give the right result. When subtracting, the carry flag must be set initially, because it provides the borrow digit (if required) in the most significant bit of the result. If the carry flag is cleared after a subtraction, it means the result was negative, because the number being subtracted was the larger. An example of this is seen later in the calculator program (Chapter 6). Taken together, the zero and carry flags allow the result of an arithmetic operation to be detected as positive, negative or zero, as shown in Table 1.4. Remember that the carry flag must be set before a subtract operation; a borrow can then be detected as C = 0.

1.3.2.3 Digit Carry (DC)

A file register can be seen as containing 8 individual bits, or 1 byte. It can also be defined as two 4-bit nibbles (a small byte!), high and low. Each nibble can be represented as 1 hex digit (0-F). The digit carry records a carry from the most significant bit of the low nibble (bit 3). Hence, the digit carry allows 4-bit hexadecimal arithmetic to be implemented in the same way that 8-bit binary arithmetic uses the carry flag C. Each nibble can also store a BCD (binary coded decimal) number, values 0–9, for BCD arithmetic (see Chapter 5).

1.3.2.4 Register Bank Select (RP1, RP0)

The PIC 16F877A file register RAM is divided into four banks of 128 locations, Banks 0–3. Only one can be selected at a time, depending on the settings of the register select bits RP0 and RP1 in the status register (Table 1.5). At power on reset, Bank 0 is selected by default; when access to a register in Bank 1, 2 or 3 is required, these bits must be changed accordingly and changed back afterwards. This can be done using BSF and BCF instructions. Only the bit/s that need changing are set or cleared to switch between banks.

The supplementary instruction BANKSEL can be used instead. The operand for BANKSEL is any register in the required bank or its label. BANKSEL detects the bank bits in the register

Table 1.4: Testing Arithmetic Results.

Flag After Operation	Zero	Carry	Result	Comment
(clear Carry)	0	0	A + B < 256	8-bit sum, no carry
ADD	1	1	A + B = 256	Exactly, carry out
A + B	0	1	A + B > 256	9-bit sum, carry out
(set Carry)	0	1	A − B < 256	8-bit difference, no borrow
SUBTRACT	1	1	A − B = 0	Numbers equal, no borrow
A − B	0	0	A − B < 0	Borrow taken, result negative

address and copies them to the status register bank select bits (see LED2 source code in the next chapter, Program 2.1).

It can be seen that some registers repeat in more than one bank, making it easier and quicker to access them, because bank switching is unnecessary. For example, the status register repeats in all banks. In addition, a block of GPRs at the end of each bank repeat, so that their data contents are available without changing banks. More recent PIC chips avoid this problem by using linear addressing of the register memory space.

1.3.2.5 Power Status Bits (PD, TO)

The Power-Down (PD) bit is cleared to zero when SLEEP mode is entered. The Time Out (TO) bit is cleared when the watchdog timer expires. These bits can then be used to trigger suitable actions when these events occur.

1.3.3 Ports

The five ports in the PIC 16F877A, labelled A—E, occupy SFR addresses 05h—09h. All pins can be used as bit or byte oriented digital input or output. Most of them also have alternate functions as summarised in Table 1.6. The 16F887 has a similar set of alternate functions, without the parallel slave port but with additional analogue inputs. In general, the physical size of a PIC chip depends on the number of port pins provided.

It can be seen that many of the port pins have two or more functions controlled by the initialisation of the relevant control registers. On power-up or reset, the port control register bits adopt a default condition (see Table 2.1 in the data sheet, right-hand columns). The TRISx (data direction) register bits in Bank 1 default to 1, setting the Ports B, C and D as

Table 1.5: Register Bank Selection.

Bank	Select Code[a]	Address$_h$	Total$_d$	Function
0	BCF 03,0	00—20	32	SFRs
	BCF 03,1	20—7F	96	GPRs
1	BSF 03,0	80—9F	32	SFRs, some repeat
	BCF 03,1	A0—EF	80	GPRs
		F0—FF	16	Repeat 70—7F
2	BCF 03,0	100—10F	16	SFRs, some repeat
	BSF 03,1	110—16F	96	GPRs
		170—17F	16	Repeat 70—7F
3	BSF 03,0	180—18F	16	SFRs, some repeat
	BSF 03,1	190—1EF	96	GPRs
		1F0—1FF	16	Repeat 70—7F
	Overall	**000—1FF**	**96**	**SFRs**
			368	**GPRs**

[a]Or use BANKSEL.

Table 1.6: 16F877A Port Alternate Functions.

	Bits	Pins	Alternate Function/s	Bit	Default
Port A[a]	6	RA0−RA5	Analogue inputs	0,1,2,3,5	Analogue Input
			Timer0 clock input	4	
			Serial port slave select input	5	
Port B[a]	8	RB0−RB7	External interrupt	0	Digital I/O
			Low-voltage programming input	3	
			Serial programming	6,7	
			In-circuit debugging	6,7	
Port C	8	RC0−RC7	Timer1 clock input/output	0,1	Digital I/O
			Capture/Compare/PWM	1,2	
			SPI, I^{2}C synchronous clock/data	3,4,5	
			USART asynchronous clock/data	6,7	
Port D[a]	8	RD0−RD7	Parallel slave port data I/O	0−7	Digital I/O
Port E	3	RE0−RE2	Analogue inputs	0,1,2	Analogue Input
			Parallel slave port control bits[a]	0,1,2	

[a]16F887 Port A: Additional comparator inputs RA0−RA5.
Port B: Additional analogue inputs RB0−RB5.
Port D: Parallel slave port not present, RD5−RD7 PWM outputs.

inputs. If this is as required, no further initialisation is needed, since other relevant control registers are generally reset to provide plain digital I/O by default.

However, there is an IMPORTANT exception. Ports A and E are set to ANALOGUE INPUT by default, because the analogue control register ADCON1 in Bank 1 defaults to 0xxx0000. To set up these ports for digital I/O, this register must be loaded with the code xxxx011x (x = don't care or undefined), say 06h. If analogue input is required only on selected pins, ADCON1 can be initialised with bit codes that give a mixture of analogue and digital I/O on Ports A and E. Note that ADCON1 is in Bank 1, so BANKSEL is needed to access it. Initialisation for analogue I/O will be explained in more detail later.

1.3.4 Timers

The simple way of creating a delay in a program is a software counting loop, as seen in Program 2.1 (LED2.ASM) in Chapter 2. However, this is an inefficient use of MCU resources, as the processer is completely occupied by the delay count. To avoid this problem, the PIC 16F877A has three hardware timers (data sheet, Sections 5, 6 and 7). These are used to carry out timing operations simultaneously with the main program execution, for faster and more efficient overall performance. A typical timer function would be to generate a regular pulse sequence at an output. A hardware timer count can be set to run and interrupt the MCU when done (see interrupts below), which can then toggle the output.

Timer0 uses an 8-bit register, TMR0, file register address 01. Its output is the overflow flag, T0IF, bit 2 in the Interrupt Control Register INTCON, address 0Bh. The timer register

Table 1.7: Typical Configurations for Timer0.

OPTION_REG	Configuration	Effect	Applications
11**011**000	Internal clock ($f_{OSC}/4$) No pre-scale	Timer mode using instruction clock	1. Preload Timer0 with initial value, and count up to 256 2. Clear Timer0 initially and read count later to measure time elapsed
110**1**0**011**	Internal clock ($f_{OSC}/4$) Pre-scale = 16	Timer mode using instruction clock with pre-scale	Extend the count period $\times$ 16 for applications 1 and 2.
11**110111**	External clock T0CKI pin	Counter mode Pre-scale = 256	Count one pulse in 256 at RA4, Max count = $256 \times 255 = 65,280$
11**111110**	Watchdog timer Pre-scale = 64	Extend watchdog reset period to $18 \times 64 = 1152$ ms	Watchdog timer checks program function every second

Relevant bits in bold.

is incremented via a clock input that is derived from either the MCU oscillator (f_{OSC}) or an external pulse train at RA4. The register counts from 0 to 255 in binary, then rolls over to 00 again, at which point T0IF is set.

If the internal clock is used, the register acts as a timer. Each instruction in the MCU takes four clock cycles to execute, so the instruction clock is $f_{OSC}/4$. The timers are driven from the instruction clock. A count of any number less than 256 can be obtained by preloading TMR0. For a count of 100, for example, it will be preloaded with 156d (d = decimal) and TMR0 will count 100 pulses until it rolls over, and T0IF is set. If the chip is driven from a crystal of 4MHz, the instruction clock will be 1 MHz, and the timer will overflow after 100µs. If this were used to toggle an output, a signal with a period of exactly $2 \times 100 = 200$µs (frequency = 5kHz) would be obtained.

The timers can also be used as counters. A sequence of external pulses can be recorded by directing the signal into the counter from the port pin. The resulting value can be read from the register when the input is finished or reaches a set value. Figure 5.1 in the data sheet shows the full block diagram of Timer0 that shows a pre-scale register and the watchdog timer, which shares the pre-scaler block.

The pre-scaler provides additional counter stages, dividing the input count by 2, 4, 8, 16, 32, 64, 128 or 256. This extends the count period or total count by the same factor, giving a greater range to the measurement but reducing its precision. The watchdog timer interval can also be extended, if this is selected as the clock source. The pre-scale select bits, and other control bits for Timer0, are found in OPTION_REG. Some typical Timer0 configurations are suggested in Table 1.7.

The larger mid-range PIC chips usually have more than one timer; the 16F877A has three. Timer1 is a 16-bit counter, consisting of TMR1H and TMR1L (file registers 0E and 0F). The count is fed to the low byte, and each time it rolls over from FF to 00, the high byte is incremented. The maximum count is therefore $2^{16}-1 = 65,535$, which allows a higher count without sacrificing accuracy.

Timer2 is an 8-bit counter (TMR2) with a 4-bit pre-scaler, 4-bit post-scaler and a comparator register, which allows the count value to be compared with a preset value, and the count terminated when they match. It can be used to generate Pulse Width Modulated (PWM) output, which provides a variable mark/space (hi/lo) ratio. This is useful for driving d.c. motors at a variable speed and digital position servos. These timers also can be used in capture and compare modes, which allow external signals to be more easily measured. There will be further consideration of these functions, with demonstration programs on timed I/O, in Chapter 6.

1.3.5 Indirect Addressing

File register 00 (INDF) is used for indirect file register addressing. The address of the register required is placed in the file select register (FSR). When data is written to or read from INDF, it is actually written to or read from the file register pointed to by FSR. This is most useful for carrying out a read or write on a block of GPRs, where FSR is simply incremented to select the next location. For example, it could be used to store a set of readings from a port over a period of time. Since 9 bits are needed to address all file registers (000−1FF), the IRP bit in the status register is used as the extra bit, acting then as a bank selection bit switching between bank pairs 0 and 1 and 2 and 3. Direct and indirect addressing of the file registers are illustrated in the data sheet (Figure 2.6).

1.3.6 Interrupt Control

Interrupts are external hardware signals which force the MCU to suspend its current process and carry out an Interrupt Service Routine (ISR). An interrupt can be generated in various ways but, in the PIC, the result is always to jump to program address 004. If more than one interrupt source is operational, then the source of the interrupt must be detected and the corresponding ISR selected. The registers involved in interrupt handling are INTCON, PIR1, PIR2, PIE1, PIE2 and PCON.

By default, interrupts are disabled, so interrupt-free programs can be loaded with their origin (first instruction) at address 0000, and the significance of address 0004 can be ignored. If interrupts are to be used, the main program start address needs to be 0005, or higher, and a 'GOTO start' (or similar label) placed at address 0000 to jump over the interrupt vector address 004. A 'GOTO ISR' instruction can then be placed at 004, using the ORG directive, which sets the address at which the instruction will be placed by the assembler. Several programs in later chapters use simple interrupts, see Program 4.3 for example.

The Global Interrupt Enable bit (INTCON,7) must be set to enable all interrupts. The individual interrupt source is then enabled. For example, the bit INTCON,T0IE is set to enable the Timer0 overflow to trigger the interrupt sequence. When the timer overflows INTCON,T0IF (Timer0 Interrupt Flag) is set, which indicates the interrupt source, and the ISR called automatically. If more than one interrupt is enabled, the relevant flags can be checked by the ISR to establish the source, and the correct ISR called. A list of interrupt sources and their control bits is given in Table 1.8.

The primary interrupt sources are Timer0 and Port B. Input RB0 is used for single interrupts, and pins RB4–RB7 can be set up so that any change on these inputs initiates an interrupt. This could be used to detect when a button on a keypad connected to Port B has been pressed, and the ISR would then process the input.

The remaining interrupt sources are enabled by the Peripheral Interrupt Enable bit (INTCON, PEIE). These are then individually enabled and flagged in PIE1, PIE2, PIR1 and PIR2. Many of these peripherals will be examined in more detail later, but the demonstration programs do not always use interrupts, to keep them as simple as possible. However, if these peripherals are used in more complex programs where multiple processes are required, interrupts are useful, even essential.

The program designer then has to decide on interrupt priority. This means selectively disabling lower priority interrupts, using the enable bits, when a more important process is

Table 1.8: Interrupt Sources and Control Bits.

Interrupt Source	Enable Bit Set	Flag Bit Set	Interrupt Trigger Event
TMR0	INTCON,5	INTCON,2	Timer0 count overflowed
RB0	INTCON,4	INTCON,1	RB0 input changed (also uses INTEDG)
RB4–7	INTCON,3	INTCON,0	Port B high nibble input changed
Peripherals	INTCON,6		*Peripheral Interrupt Enable bit*
TMR1	PIE1,0	PIR1,0	Timer1 count overflowed
TMR2	PIE1,1	PIR1,1	Timer2 count matched period register PR2
CCP1	PIE1,2	PIR1,2	Timer1 count captured in or matched CCPR1
SSP	PIE1,3	PIR1,3	Data transmitted or received in Synchronous Serial Port
TX	PIE1,4	PIR1,4	Transmit buffer empty in Asynchronous Serial Port
RC	PIE1,5	PIR1,5	Receive buffer full in Asynchronous Serial Port
AD	PIE1,6	PIR1,6	Analogue to Digital Conversion completed
PSP	PIE1,7	PIR1,7	A read or write has occurred in the Parallel Slave Port
CCP2	PIE2,0	PIR2,0	Timer2 count captured in or matched CCPR2
BCL	PIE2,3	PIR2,3	Bus collision detected in SSP (I^2C mode)
EE	PIE2,4	PIR2,4	Write to EEPROM memory completed

in progress. For example, when reading a serial port, the data has to be picked up from the port before being overwritten by the next data to arrive, so this should have priority. The limited stack depth (8 return addresses) in the PIC must be taken into account in designing interrupt-driven applications, especially if several levels of subroutine are implemented as well as multiple interrupts.

1.3.7 Peripheral Control

The remainder of the SFRs are used to control various peripheral functions. Their set-up will be explained as each is examined in turn with sample programs. The only peripheral which does not require external connections is the Electrically Erasable Programmable Read Only Memory. This is a block of non-volatile read-and-write memory that stores data during power down, such as a security code or combination for an electronic lock. A set of registers in Banks 2 and 3 is used to access this memory, as well as a special EEPROM write sequence designed to prevent accidental overwriting of the secure data. See Section 4 of the data sheet for details.

1.4 Application LED1

At this stage, it would be useful to look at some hardware in which we can demonstrate basic PIC program operation. The application LED1 is designed to be as simple as possible, simply outputting a binary count by incrementing an 8-bit port. This will also allow us to start using the development system and check that the program downloading to hardware works correctly.

1.4.1 LED1 Hardware

The hardware consists of a PIC 16F877A, CR clock components and a set of LEDs connected to Port B via current limiting resistors to display the output. The output can be viewed as a visible binary count, or monitored on a multi-channel oscilloscope or logic analyser, where it can be viewed as a set of output square waves that double in frequency from one output to the next. By default, the chip uses a CR network connected to input CLKIN to control the clock speed, which in this case uses a variable resistance so the operating frequency can be adjusted to a convenient value (40kHz). This gives an instruction cycle time of 100μs. A schematic of the circuit is reproduced in Figure 1.6.

The power supply pins are not seen on the MCU schematic component but are implicit in the design. V_{SS} (pins 12 and 31) is normally connected to 0V, and V_{DD} (pins 11 and 32) is connected to a 5V supply. Obviously, they do need to be connected in any final hardware layout. MCLR (Master Clear) also needs to be connected to a logic high (V_{DD}) or the program will not run. Unused pins can be left open circuit, as they default to inputs.

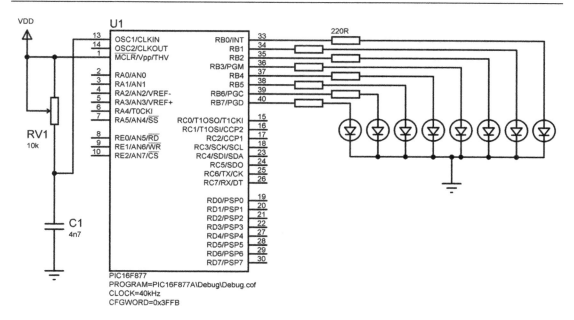

Figure 1.6
LED1 schematic.

1.4.2 Simple PIC Program

The first test program, LED1, will output a binary count sequence by incrementing the Port B data register. The clock rate is set to 40kHz by the component values $C = 4.7\text{nF}$, $R = 5\text{k}\Omega$ (10k preset), giving $C \times R \sim 25\mu\text{s}$. This is the rise time of the clock network, which roughly corresponds to the clock period, giving an instruction execution time of approximately100μs. This can be adjusted on test in the hardware.

The source code and list file for a simple program are shown in Program 1.1. It has been kept as simple as possible to highlight the essential elements, while still producing a useful output. The last statement END is not part of the program; it is required to terminate the code assembly process.

The meaning of the program instructions in the source code are as follows:

1. Load W with literal value 00h (hex)
2. Store this value in Port B (06h) data direction register
3. Clear Port B data register to 00h to switch off LEDs
4. Increment Port B (line labelled 'again') to count on LEDs
5. Jump back to previous instruction using target label 'again'

The program has the basic structure of a real-time application, with an initialisation phase and a loop sequence. We wish to output from Port B, so its data direction register must be

(a)

```
        MOVLW  00
        TRIS   06
        CLRF   06
again   INCF   06
        GOTO   again

        END
```

(b)

```
MPASM   5.46      LED1.ASM   9-18-2012   9:11:12     PAGE   1


LOC   OBJECT CODE     LINE SOURCE TEXT
   VALUE

0000   3000             00001                   MOVLW    00
Warning[224]: Use of this instruction is not recommended.
0001   0066             00002                   TRIS     06
0002   0186             00003                   CLRF     06
Message[305]: Using default destination of 1 (file).
0003   0A86             00004 again   INCF   06
0004   2803             00005                   GOTO     again
                        00006                   END



SYMBOL TABLE
   LABEL                          VALUE

__16F877A                         00000001
again                             00000003


MEMORY USAGE MAP ('X' = Used,  '-' = Unused)

0000 : XXXXX---------- ---------------- ----------------

All other memory blocks unused.

Program Memory Words Used:     5
Program Memory Words Free:  8187


Errors   :     0
Warnings :     1 reported,     0 suppressed
Messages :     1 reported,     0 suppressed
```

Program 1.1
LED1 application files: (a) source code LED1.ASM and (b) list file LED1.LST.

set up accordingly. All ports have two basic 8-bit registers for simple digital I/O, the data register (DR) itself and a data direction register (DDR). A zero in a particular bit in the DDR makes the corresponding bit of the DR operate as output. On power-up, the data direction register bits default high, so the ports become input unless initialised as output. Recall, if the pin has an alternate function as an analogue input, this is the default condition, so additional initialisation is needed to make it work as a digital I/O.

The first two instructions initialise Port B for output by loading W with a DDR code with all zero bits and copying it to DDRD using the TRIS instruction. The operand of this instruction is the port data register address 08h, although the actual destination of the code is register 88h, DDRD. The clear instruction then sets all these output bits low, turning off the LEDs. The following instruction increments Port B, showing the value 0000 0001 on the LEDs, and the last causes a jump back to the previous instruction, repeating the increment operation endlessly (until the MCU is reset or switched off). The result is a binary count on the LEDs, from 0000 0000 to 1111 1111, at which point it restarts at zero again.

1.4.3 Writing the Program

PIC application programs may be written using MPLAB, the free development system from Microchip. The process outlined here is explained from first principles in 'PIC Microcontrollers' by the author. MPLAB8 is used for this simple example, since it does not require a project structure to be created. For professional work, it has been superseded by MPLABX, which will be considered further in Chapter 2.

Assuming MPLAB8 has been downloaded from www.microchip.com, installed and started, click on the Configure menu and select the target device as 16F887A, then open the Configuration Bits dialogue. Uncheck the 'Configuration bits set in code' option because the processor configuration bits will set in this dialogue, not in the source code. Options can now be changed by clicking an item in the settings column. Set the Oscillator to RC, disable the watchdog timer (WDT), enable the power-up timer (PWRT), disable the brown-out reset and low-voltage programming and disable code protection. Note that the configuration code that will be downloaded to the chip is 3F33h.

A new file can now be created and the source code typed in, as shown in Program 1.1(a), and saved as LED1.ASM in a folder named LED1. The instruction mnemonics must be tabbed (or spaced) in from the left margin to be correctly recognised by the assembler, while labels such as 'again' are not. It is recommended that the labels, mnemonics and operands are arranged in columns to aid clarity for program analysis.

When the text has been entered, select Project, Quickbuild from the menus; the source code is then assembled, i.e. converted to machine code, file name LED1.HEX. A successful build should be confirmed in the output window (ignore the warning given about the TRIS instruction). The source and machine code can be viewed together in the (plain text) list file LED1.LST that is created in the application folder (Program 1.1(b)).

The list file includes the source code at the right, with, from the left, the memory location where each instruction is stored (0000−0004), the hex machine code and source line numbers. Warnings and messages generated by the assembler and inserted in the list file can be switched off if preferred. Note that the terminal directive END is not converted to machine code. Comments will be added later in the last column. The list file also contains warning messages and information about the memory usage generated by the assembler.

1.4.4 Simulation of LED1

Applications can be tested by simulation, either in MPLAB or Proteus VSM. In MPLAB, no external hardware is simulated, only the operation of the MCU itself. MPLAB is free of charge and also provides the tools for program downloading to the target hardware. For the basic test programs which do have specific devices attached to the I/O pins, MPLAB is sufficient. In professional practice, it may not be possible to simulate complex peripherals in VSM, and the project management tools in MPLAB may be required.

Assuming the program has been built (assembled) as above, it can be tested by selecting Debugger, Select Tool, MPLAB SIM. A debugging toolbar will appear with buttons to run, pause and single step the program. If run, nothing much appears to happen, but if paused, the current execution point in the program is displayed as an arrow in the source code window. The program can now be single stepped to check its sequence.

To see the effect of the program on the SFRs, select them in the View menu and confirm that Port B is incrementing. The output can also be viewed in the virtual Logic Analyzer by selecting the Port B pins via the Channels button. The program can also be single stepped or animated and break points used to assist with debugging. When the windows have been conveniently arranged, the workspace can be saved for future recovery (Figure 1.7).

If we wish to study the program timing, the MCU clock speed must be set in the simulator. With the debugger MPSIM enabled, go to the Debugger, Settings dialogue and change the Processor Frequency to 40kHz. The Stopwatch window can now be opened and the step time (one instruction) confirmed to be 100μs. Note that the GOTO takes two cycles. If a break point is set at the start of the output loop (just double left click on the left margin of the source code line required), the stopwatch can be used to check the output timing,

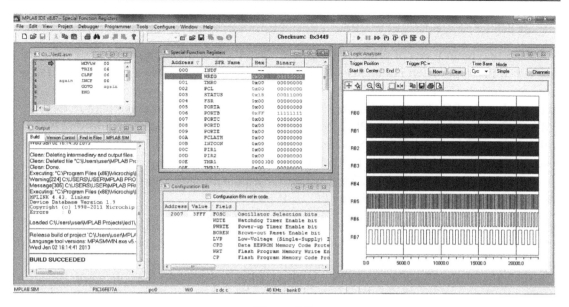

Figure 1.7
MPLAB8 screenshot of LED1 simulation.

Table 1.9: MPLAB8 Quickbuild File Set.

File Name	Function
LED1.ASM	Source text file
LED1.HEX	Binary machine code program for downloading to chip
LED1.LST	List file contains source code and machine code
LED1.ERR	Error messages as seen in output window
LED1.COF	MPLAB debugging file
LED1.MAP	Map file shows memory usage
LED1.O	Object file can be used as library file

and the duration of the loop (300μs). This means that RB0 will toggle with a period of 600μs and RB7 with a period of $0.6 \times 128 = 76.8$ms (768 instruction cycles).

Debugging will be covered in more detail in Chapter 3, using similar tools in Proteus VSM. Since MPLABX will be needed for program downloading and hardware debugging later, it will be used in the next chapter for further assembler programming. A project file set for LED1 which can be tried out in MPLAB8 is included in the demo downloads for comparison with MPLABX. The file set generated by the Quickbuild option contains those listed in Table 1.9. This allows the individual files to be examined, which is not so easy in MPLABX.

1.5 Downloading and Testing

There are two main types of programming hardware. Pre-programming (before the chip is inserted in the application circuit) is shown in Figure 1.8(a), with a host PC connected to a programming unit via a serial link. The target chip is inserted into the programming unit connected to the PC COM (RS232) port. The program is written and assembled in the PC and is then downloaded using MPLAB or the programming application module supplied with the programmer unit. The RS232 protocol, the simplest serial data format, will be described later as a standard port in the PIC 16F877A itself. The chip is programmed via pins RB6 (clock) and RB7 (data). This method is principally now used for production programming of multiple chips via the PC USB port.

The basic programming unit of this type supplied by Microchip is PICSTART Plus. It has a zero insertion force (ZIF) socket in which the target chip is placed and contains another PIC within to handle the programming. Since the COM port is no longer fitted as standard to most PCs, this programming system is largely obsolete for training and development work. However, a bulk programming unit PM3, which can operate in stand-alone mode once the application code has been downloaded, is still a current product for supporting commercial designs.

1.5.1 In-Circuit Programming and Debugging

A more versatile method of program downloading provides in-circuit programming and debugging (Figure 2.1(b)). The PIC chip remains in the target board after construction and

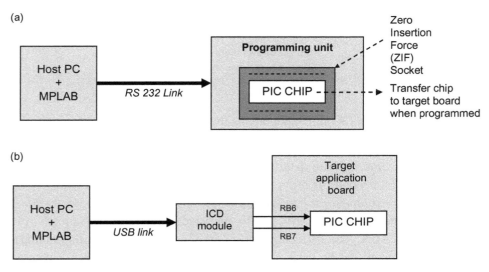

Figure 1.8
PIC development systems: (a) programming unit and (b) ICD system.

is programmed in circuit. The user program can then also (sometimes) be debugged in circuit, operating with the real hardware. This provides the final fault-finding step and hardware testing. The in-circuit programming and debugging (ICPD) connections are shown in Figure 1.9. These are included in the hardware schematic shown in Figure 3.1.

Generally, ICPD modules are fully integrated into the MPLAB system and, once connected, are selected for use from the menus. When the application program has been satisfactorily tested in simulation, the programmer is plugged into the host USB port and the target hardware. In MPLAB8, the programmer type is selected from the Programmer menu, and the relevant toolbar is enabled if successful communication with the programmer has been established. When the application program has been downloaded, the same device has to be selected from the debugger menu to enable in-circuit testing. In MPLABX, the hardware tool only has to be selected once in project properties dialogue from the File menu.

The program under test is controlled via the programmer/debugger from within MPLAB, where the source code is displayed in debug mode. The program can be run, stopped, paused and restarted under manual control or with break points, just as in simulation mode. The current execution point is highlighted, and the state of the file registers displayed and traced.

To achieve this, the MCU must have special debugging features built in, so that the program can be interrupted as required and the processor status reported. A block of debug code will be loaded into a reserved area of memory, and an NOP (No Operation) must be placed at program location 0000 by the user to allow for a call to the debug code before the program starts executing. Specific SFRs must also be reserved for ICD use (see Table 1.3).

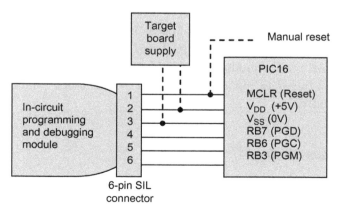

Figure 1.9
In-circuit programming and debugging connections.

(a)   (b)

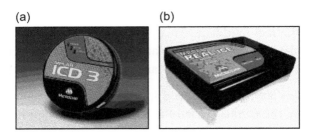

Figure 1.10
Microchip programmer/debugger modules: (a) ICD3 and (b) Real ICE. *(Courtesy of Microchip Inc.)*

1.5.2 ICPD Hardware

Microchip provides a variety of programming and in-circuit testing modules, and there are, at the time of writing, three main products available of increasing power and cost. The PICkit3 is the entry level device (Figure I.1) that nevertheless provides a debugging tool that is more than adequate for beginners. ICD3 (Figure 1.10(a)) is a slightly more advanced version of the basic programmer/debugger. It is faster and has additional debugging features. ICDx hardware generally uses a phone jack style modular connector for interfacing to the target, rather than the in-line connector associated with PICkitX. Real ICE (Figure 1.10(b)) is designed for high volume or commercial work; it offers target hardware system monitoring using additional logic probes and more powerful data trace, capture and display modes for complete system debugging.

Note that in-circuit programming and debugging is not necessarily built into some smaller, low pin-count PIC chips, because the requisite on-chip ICPD hardware resources (i.e. extra pins and internal circuits) would represent a significant extra cost. In this case, a header module is used which carries a version of the chip that has these on-chip resources. This is used to debug the target board, as a pin compatible replacement with ICPD connector, and a pre-programmed chip fitted after testing. These ICPD headers are usually compatible with ICDx modules. The PIC 16F690, fitted in the low pin-count demo board, requires this ICD support. The 16F877A was among the first PIC chips to support ICPD internally.

1.5.3 LED1 Program Testing

LED1 is not really intended for hardware implementation, only to introduce MCU principles. However, if required, LED1 can be downloaded and tested using the PICkit2 or 3 programmer. A simple prototype set-up is seen in Figure 1.11 which was used to check these functions using PICkit2.

Figure 1.11
Prototype programming test.

When the programmer is selected from the menus, a programming toolbar appears with buttons to download and run. The MCU is set to run by setting MCLR. An external power supply may be needed for the target system, as the programmer can only supply limited current. The program LED1 can be single stepped in hardware, with a test LED connected to RB0. The output with a 40kHz clock is a little too fast to be visible but can be viewed on a logic analyser or multi-channel oscilloscope as a set of 8 pulse waveforms of decreasing frequency.

When an application is complete, the chip is reprogrammed with ICPD disabled. MCLR must be pulled up to V_{DD} for the chip to run, via a manual reset button if required. An application circuit is described in Figure 3.1 (LED2) that includes programming connections. The program has the same output function but includes a delay routine to render the output pulses visible and inputs to run and reset the count.

1.5.4 Development Steps

The application development procedure can be summarised as follows:

- *Design and construct application hardware*
- *Design application firmware*
- *Edit source code (MPLAB)*
- *Assemble (or compile) the source code (MPASM)*
- *Test it in simulation mode (MPSIM)*
- *Download and test in target hardware (ICPD)*
- *Modify and repeat above stages if necessary*
- *Produce final prototype and production version*

Chapter 2 will cover program design and implementation, and Chapter 3 will describe the use of Proteus VSM for circuit design, schematic capture and interactive simulation. It is therefore recommended that any circuit construction work and hardware testing be deferred until these chapters have been studied. Prototype hardware for LED2 can then be attempted if desired. Further information on prototyping and other implementation issues can be found in 'PIC Microcontrollers' by the author.

1.6 Conversion to 16F887 and Other Chips

It is necessary to select the most appropriate chip for a particular design. An application can be developed on a chip such as the 16F877A that has a comprehensive range of features but, in the end, there may well be unused ports and interfaces, and the final program may not need so much memory. The application may therefore best be 'ported' to a smaller, cheaper chip whose features more closely match the requirements of the application in question.

In addition, a particular device may be superseded by an improved equivalent, as in the case of the '877A', requiring its replacement in an existing design by a more recent product. Cost is also a factor; the guide price at the time of writing for the 16F877A is $4.94 but for the 16F887 is $1.78, so even if the original chip is not out of production, cost is an incentive to update, particularly if the application is produced in volume.

The chip data sheet is the most important reference for any microcontroller application development work, so this must be consulted carefully when updating designs for a replacement chip. Microchip provides an application note DS41305A which covers the differences between the 16F877A and the 16F887. Strangely, the most significant difference is not mentioned explicitly − the availability of an internal clock oscillator in the more recent chip. This means that the external clock components are not required unless a more precise crystal clock is needed. Conveniently, it makes no difference in simulation, since the MCU clock is configured in software anyway.

Other hardware differences are that the '887' has two independent analogue comparators and does not have a parallel slave port. The former provides additional analogue interfacing options (see Chapter 7) and a parallel data connection would just have to use the general features of the digital ports to manage a bus system, should that be required. The other updates are mainly enhancements of existing features, in particular additional analogue inputs, giving 14 in all (AN0−AN13), and three PWM outputs on Port D. The 16F887 also has additional low-power operating features. The chips are otherwise pin compatible, so the same circuit can be used with suitable configuration and program changes. Obviously, the correct initialisation file must also be included in the program header (see Chapter 2).

The pin out for the 16F887 shown in Figure 1.12 may be compared with that of the 16F877A shown in Figure 1.4 (alternate pin functions are shown in brackets). This gives

Alternate functions				Alternate functions
Vpp / RE3	**MCLR**	1	40 **RB7**	Program Data / Interrupt
AN0 / ULPWU / C12IN0-	**RA0**	2	39 **RB6**	Program Clock / Interrupt
AN1 / C12IN0+	**RA1**	3	38 **RB5**	AN13 / Timer 1 Input / Interrupt
AN2 / VREF- / CVREF / C2IN+	**RA2**	4	37 **RB4**	AN11 / Interrupt
AN3 / VREF+ / C1IN+	**RA3**	5	36 **RB3**	AN9 / C12IN2 / LV Program
Timer 0 Clock Input / C1OUT	**RA4**	6	35 **RB2**	AN8
AN4 / Slave Select / C2OUT	**RA5**	7	34 **RB1**	AN10 / C12IN3-
Analogue Input AN5	**RE0**	8	33 **RB0**	AN12 / Interrupt
Analogue Input AN6	**RE1**	9	32 **V$_{DD}$**	+5V Power Supply
Analogue Input AN7	**RE2**	10	31 **Vss**	0V Power Supply
+5V Power Supply	**V$_{DD}$**	11	30 **RD7**	PWM Output D
0V Power Supply	**Vss**	12	29 **RD6**	PWM Output C
RA6 / CR clock / XTAL	**CLKIN**	13	28 **RD5**	PWM Output B
RA7 / XTAL	**CLKOUT**	14	27 **RD4**	
Timer 1 Input / Output	**RC0**	15	26 **RC7**	USART Receive / Data
Timer 1 Input / Output	**RC1**	16	25 **RC6**	USART Transmit / Clock
Timer 1 Input / PWM Output A	**RC2**	17	24 **RC5**	I²C Data Out
SPI Clock	**RC3**	18	23 **RC4**	I²C Data In / SPI Data
	RD0	19	22 **RD3**	
	RD1	20	21 **RD2**	

ULPWU = Ultra Low Power Wake Up input; C1, C2, C12 = Comparator inputs and outputs.

Figure 1.12
PIC 16F887 pin out.

some further indication of the additional features of the later chip. It should be substituted in hardware implementations of the designs in this book, but the original MCU can be used in simulation since its model licence is cheaper.

Questions 1

Refer to the 16F877A data sheet as required.

1. State the three main elements in any microprocessor system. (3)
2. State the difference between a microprocessor and a microcontroller. (3)
3. Describe briefly the process of fetching an instruction in a microcontroller. (3)
4. State the advantages of flash ROM, compared with other memory types. (3)
5. Explain briefly why serial data communication is generally slower than parallel. (3)
6. Explain briefly why Ports A and E in the PIC 16F877A cannot be used for digital input without initialisation. (3)
7. Explain briefly the difference between an SFR and a GPR. (3)
8. Show which bits in the binary instruction 'MOVWF 0C' are allocated to the op-code and the operand. (3)
9. Explain briefly why bank selection is necessary when initialising a parallel port for output in the PIC, and how this is achieved. (3)
10. Explain briefly a major advantage of using a hardware timer rather than a delay loop. (3)
11. Explain briefly the contents of each column in the program list file. (5)
12. Outline the procedure for creating and downloading a program to the PIC MCU using an ICPD programmer. (5)

Assignments 1

1.1 Program Execution

Describe the process of program instruction execution in a PIC MCU by reference to the data sheet. Explain the role of each block, and how the instructions and data are moved around. Use the instructions MOVLW XX, ADDWF XX and CALL XXX as examples to explain the execution sequence. Identify the binary codes that will appear on the internal busses when these instructions are executed. Refer to 'PIC Microcontrollers (Ed 3)' if necessary.

1.2 MPLAB8 Test

Download and install the MPLAB8 development system (if available). Enter and save the program LED1 source code in a suitable folder. Assemble (Quickbuild) and run the program in simulation mode. Set the MCU clock to 40kHz and disable the watchdog timer. Display the SFRs and confirm that PCL tracks the execution point and Port B increments from zero. Note the effect on the Z flag when the port register is cleared, then incremented. Use a break point and the stopwatch to measure the loop execution period (600μs).

CHAPTER 2

PIC Programming

Summary

- The development system has an editor, assembler, simulator and programmer
- Assembly language uses instruction set mnemonics to create source code
- The configuration code sets up MCU clock, start-up and memory options
- Programs should be well commented and structured for analysis and debugging
- Assembler directives can be used to improve the efficiency and flexibility of code
- Programs are designed using flowcharts, structure charts or pseudocode
- C programming is more user friendly but uses more memory

PIC microcontroller architecture has been introduced in Chapter 1, so we now turn to the software (or firmware, since it is stored in non-volatile memory) required to make it run. The source code is written on a PC host computer in the edit window of MPLAB (the manufacturer's development system), assembled and downloaded to the chip. A demo program LED2 will illustrate some of the main principles of assembler programming for the PIC.

The application firmware is created when source code is assembled in MPLAB using the MPASM assembler utility. The Microchip development environment and the MPASM User Guide are downloadable from www.microchip.com. We have used MPLAB8 in the last chapter because it is simpler to create a stand-alone program. In this chapter we will use the professional version available at the time of writing, MPLABX v1.41, where a project must be created for all applications. The application schematic for LED2 can be seen in Figure 3.1, where the procedure for creating the drawing and testing the circuit by interactive simulation will be outlined after we have looked at software development.

2.1 Application LED2

The application LED2 will have the same output function as LED1 but introduce control inputs to provide run/stop/restart functions and a delay routine to slow down the output rate so that it is more easily visible to the user. The program will also use basic source code features such as register labels and assembler directives. A block diagram can be used to outline a hardware design, as seen in Figure 2.1(a), although it is fairly trivial in this case.

Interfacing PIC Microcontrollers.
DOI: http://dx.doi.org/10.1016/B978-0-08-099363-8.00002-9

(a)

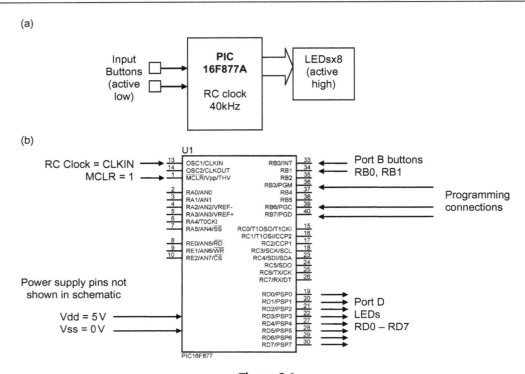

(b)

Figure 2.1
LED2 hardware: (a) block diagram and (b) MCU connections.

This must then be translated into corresponding hardware connections; the pin allocation is shown in Figure 2.1(b).

Control switches will be connected to RB1 and RB2 using pull-up resistors, so they will generate default logic high inputs, going low when pressed. The output has been moved to Port D, so that RB3, RB6 and RB7 can be used exclusively for the programming connections. Program execution and hardware debugging will be controlled via an ICPD module plugged into the 6-pin programming connector.

The complete project fileset can be found in the demo downloads in the folder 'led2.x'. If the project were created from scratch in MPLABX, New Project and the options Microchip Embedded, Standalone Project would be selected, then MCU (16F877A), simulator and assembler (mpasm). When the project is named, a folder 'led2.x' containing a default fileset is automatically created. A new source code file 'led2.asm' needs to be created by clicking on the Source File in the project window, and right clicking for New, Assembly File. The file must be named 'led2' and the source code edit window is then activated. The source code to be entered would be as seen in the list file to the right of the source file line numbers. Save and build creates machine code (.hex) and debug (.cof) files in the folder led2.x/dist/default/production. The error (.err), list (.lst) and object (.o) files are created in a build/default/production folder.

The led2 list file shows the memory address, machine code, source line number and source code in columns. There are also detailed comments to aid program debugging and later maintenance. Note that two types of labels are used in program to represent numbers: label equates are used at the top of the program to declare labels for the file registers that will be used in the program, and address labels are placed in the first column to mark the destination for GOTO and CALL instructions. Various directives (instructions to the assembler) are used, but CODE, which indicates the start of the program code, and END to terminate it, are the only ones that are essential (Program 2.1).

Pre-defined terms such as the assembler directives, instruction mnemonics (e.g. MOVLW), and standard port labels (e.g. PORTB) are in upper case because that is how they are defined in the assembler specification, data sheet for the MCU and standard MCU include file respectively. In fact, only the address labels are case sensitive by default, so the register labels may be seen in lower case. Here, lower case and mixed case are used for locally defined address and register labels respectively. The content of each column of the list file is detailed in Table 2.1.

If the simulator is selected as the test tool during project creation, the program can be debugged in software prior to downloading to hardware (Figure 2.2). When the Debug Project tool button is pressed, the project is rebuilt and the simulator activated. To test LED2, the inputs need to be simulated via a stimulus dialogue opened by selecting Window, Simulator, Stimulus. In this case, RB1 and RB2 operate as a reset and run input respectively. Initially, they must be set high, and taken low to trigger their action. When running, Port D will increment continuously. This can be observed in the SFR window (Window, PIC Memory Views, SFRs) or in the logic analyser (Window, Simulator, Analyzer). In the latter case, the required outputs are selected via the 'Edit pin channel definitions' button. Note that the 'Reset Zoom' button must be pressed after a run to see the resultant waveforms. The Dashboard window is also useful, showing a summary of the project resources and status.

2.2 Assembly Language

The assembly language program is based on the instruction set defined for a specific MCU, plus assembler directives which control the conversion to machine code, but are not part of the final program. The instruction syntax is similar for all PIC chips, but the larger, more powerful chips tend to have more instructions. The '877A' and 887 have a minimal set of 35 instructions that is common to all mid-range PIC chips.

```
MPASM  5.46                    LED2.ASM   1-2-2013   14:44:33        PAGE   1


LOC   OBJECT CODE     LINE SOURCE TEXT
  VALUE

                      00001 ;;;;;;;;;;;;;;;;;;;;;;;;;;;;;;;;;;;;;;;;;;;;;;;;;;;;;;;;;;;;
                      00002 ;
                      00003 ;      Source File:    LED2.ASM
                      00004 ;      Author:         MPB
                      00005 ;      Date:           2-1-13
                      00006 ;
                      00007 ;;;;;;;;;;;;;;;;;;;;;;;;;;;;;;;;;;;;;;;;;;;;;;;;;;;;;;;;;;;;
                      00008 ;
                      00009 ;      Slow output binary count is stopped, started
                      00010 ;      and reset with push buttons.
                      00011 ;
                      00012 ;      Processor:      PIC 16F877A
                      00013 ;
                      00014 ;      Hardware:       Prototype
                      00015 ;      Clock:          RC = 40kHz
                      00016 ;      Inputs:         Port B: Push Buttons
                      00017 ;                      RB1, RB2 (active low)
                      00018 ;      Outputs:        Port D: LEDs (active high)
                      00019 ;
                      00020 ;      WDTimer:        Disabled
                      00021 ;      PUTimer:        Enabled
                      00022 ;      Interrupts:     Disabled
                      00023 ;      Code Protect:   Disabled
                      00024 ;
                      00025 ;;;;;;;;;;;;;;;;;;;;;;;;;;;;;;;;;;;;;;;;;;;;;;;;;;;;;;;;;;;;
                      00026
                      00027        PROCESSOR 16F877      ; Define MCU type
2007   3733           00028        __CONFIG 0x3733       ; Set config fuses
                      00029
                      00030 ; Register Label Equates...................................
                      00031
  00000006            00032 PORTB   EQU    06    ; Port B Data Register
  00000008            00033 PORTD   EQU    08    ; Port D Data Register
  00000088            00034 TRISD   EQU    88    ; Port B Direction Register
  00000020            00035 Timer   EQU    20    ; GPR used as delay counter
                      00036
                      00037 ; Input Bit Label Equates .................................
                      00038
  00000001            00039 Inres   EQU    1     ; 'Reset' input button = RD0
  00000002            00040 Inrun   EQU    2     ; 'Run' input button = RD1
                      00041
                      00042 ;;;;;;;;;;;;;;;;;;;;;;;;;;;;;;;;;;;;;;;;;;;;;;;;;;;;;;;;;;;;
                      00043
                      00044        CODE   0      ; Program code start address
                      00045
                      00046 ; Initialise Port B (Port A defaults to inputs)............
                      00047
0000   1683 1303      00048        BANKSEL TRISD         ; Select bank 1
0002   3000           00049        MOVLW  b'00000000'    ; Port B Direction Code
0003   0088           00050        MOVWF  TRISD          ; Load the DDR code into F86
0004   1283 1303      00051        BANKSEL PORTD         ; Select bank 0
0006   2???           00052        GOTO   reset          ; Jump to main loop
                      00053
                      00054
```

Program 2.1
LED2 list file.

```
                        00055 ; 'delay' subroutine ........................................
                        00056
0007    00A0            00057 delay   MOVWF   Timer       ; Copy W to timer register
0008    0BA0            00058 down    DECFSZ  Timer       ; Decrement timer register
0009    2???            00059         GOTO    down        ; and repeat until zero
000A    0008            00060         RETURN              ; Jump back to main program
                        00061
                        00062
                        00063 ; Start main loop ..........................................
                        00064
000B    0188            00065 reset   CLRF    PORTD       ; Clear Port B Data
                        00066
000C    1C86            00067 start   BTFSS   PORTB,Inres ; Test reset button
000D    2???            00068         GOTO    reset       ; and reset Port B if pressed
000E    1906            00069         BTFSC   PORTB,Inrun ; Test run button
000F    2???            00070         GOTO    start       ; and repeat if n pressed
                        00071
0010    0A88            00072         INCF    PORTD       ; Increment output at Port B
0011    30FF            00073         MOVLW   0FF         ; Delay count literal
0012    2???            00074         CALL    delay       ; Jump to subroutine 'delay'
0013    2???            00075         GOTO    start       ; Repeat main loop always
                        00076
3733                    00077         END                 ; Terminate source code
```

Program 2.1
(Continued)

Table 2.1: List File Elements.

List File Column	Example Content	Meaning
0	000C	Memory location at which machine code instruction is stored
1	1C08	Machine code instruction, including op-code and operands
2	00065	Source code line number
3	start	Address label marking jump destination
4	BTFSS	Instruction mnemonic or assembler directive
5	PORTD,Inres	Instruction operand labels
6	; Test reset	Comment delimited by semicolon

2.2.1 Assembler Code

The machine code instruction provides the execution unit of the MCU with a code to carry out a particular operation (move, calculate, test, etc.). It could be entered in plain binary, but this would require us to look up the code each time; a more useful option is the instruction mnemonic. Labels are used to represent the op-code and operands, and these are replaced by the assembler program with the corresponding binary codes, to produce the machine code program (hex file). The sample instruction found at address label 'start' is dissected in Table 2.2.

It can be seen that in this case the 14-bit instruction code can be broken into three functional parts. The first two bits are always zero; the unused bits are 14 and 15. The next

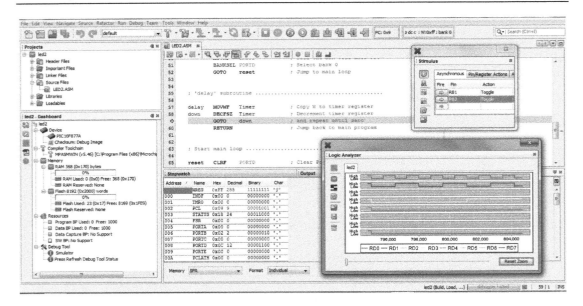

Figure 2.2
MPLABX simulation of LED2.

Table 2.2: Instruction Analysis.

Label	Hex	Binary	Meaning	Range
start	000C	0000 0000 0000 1100	Program memory address label	0000—1FFF (8K)
BTFSS	1C08	00 01 11-- ---- ----	Op-code (bits 13, 12, 11, 10 only)	—
PORTD		-- -- ---- -000 1000	File register address = 08	00—7F (128)
Inres		-- -- --00 0--- ----	File register bit = 0	0—7

four give the code for the instruction BTFSS (Bit Test File register and Skip if Set); the next three identify the bit to be tested (0) and the last seven the file register address (08) in which this bit is located. A similar structure can be seen in the code for each instruction in the data sheet (Table 15.2). The number of bits used to define the instruction itself varies.

Other instructions do not necessarily have all these operands. For example, CLRW has no operands, since the target register identity is implicit in the instruction. CLRF has one, the file register address. In most of the byte-oriented file register operations, the default destination of the modified data is the file register (bit 7 = 1) but is switched to W by setting bit 7 = 0.

Note that an unused memory location normally contains all ones (3FFF). In the instruction set code 3FFF = ADDLW 0FF, meaning add the literal FF to W. Blank locations will repeat this operation until the program counter rolls over to zero and the program restarts.

Therefore, if the program does not loop continuously, it should be terminated with a SLEEP instruction to stop the program at that point. Note that code 0000 = NOP, no operation.

In the definitive instruction set in the data sheet, the elements of each are identified as follows:

```
f = file register (00-7F)
d = destination (1 = file register, 0 = W)
k = literal (00-FF)
b = bit number (0-7)
```

The effect of each instruction on the status flags is also defined in the full instruction set. In general, the arithmetic instructions affect the Carry (C), Digit Carry (DC) and Zero (Z) flags, while most of the logic instructions affect Z only. The rotate instructions affect C only, while SLEEP and CLRWDT (Clear Watchdog Timer) affect the Time Out (TO) and Power Down (PD) flags.

The assembly language program is therefore written using pre-defined labels for the instruction (mnemonics) with user-defined labels for destination addresses, registers, bits and literals. In general, any number that appears in the program can be replaced by a label.

2.2.2 Assembler Syntax

The essentials of program syntax (grammar and spelling) are illustrated in LED2. The instructions themselves are placed in the second and third columns; they are separated by a tab for clarity, but a space is also acceptable to the assembler. The line comments are delimited by a semicolon. The comment should describe the effect of the instruction, rather than simply repeating the meaning of the mnemonics and labels.

A detailed comment block at the top of the program is desirable but should be in proportion to the complexity of the program. The source code file name, author and date, and/or version number are a minimum recommendation. A program description should then follow, and development system details. Information on the intended target hardware is useful, especially the MCU I/O allocation.

This is then followed by the MCU selection and configuration word. These are assembler directives and are not converted to machine code, which can be confirmed by the absence of any corresponding code in the left-hand column of the list file. The PROCESSOR directive tells the assembler which MCU will be used, because there is significant variation between them, e.g. memory size and the type and number of ports available. It can then throw up an error if the program tries to use non-existent resources.

The __CONFIG directive sets the programmable fuses in the chip which cannot be changed except by reprogramming. These are detailed below and include the clock type selection, code protection options, and watchdog and power-up timer (PUT) enable (see Chapter 1 for more details). The configuration code (3733h) and its location (2007h) appear in the left column of the list file.

Another commonly used directive is EQU. This allows any number in the program to be represented by a label, notifying the assembler that the label given should be replaced by the number to which it is equated. File register addresses (PORTB, TRISB, PORTD, Timer), bit numbers (Inrun, Inres) and literals may be represented in this way.

The directive CODE 0 precedes the actual program code. This is required in current assemblers to indicate the start of the instruction list and locates the start of the code at program memory address 0000. The only other directive that must be included is END, which indicates the end of the source code to the assembler.

Within the code section, program memory address labels (delay, down, reset, start) are declared implicitly by their placement at the beginning of a source code line. Short labels are used here so they fit into an 8-character column in the source code. Longer labels (e.g. start_of_main_program) may be used, in which case the label can be one line and the associated code on the next − the line return is not significant to the assembler. However, spaces should not be used within the label; underscore is usually used instead. Avoid characters other than letters, numbers and underscore in labels, since they may have special significance to the assembler. Similarly, assembler directives and other pre-defined keywords must be not be used as labels; an error message on assembly should identify this problem.

The special function register names are usually defined by including a standard header file that is supplied with MPLAB for each PIC chip. The header file P16F877A.INC contains a complete list of the SFRs and their addresses, and this will be used in later examples when a longer list of file registers is needed. These provide a standard labeling system which aids consistency between application developers. The standard labels are all in upper case, e.g. PORTA. The use of the standard SFR label file is recommended for all PIC programs. In MPLABX, the standard labels definitions are included implicitly when the MCU is selected for the project. The label list can be studied by opening the INC file in the Microchip/ MPLABX/mpasmx folder of the development system fileset in Program Files.

2.3 MCU Configuration

The target MCU needs to be specified at the top of the program, and the operational options selected by including a configuration code in the program header block.

2.3.1 PROCESSOR Directive

The assembler will create the machine code for the MCU specified and will be able to check that the assembler code is correct for this processor. For example, it will confirm that only those ports available in hardware are used in the program. In MPLABX, it is not essential as the MCU has already been defined in the project set-up but will be needed when testing in Proteus VSM. The LIST directive can also include the MCU selection, as well as many other useful enhancements, but has not been used here for the sake of simplicity.

2.3.2 CONFIG Directive

The assembler directive __CONFIG (double underscore prefix), which allows the chip configuration word to be specified, is included at the top of most PIC programs. A special area of program memory outside the normal range (address 2007h) stores this code. The configuration bits specify the clock type and other MCU options for that application. If undefined, as in LED1, the default code 3FFFh will be used, where the RC clock is selected, watchdog timer (WDT) enabled and all other options disabled.

The function of each bit is given in Table 2.3, along with some typical configuration settings. Further details can be found in the data sheet, Section 12. In simulation mode (MPSIM and VSM), the configuration word can also be set in the MCU properties dialogue

Table 2.3: 16F877A Configuration Bits.

Bit	Label	Function	Disabled	Enabled	Default	Typical
15	—	None	x	x	0	0
14	—	None	x	x	0	0
13	CP	All code protection enable	1	0	1	1
12	—		x	x	1	1
11	DEBUG	ICD enable	1	0	1	0
10	WRT1	Program memory write protect enable	1	0 0 1	1	1
9	WRT0	(selected blocks)	1	0 1 0	1	1
8	CPD	EEPROM write protect enable	1	0	1	1
7	LVP	LVP enable	0	1	1	0
6	BOREN	Brown-out reset (BoR) enable	0	1	1	0
5	—	None	x	x	1	1
4	—	None	x	x	1	1
3	PWRTEN	PUT enable	1	0	1	0
2	WDTEN	WDT enable	0	1	1	0
1	FOSC1	Oscillator type select	—	0 1 0 1	1	0
0	FOSC0	RC = 11, HS = 10, XT = 01, LP = 00	—	0 0 1 1	1	1

Default = 3FFF (RC clock, PuT disabled, WdT enabled).
Typical RC clock = 3FF3 (RC clock, ICD disabled, PuT enabled, WdT disabled).
Typical XT clock = 3731 (XT clock, ICD enabled, PuT enabled, WdT disabled).

prior to testing the program, for experimental purposes, but must be defined in the source code for final downloading.

The configuration options in the 16F877A are explained below, using the standard label for each bit in the configuration code.

2.3.2.1 Code Protection (CP, WRT, CPD)

Normally, the program machine code can be read back to the programming host computer via the programmer. It can then be disassembled in the host PC and a source program recovered. This may need to be prevented for commercial or security reasons, so CP protects all program flash memory from being read. The code protection bits (WRT1: WRT0) disable writes to selected program ranges. Data EEPROM may also be protected from external reads using the CPD bit, while internal read and write operations are still allowed, regardless of the state of the code protection bits.

2.3.2.2 In-Circuit Debugging

First generation PIC chips had to be removed from circuit for programming in a separate programming unit. With in-circuit debugging (ICD), the chip can be initially programmed, or reprogrammed after debugging, while remaining in the target board. The normal debugging techniques of single stepping, break points and tracing are available in ICD mode, allowing a final stage of debugging in the prototype hardware, where problems with the interaction of the MCU with the real circuit can be resolved. ICD is enabled in the configuration word but disabled in the final working version of the downloaded program, by selecting the appropriate programmer/debugger option. If ICD is disabled, RB6 and RB7 can be used for general I/O.

2.3.2.3 Low-Voltage Programming

Normally, when the chip is programmed, a high voltage (12−14V) is applied to the PGM pin (RB3). To avoid the need to supply this voltage during in-circuit programming, a low-voltage programming (LVP) mode is available. However, using this option means that RB3 is not then available for general I/O functions during normal operation. The standard programming tools use high-voltage programming.

2.3.2.4 Power-Up Timer

When the supply power is applied to the programmed MCU, the start of program execution should be delayed until the power supply and clock are stable. The PUT should therefore be enabled (PWRTE = 0) as a matter of routine. It avoids the need to reset the MCU manually at start-up, or connect an external reset circuit, as was necessary with some microprocessors. An internal oscillator provides a delay between the power coming on and an internal MCU reset of about 72ms. This is followed by an oscillator start-up delay of 1024 cycles of the clock before program execution starts. At a clock frequency of 4MHz, this works out to 256µs. Clearly, the start-up delay is unnecessary in simulation mode.

2.3.2.5 Brown-Out Reset

Brown-out refers to a short-term dip in the power supply voltage, caused by mains supply fluctuation or other supply fault, which might disrupt the program execution. If the Brown-Out Reset Enable bit (BOREN) is set, a PSU glitch of longer than about 100µs will cause the device to be held in reset until the supply recovers and then wait for the PUT to time out, before restarting.

2.3.2.6 Watchdog Timer

The WDT is designed to automatically reset the MCU if the program malfunctions. This could be caused by an undetected bug in the program, or an unplanned sequence of inputs, causing the program to get stuck in a loop. A dedicated internal oscillator and counter automatically generate a reset about every 18ms. If the WDT is enabled, it must be regularly reset by an instruction in the program loop (CLRWDT) to prevent the auto-reset. If the program hangs, and the WDT reset instruction not executed, the MCU will restart, and (possibly) continue correctly, depending on the nature of the fault and precautionary routines included in the program.

Note that the WDT is enabled by default so will operate if no configuration word is defined. This applies to the test program LED1, so it restarts after each WDT timeout. This can be seen in the messages generated in the simulator. WDT wakes up the MCU from sleep mode (after the execution of the SLEEP instruction) so can be used to implement intermittent operation which saves battery power, for example, when inputs only need to be processed at intervals.

2.3.2.7 RC Oscillator (RC)

The MCU clock drives the program along, providing the timing signals for internal program execution. The RC (Resistor–Capacitor) clock is a low-cost method of controlling execution speed, requiring only two inexpensive external components. The product $R \times C$ sets the time constant for an internal driver circuit, hence the clock frequency. It has the advantage that the clock rate can be adjusted by using variable pot, as in the LEDx applications. However, it is not very stable or precise, maybe $\pm 3\%$, depending on the external component characteristics. The RC clock will operate up to about 50 kHz.

2.3.2.8 Crystal Oscillators (LP, XT, HS)

If greater precision and stability is required, a crystal oscillator is needed. This option is required if the program uses the hardware timers to make accurate measurements or generate precise output signals. It uses a quartz crystal connected across the clock pins, with a small capacitor (15pF) to ground on each pin to provide stability. The crystal acts as a self-contained resonant circuit, where the quartz or ceramic crystal vibrates at a precise frequency when placed in a suitable driver circuit.

The crystal oscillator typically has an accuracy of better than 50 parts per million (PPM), equivalent to $\pm 0.005\%$. For low-power operation in the 16F877A, an LP crystal will provide frequencies from 32 to 200kHz. The XT crystal operates up to 4MHz, with an HS type required for the maximum clock speed of 20MHz. Each of these has its own drive requirements, selected using FOSC1 and FOSC0 bits. Power consumption is proportional to clock speed, so a trade-off must be decided.

Four megahertz is a convenient value, giving an instruction cycle time of 1 μs, making timing calculations a little simpler. The 16F887 and many other PICs now incorporate an internal oscillator option which needs no external components and can be trimmed via a calibration register to a precise frequency but is typically only accurate to about $\pm 1\%$. The 'A' suffix at the end of the 16F877A chip number indicates that the maximum clock rate is increased from 10 MHz in the original 16F877 to 20 MHz.

2.3.3 Typical MCU Configurations

The default setting for the configuration bits is 3FFFh, which means that the code protection is off, ICD disabled, program write enabled, LVP enabled, brown-out reset enabled, PUT disabled, WDT enabled, and RC oscillator selected. A typical setting for basic development work will enable ICD, enable the power-up timer for reliable starting, disable all other options, and use the XT oscillator type. Configuration code 3731h will select these options.

2.4 PIC Instruction Set

Each microcontroller family has its own set of instructions, which carry out a similar set of operations but using different syntax. This reflects the variation in internal architecture of different types of MCU. More complex processors such as those in the Intel series have a more extensive instruction set, with more options within each instruction, but this tends to cause slower instruction execution. The PIC uses a minimal set of uniform instructions, which makes it a good choice for learning, as well as providing a speed advantage over other types in high performance systems.

The definitive instruction set is listed in the 16F877A data sheet as Table 15.2. A simplified version of this organised by functional groups is given in Table 2.4. It consists of 35 separate instructions, some with alternate result destinations. The default destination for the result of an operation is the file register, but the working register W is sometimes an option.

Table 2.4: PIC Instruction Set by Functional Groups.

PIC Mid-Range Instruction Set	
Operation	**Example**
Move	
Move data from F to W	MOVF GPR1,W
Move data from W to F	MOVWF GPR1
Move literal into W	MOVLW num1
Test the register data	MOVF GPR1,F
Register	
Clear W (reset all bits and value to 0)	CLRW
Clear F (reset all bits and value to 0)	CLRF GPR1
Decrement F (reduce by 1)	DECF GPR1
Increment F (increase by 1)	INCF GPR1
Swap the upper and lower four bits in F	SWAPF GPR1
Complement F value (invert all bits)	COMF GPR1
Rotate bits Left through carry flag	RLF GPR1
Rotate bits Right through carry flag	RRF GPR1
Clear (=0) the bit specified	BCF GPR1,but1
Set (=1) the bit specified	BSF GPR1,but1
Arithmetic	
Add W to F, with carry out	ADDWF GPR1
Add F to W, with carry out	ADDWF GPR1,W
Add L to W, with carry out	ADDLW num1
Subtract W from F, using borrow	SUBWF GPR1
Subtract W from F, placing result in W	SUBWF GPR1,W
Subtract W from L, placing result in W	SUBLW num1
Logic	
AND the bits of W and F, result in F	ANDWF GPR1
AND the bits of W and F, result in W	ANDWF GPR1,W
AND the bits of L and W, result in W	ANDLW num1
Inclusive OR the bits of W and F, result in F	IORWF GPR1
Inclusive OR the bits of W and F, result in W	IORWF GPR1,W
Inclusive OR the bits of L and W, result in W	IORLW num1
Exclusive OR the bits of W and F, result in F	XORWF GPR1
Exclusive OR the bits of W and F, result in W	XORWF GPR1,W
Exclusive OR the bits of L and W	XORLW num1
Test & Skip	
Test a bit in F and Skip next instruction if it is Clear (=0)	BTFSC GPR1,but1
Test a bit in F and Skip next instruction if it is Set (=1)	BTFSS GPR1,but1
Decrement F and Skip next instruction if F = 0	DECFSZ GPR1
Increment F and Skip next instruction if F = 0	INCFSZ GPR1
Jump	
Go to a labelled line in the program	GOTO start
Jump to the label at the start of a subroutine	CALL delay
Return at the end of a subroutine to the next instruction	RETURN
Return at the end of a subroutine with L in W	RETLW num1
Return from interrupt service routine	RETFIE
Control	
No Operation — delay for 1 cycle	NOP
Go into standby mode to save power	SLEEP
Clear watchdog timer to prevent automatic reset	CLRWDT

Note 1: For MOVE instructions data is copied to the destination but retained in the source register.

Note 2: General Purpose Register 1, labelled 'GPR1', represents all file registers (00—4F). Literal value 'num1' represents all 8-bit values 00 − FF. File register bits 0—7 are represented by the label 'but1'.

Note 3: The result of arithmetic and logic operations can generally be stored in W instead of the file register by adding 'W' to the instruction. The full syntax for register operations with the result remaining in the file register F is ADDWF GPR1, F, etc. F is the default destination and W is the alternative, so the instructions above are shortened to ADDWF GPR1, etc. This will generate a message from the assembler that the default destination will be used.

Note 4: Instructions TRIS and OPTION are no longer recommended but are supported by MPASM.

As recognised by the Microchip PIC MCU assembler MPASM.

F = Any file register (specified by address or label), example used is labelled GPR1.

L = Literal value (follows instruction), example used is labelled num1.

W = Working register, W (default label).

Labels: Register labels must be declared in include file or by register label equate (e.g. GPR1 EQU 0C).

Bit labels must be declared in include file or by bit label equate (e.g. bit1 EQU 3).

Address labels must be placed at the left margin of the source code file (e.g. start, delay).

The effect of each instruction is described briefly in this table and in detail in the MCU data sheet, Section 15.2. All PIC instruction sets use the same basic syntax, with the higher performance chips featuring additional instructions.

2.4.1 Instruction Types

The functional groups of instructions, and how they work, are described below. The use of most of these instructions will be illustrated in due course within the demonstration programs for each type of peripheral interface.

2.4.1.1 Move

The contents of a register are copied to another. Note that we cannot move a byte directly from one file register to another in the mid-range PIC MCU; it has to go via the working register. To put an arbitrary data byte (a literal) into a register, we must use MOVLW to put it into W initially. It can then be moved to another register, or processed, as required. This is one penalty for the simplified instruction set and a streamlined PIC internal architecture.

The syntax is also not symmetrical; to move a byte from W to a file register, MOVWF is used. To move it the other way, MOVF F,W is used, where F is any file register address. This means that MOVF F,F is also available. This may seem pointless but in fact can be used to test a register without changing it, because it will affect the status bits (see below).

2.4.1.2 Register

Register operations affect only a single register, and all except CLRW (clear W) operate on file registers. Clear sets all bits to zero (00h), decrement decreases the value by 1 and increment increases it by 1. Swap exchanges the upper and lower four bits (nibbles). Complement inverts all the bits, which in effect negates the number (see 2s complement arithmetic in Chapter 5). Rotate moves all bits left or right, including the carry flag in this process (see below for flags). Clear and set a bit to operate on a selected bit, where the register and bit need to be specified in the instruction.

2.4.1.3 Arithmetic

Arithmetic operations (add and subtract) are applied to pairs of 8-bit binary numbers. One must be placed in the working register, the other in a file register. For example, ADDWF GPR1 adds the contents of W to the specified file register (W remains unchanged). Addition and subtraction in binary give the same result as would be obtained by working in decimal or hex. If the result generates an extra binary bit (e.g. FF + 03 = 102h) or requires a borrow digit (e.g. 103−05 = FE), the carry flag is used. Arithmetic operations are described in more detail in Chapter 5.

2.4.1.4 Logic

Logic operations are carried out on the corresponding bit pairs within the two binary numbers being operated on, giving the result which would be obtained if they were fed to the relevant logic gate, for example 00001111 AND 01010101 = 00000101. The instruction ANDWF GPR1 will AND the contents of W with the specified register, leaving the result in GPR1. The bit logic operations are as follows:

```
0 AND 0 = 0    0 IOR 0 = 0    0 XOR 0 = 0
1 AND 0 = 0    0 IOR 1 = 1    0 XOR 1 = 1
1 AND 1 = 1    1 IOR 1 = 1    1 XOR 1 = 0

IOR = Inclusive OR
XOR = Exclusive OR
```

Taken with the logic inversion operation (bit complement), these options are sufficient to implement all possible logic and arithmetic operations. If necessary, reference should be made to an introductory text for further details of arithmetic and logical operations, such as 'PIC Microcontrollers' by the author.

2.4.1.5 Test, Skip and Jump

A mechanism is needed to make decisions (conditional program branches) which depend on some input condition or the result of a calculation. Programmed jumps are initiated using a bit test and conditional skip, followed by a GOTO or CALL. The bit test can be made on any file register bit. This could be a port bit, to check if an input has changed, or a status bit in a control register.

BTFSC (Bit Test and Skip if Clear) and BTFSS (Bit Test and Skip if Set) are used to test the bit and skip the next instruction, or not, according to the state of the bit tested. DECFSZ and INCFSZ implement the most common type of branch operation − decrement or increment a register and jump depending on the effect of the result on the zero flag (Z is set if the result in the destination register is zero). Decrement is probably used more often (see LED2 delay routine), but increment also works because when a register is incremented from the maximum value (FFh) it goes back to zero (00h).

The bit test and skip may be followed by a single instruction to be carried out conditionally, but GOTO and CALL allow a block of conditional code. Using the GOTO *label* simply transfers the program execution point to some other point in the program indicated by a label in the first column of the source code line, but the CALL *label* means that the program returns to the instruction following the CALL when RETURN is encountered at the end of the subroutine (see below).

RETLW (Return with Literal in W) is useful for making program data tables (see the keypad program in Chapter 4). RETFIE (return from interrupt) is explained below.

2.4.1.6 Control

NOP simply does nothing for one instruction cycle (four clock cycles). This may seem pointless but is in fact very useful for putting short delays in the program so that, for example, external hardware can be synchronised or a delay loop adjusted for an exact time interval. In the demo programs seen later, NOP is used to allow an ICD operation to be inserted when the program is downloaded, or to pad a timing loop so that it is exactly 1 ms.

SLEEP stops the program and forces the MCU to wait for a reset or other interrupt. It should be used at the end of any program that does not loop back continuously, to prevent the program execution continuing into unused locations. If the program is not stopped, it will run through program memory and restart from the first instruction when the program counter rolls over to 0000.

CLRWDT means clear the WDT (watchdog timer). If the program gets stuck in a loop or stops for any other reason, it will be restarted automatically by the WDT. To stop this happening when the program is operating normally, the WDT must be reset at regular intervals of less than, say, 10ms, within the program loop, using CLRWDT. The WDT will wake up the MCU from SLEEP mode and therefore can be used to implement intermittent power saving operation.

2.4.2 Obsolete Instructions

There are a few instructions that are no longer recommended but are retained for backward compatibility and which are preferred sometimes for simplicity.

2.4.2.1 TRIS

This was an instruction originally provided to make port initialisation simpler (see program LED1). It selects register bank 1 so that the TRIS data direction registers (TRISA, TRISB, etc.) can be loaded with a data direction code (0 = output). The manufacturer no longer recommends use of this instruction, although it is still supported by the current assembler versions and is useful when learning with very simple programs. The assembler directive BANKSEL should be used in most programs, because it gives access to all the registers in banks 1, 2 and 3. The other option is to change the bank select bits RP0 and RP1 explicitly in the STATUS register directly, using BSF and BCF.

2.4.2.2 OPTION

This provides direct access to the control register OPTION_REG which controls Timer0, the first hardware counter, and, like TRIS, it is also no longer recommended. BANKSEL may be used to select bank 1 that contains OPTION_REG, which can then be accessed directly.

2.5 Program Execution

The PIC instruction contains both the op-code and operand. When the program executes, the instructions are copied to the instruction register in sequence, and the upper bits, containing the op-code, are decoded and used to set up the operation within the MCU (see Figure 1.5). The program counter keeps track of program execution, clearing to zero on power-up or reset. With 8k of program memory, a counter ranging from 0000 to 1FFF (8191) is required (13 bits) to generate all the necessary addresses. The PCL (program counter low) register (SFR 02) contains the low byte, and this can be read or written like any other file register. The high byte is only indirectly accessible via PCLATH (Program Counter Latch High, SFR 0A).

2.5.1 Subroutines

Subroutines are used to create functional blocks of code and provide good program structure. This makes it easier for the program to be understood, allows a block of code to be reused and ultimately allows ready-made library routines to be created for future use.

A label is used at the start of the subroutine, which the assembler then replaces with the actual program memory address. When a subroutine is called (CALL addlab), this destination address is copied into the program counter, and the program continues from the new address. At the same time, the return address (the one following the subroutine CALL) is pushed onto the stack, a block of memory dedicated to this purpose. In the PIC16, there are eight stack address storage levels, which are used in turn.

The subroutine is terminated with a RETURN instruction that causes the program to go back to the original position and continue. This is achieved by pulling the address from the top of the stack and replacing it in the program counter. It should be clear that CALL and RETURN must always be used in sequence in order to avoid a runtime stack error. In the PIC, the stack is not directly accessible.

A delay subroutine is included in the program LED2 that demonstrates the syntax required. The stack mechanism and program memory arrangement is shown in Figure 2.1 in the data sheet, and a somewhat simplified version is shown here in Figure 2.3. We can see that the

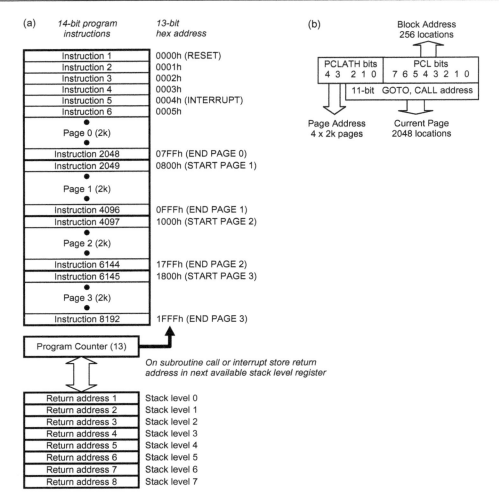

Figure 2.3
P16F877A program memory and stack: (a) program memory addressing and stack operation, and (b) block and page addressing by the program counter.

program memory space is divided into pages of 2048 locations by the fact that jumps only use 11 bits of the program counter.

2.5.2 Page Boundaries

The 16F877A and 16F887 chips have 8k of program memory in all, $8192 = 2^{13}$ locations, needing a 13-bit address. The 8-bit PCL register PCL can address a block of $2^8 = 256$ locations (Figure 2.3(b)). The address high bits are not directly accessible but can be written indirectly via the low five bits of PCLATH register (PCLATH < 4:0>). Jump instructions (CALL or

GOTO) provide an 11-bit destination addresses. This can only address $2^{11} = 2048$ locations, effectively dividing address space into 2k pages.

Page selection is thus controlled by bits 3 and 4 of PCLATH, so a jump across the program memory page boundary may require the page selection bits (PCLATH <4:3>) to be modified by the user program to select page 0, 1, 2 or 3 as necessary. Sections 2.3 and 2.4 in the 16F877A data sheet detail how to handle these problems.

The stack stores all 13 bits, so in normal subroutine and interrupt operations, there should be no problem with page boundaries, but if the PCL is modified directly, as in a table read, the relevant PCLATH bits may need to be modified explicitly if the table crosses a page or block boundary.

2.5.3 Interrupts

The stack is also used when an interrupt is processed. This is effectively a call and return that is initiated by an external hardware signal, which forces the processor to immediately jump to a specified instruction sequence, an Interrupt Service Routine (ISR). For example, the MCU can be set up so that when a hardware timer times out (finishes its count), the process required at that time is called via a timer interrupt.

When an interrupt input is received, the current instruction is completed and the address of the next instruction (the return address) is pushed into the first available stack location. The ISR is terminated with the instruction RETFIE, which causes the return address to be pulled from the stack. Program execution then restarts at the original location. However, we have to remember to take into account any changes in the registers which may have happened during the ISR. If necessary, the registers must be saved at the beginning of the ISR in a spare set of file registers and recovered afterwards (context saving). A simple example using a timer interrupt is seen later in a test program which generates a pulse output (Chapter 6).

2.6 Program Structure

The source code shown in LED2 is organised in blocks, to make it easier to understand. Good layout and readability are important as it is quite possible that another software engineer will want to repair, modify or otherwise update your program. Each block is described in a full line block comment, and each instruction explained in a line comment. Functional parts of the code are separated into the following blocks:

1. *Header comment*
2. *MCU configuration*
3. *Label equates*

4. *Port initialisation*
5. *Subroutines/macros*
6. *Main program*

The subroutines are placed before the main program so that the destination address labels are declared before being encountered as operands in subsequent blocks. However, the assembler does allow us to place the main routine first, followed by the subroutines, and this may be considered more logical.

The use of subroutines encourages structured programming, where distinct operations are created in separate blocks and then called as necessary. The subroutine may be called as many times as required but only needs to be written once. If convenient, these blocks can be converted to separate source code files (include files), or pre-compiled blocks (library files), as used in a higher level language such as 'C' (see below). However, remember that each subroutine call uses one stack level; the size of the stack therefore limits the number of levels of subroutine (eight in the 16F877A and 16F887), so the program cannot be too highly structured. Also, a subroutine call takes extra time because CALL and RETURN need two instructions cycles, the second for updating the program counter, so a program with fewer subroutines will be faster.

2.7 Assembler Directives

The main function of the assembler is to convert source code, written using the defined instruction set, into machine code for downloading to the MCU. Assembler directives are provided in addition to the instruction set to improve the speed, efficiency and flexibility of the programming process for the more experienced developer. A few will be described here and a new version of the LED output program (Program 2.2, LED3) used to demonstrate the syntax.

2.7.1 Common Directives

Some of the more commonly used directives are listed in Table 2.5. Only two are essential, CODE at the beginning of the actual program code, and END which terminates the assembler. CODE has only become essential in recent versions of the assembler/linker package. Previously, ORG was sufficient to locate the start of the program code in memory. CODE also differentiates the program code from data blocks elsewhere in memory that are defined using the DATA directive. CODE nnnn places the first instruction of the program code at location nnnn; CODE 0 is usually used in this book. EQU is also used throughout this book to declare constant labels, such as those representing port register addresses. SET is available for assembler variables that may be changed in the course of the program or used to control assembler options.

```
MPASM  5.46                  LED3.ASM   1-2-2013  17:21:19        PAGE   1

LOC  OBJECT CODE    LINE SOURCE TEXT
     VALUE

                    00001 ;;;;;;;;;;;;;;;;;;;;;;;;;;;;;;;;;;;;;;;;;;;;;;;;;;;;;;
                    00002 ;
                    00003 ;        Source File:    LED3.ASM
                    00004 ;        Author:         MPB
                    00005 ;        Date:           2-1-13
                    00006 ;        Dev.System:     MPLABX
                    00007 ;
                    00008 ;;;;;;;;;;;;;;;;;;;;;;;;;;;;;;;;;;;;;;;;;;;;;;;;;;;;;;
                    00009 ;
                    00010 ;        Slow output binary count is stopped, started
                    00011 ;        and reset with push buttons.
                    00012 ;        Modified with extra directives
                    00013 ;
                    00014 ;;;;;;;;;;;;;;;;;;;;;;;;;;;;;;;;;;;;;;;;;;;;;;;;;;;;;;
                    00015
                    00016 ;        Declare processor, supress messages and warnings,
                    00017 ;        do not print symbol table, configure (RC clock)
                    00018
                    00019        LIST p=16f877a, w=2, st=off, mm=off
2007    3733        00020        __CONFIG 0x3733
                    00021
                    00022 ;        Declare GPR label and literal constant
                    00023 ;        Define input labels
                    00024 ;        Include standard SFR label file
                    00025 ;        Include PortB initialisation file
                    00026
  00000020          00027 Timer EQU 20
  00000000          00028 DDCodeD SET b'00000000'
  0003              00029 CONSTANT Count=003
                    00030 #DEFINE RunBut PORTB,2
                    00031 #DEFINE ResBut PORTB,1
                    00032
                    00033 ; Program code ;;;;;;;;;;;;;;;;;;;;;;;;;;;;;;;;;;;;;;;;
                    00034
                    00035 CODE    0                   ; define code segment
                    00036
                    00037 #INCLUDE "P16F877A.INC"
                    00001        LIST
                    00002
                    00003 ;=======================================================
                    00004 ;  MPASM PIC16F877A processor include
                    00005 ;
                    00006 ;  (c) Copyright 1999-2012 Microchip Technology
                    00007 ;=======================================================
                    00008
                    00566        LIST
                    00038 #INCLUDE "DOUT.INI"
0000  1683 1303     00001        BANKSEL TRISD       ; Select bank 1
0002  3000          00002        MOVLW   DDCodeD     ; Port D Direction Code
0003  0088          00003        MOVWF   TRISD       ; Load DDR code into F88
0004  1283 1303     00004        BANKSEL PORTD       ; Select bank 0
                    00039
                    00040
```

Program 2.2
LED3 list file with assembler directives.

```
                             00041 ; 'delay' macro .......................................
                             00042
                             00043 delay   MACRO                  ; macro definition starts
                             00044         MOVWF   Timer          ; Copy W to timer register
                             00045 down    DECF    Timer          ; Decrement timer register
                             00046         BNZ     down           ; and repeat until zero
                             00047         ENDM                   ; macro definition ends
                             00048
                             00049 ; Main loop .............................................
                             00050
0006   0188                  00051 reset   CLRF    PORTD          ; Clear Port D Data
                             00052
0007   1C86                  00053 start   BTFSS   ResBut         ; Test reset button
0008   2???                  00054         GOTO    reset          ; and reset if pressed
0009   1906                  00055         BTFSC   RunBut         ; Test run button
000A   2???                  00056         GOTO    start          ; and repeat if n pressed
                             00057
000B   0A88                  00058         INCF    PORTD          ; Increment output Port B
000C   3003                  00059         MOVLW   Count          ; Delay count literal
                             00060         delay                  ; Insert macro 'delay'
000D   00A0              M           MOVWF   Timer          ; Copy W to timer register
000E   03A0              M down      DECF    Timer          ; Decrement timer register
000F   1D03  2???        M           BNZ     down           ; and repeat until zero
0011   2???                  00061         GOTO    start          ; Repeat main loop always
                             00062
3733                         00063         END                    ; Terminate source code
```

Program 2.2
(Continued)

Table 2.5: Selected Assembler Directives.

Directive Example	Meaning
CODE 0	Start of actual program source code
DATA	Start of data block separate from the program code
LIST p = 16f877a, w = 2, st = off	Listing options: e.g. select MCU, print errors only, no symbol table
ORG 05	Set the first program memory address for the code that follows
END	End of program source code
PORTA EQU 05	Declare a label (assembler constant)
Max SET 200	Declare a label value which may be changed later (assembler variable)
PROCESSOR 16F877A	Select the MCU type
CONSTANT Hours_in_day = 24	Declare a constant
__CONFIG 0 × 3731	Set processor configuration word
BANKSEL TRISC	Select file register bank containing the register specified
#INCLUDE "C:\PIC\P16F877A.INC"	Include additional source file from directory specified
#DEFINE Cflag 3,0	Substitute text 'Cflag' with '3,0', e.g. 'BSF Cflag'
pulse MACRO	Declare macro definition with address label
ENDM	End a macro definition
NOEXPAND	Do not print macro each time used
END	End of source code

LIST is a commonly used directive as it can be used to control several aspects of the list file output. In LED3, it is used to specify the MCU, to disable messages and warnings (w = 2), turn off the symbol table (st = off) and memory map (mm = off) in the listing. NOLIST turns off the list from that point on. # DEFINE allows one text string to be substituted for another, to simplify operands. # INCLUDE inserts the specified text file as source code. This can be used for standard MCU header files and user-generated source code. The hash prefix is used for compatibility with C directive syntax (see below).

2.7.2 Macros

The macro provides an alternative to the subroutine that eliminates its disadvantages. The block of code for the same routine is defined once, as in the subroutine, but then inserted as source code by the assembler whenever required, so CALL, RETURN and the stack operation are avoided. This increases the source code size and overall program memory requirement, but reduces the program execution time by eliminating the time taken to jump to the subroutine, and back, when CALL and RETURN are executed. Program 2.2 demonstrates the required syntax by using a macro for the delay routine. The macro definition block starts with the label and directive 'delay MACRO' and is inserted by simply using the label as an instruction. Thus, the macro effectively allows the creation of user-defined instructions.

2.7.3 LED3 List File

Note the following features of the LED3 list file (Program 2.2):

- *List file options allow unwanted elements of printout to be suppressed*
- *Configuration settings are detailed in Chapter 1*
- *Equate is used for file register addresses*
- *Data Direction code is SET and may be changed later*
- *Timer count value is defined as a constant*
- *Text substitution is used for input bits*
- *File path for standard P16F877A include file is specified in double quotes*
- *Microchip header file title line is too long in this format*
- *List file print for standard P16F877A include file is suppressed (398 lines)*
- *Port B initialisation include file path is not specified, uses default (<>)*
- *Delay macro does not need to be skipped in program execution*
- *Special instruction BNZ is used in delay macro*
- Text substitution is used for input testing
- *Macro expansion is indicated by 'M' instead of line number*
- *END is the only essential assembler directive*
- *Warnings and messages are suppressed*
- *Memory map and symbol list are suppressed*

In order to keep the sample programs provided later as easy to understand as possible, the use of these directives will be minimal. However, once the essentials of the assembly language have been mastered, the more powerful features of the assembler can be incorporated in your applications, based on a close study of the help files provided with MPLAB and other relevant sources.

2.8 Software Design

When the principles of assembly language programming are reasonably well understood, methods of software design need to be considered. This involves taking the program specification and working out how to construct the program. A design method is needed to outline the program structure and logic, which can then be applied before conversion into actual program code.

The language to be used to write the program will determine which are the most useful design methods. Traditionally, assembly language program design has been illustrated with flowcharts, which provide a pictorial representation that is useful when learning. However, they are cumbersome when the programs get more complicated. Structure charts may then be helpful, to show the program as a hierarchy of functional blocks.

The main alternative to assembly language is 'C', which has a slightly more user-friendly syntax than the assembler, but still provides direct access to the MCU operations. The C compiler converts source code functions and statements to the pre-defined machine code blocks. The component functions are pre-compiled in a library that may contain standard functions or specific ones written by the user. C programs are often outlined using pseudocode using structured text statements and is introduced briefly in the next section.

This book generally uses program outlines that provide a text-based representation of the program structure that is applicable to all programming languages. This is a flexible technique that allows statements to be translated directly into assembler routines but has no agreed standards.

2.8.1 Application Specification

The first step in program design is to write a specification. This is essential in commercial work, since there must be clear agreement between client and developer, or manager and software engineer, what the precise function will be. A basic specification for the application LED2 is suggested as follows (a commercial specification would be much more detailed). It also specifies the hardware that has already been described.

A portable, self-contained unit is required for educational purposes, at minimum cost, which displays a repeating 8-bit binary count on a set of LEDs. The count should start at zero and be displayed when a non-latching push button (RUN) is held down. The count should stop when RUN is released and reset to zero with another push button (RESET). The output sequence should be easily visible, with each full count cycle taking at least 10 s.

2.8.2 Flowcharts

There are two main forms of flowchart. Data flowcharts may be used to represent complex data processing systems, but here we will use a minimal set of symbols to represent an assembly language program. Program flowcharts may be used to represent overall program structure and sequence, but not the details. An example is shown in Figure 2.4 that represents the program LED2.

The name of the program or project is given in the *start* symbol at the top of the flowchart. This is followed by the initialisation sequence in a plain rectangular *process* symbol and an *input/output* operation (clear the LEDs) in the parallelogram-shaped symbol. A program *decision* (button pressed?) is enclosed in a diamond shape, with two possible outputs. The

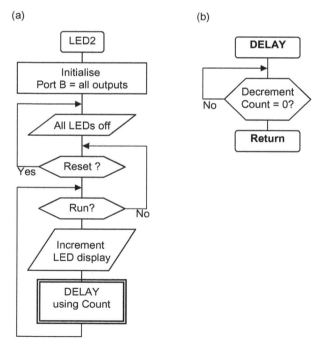

Figure 2.4
LED2 flowcharts: (a) main routine and (b) delay subroutine.

selection test is expressed as a question; the active decision is labeled yes or no, so it is unnecessary to label the default path.

The flow is implicitly down the page, so plain connecting lines may be used, with the branch forward or back using an arrow line style. The subroutine name is enclosed in a double line box and expanded into a separate flowchart below. A parameter (the delay time) is passed to the subroutine as the register variable 'count'.

These operations can be translated into PIC assembler as given in Table 2.6. Obviously, the precise implementation will depend on the exact sequence required, but generally:

- *A process is a sequence with no external branches*
- *An I/O operation uses the ports*
- *A branch will use bit test and skip*
- *A subroutine uses CALL and RETURN*

Flowcharts are useful for providing a graphical representation of the program, e.g. for a presentation, but they are time consuming to create. The flowcharts shown here were drawn just using the drawing tools in a word processor, so the creation of flowcharts to a reasonable standard is not difficult for the occasional user. Specialist drawing packages are also available, which make the process quicker and easier for the professional software engineer. In most of the programs in this book, 'End' is not needed in the flowchart, as the main sequence loops continuously until reset. If this is not the case, SLEEP should be used as the final instruction. This assumes the MCU will restart from the reset location 0000 after a hardware reset input or a watchdog timeout.

Table 2.6: Flowchart Implementation.

Operation	Symbol	Assembler
Start or end	LED2	CODE 0 SLEEP
Process sequence	Initialise Port B = all outputs	BANKSEL TRISB MOVLW B'00000000' MOVWF PORTB
Input or output	All LEDs off	CLRF PORTB
Branch selection	Reset ?	BTFSS PORTD,Inres GOTO reset
Subroutine or function	DELAY using Count	MOVLW OFF CALL delay

2.8.3 Pseudocode

Pseudocode shows the program as a text outline, using higher level language constructs to represent the basic processes of sequential processing, selection and repetition. LED2 is represented in Figure 2.5, although it is a very trivial example.

The program outline uses high level key words such as IF and DO...WHILE to control the sequence. It is not an ideal method for a very simple program like this but is useful for more complex programs. In particular, it translates directly into 'C', if the high level language is preferred. Note that in this case, the program outline does not make any assumptions about the hardware implementation.

2.8.4 Structure Charts

Structure charts are most useful in more complex programs, but the concept can be illustrated as shown in Figure 2.6. Each program component is included under standard

```
Project:        LED2   MPB    12-0213     Ver 2.0
Hardware:       LED2   MCU = P16F877A   RC clock = 40KHz
Description:    LED binary counter with stop and reset buttons
------------------------------------------------------------------

    Declare
        Registers        Input, Output, Count
        Bits             Reset, Run

    Initialise
        Inputs           Reset, Run
        Outputs          LEDS

    Main
        DO
                IF Reset pressed
                    Switch off LEDs
                DO
                    Increment LEDS
                    Load Count
                    DELAY using Count
                WHILE Run pressed
        ALWAYS

    Subroutine
        DELAY
                DO
                    Decrement Count
                WHILE Count not zero
        RETURN
```

Figure 2.5
LED2 program outline.

headings: inputs, processes and outputs, and can be broken down further in more complex programs, so that components can be created independently and then integrated.

2.9 'C' Programming

Assembly language is unique to each type of processor, while C provides a common programming method for all MCU types. C uses syntax that is closer to spoken English and easier to understand than assembly code. C is therefore the high level language of choice for the more complex microcontrollers. C programs are generally converted to assembly language by a complier and then assembled into machine code. A range of different development systems and compliers are available, but all use the same basic syntax defined as ANSI (American National Standards Institute) C. A simple example is outlined below so that it can be compared with the assembly language equivalent.

2.9.1 LEDC Program

The source code for the C program, LEDC, is listed as given in Program 2.3. It outputs a binary count at Port B, controlled by buttons at RD0 and RD1. The header file containing the standard register labels for the 16F877A is included in a similar way to the assembler equivalent.

The output port is declared as an 8-bit variable (PortB) and its address assigned (6). The main program block starts with the statement 'void main()' and is enclosed in braces (curly brackets). The output port is then initialised using a library function provided with the compiler 'set_tris_b(0)', where 0 is the data direction code in decimal form.

An initial value of zero is output to switch off the LEDs. The control loop starts with the loop condition statement 'while(1)', which means repeat the statements between the braces endlessly. The buttons are tested using 'if (condition)' statements, and the actions following

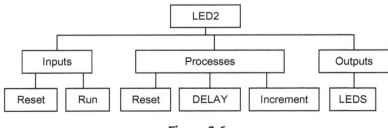

Figure 2.6
Structure chart.

```
//    LEDC.C      ***********************    MPB    19-11-05

#include <16F877.h>          // Include standard MCU labels
#byte PortB=6                // Output port data type and address


void main()                  // Start of program
{
   set_tris_b(0);            // Initialise output port
   PortB=0;                  // Initial output value

   while(1)                  // Endless loop between braces
   {
      if (!input(PIN_D0))    // Reset button pressed?
            PortB=0;         // if so, switch off LEDs

      if (!input(PIN_D1))    // Run button pressed?
            PortB++;         // if so, increment binary display
   }
}                            // End of program
```

Program 2.3
LEDC source code.

carried out if the condition is true. The condition is that the input is low (! = not), and pin labels as defined in the header file are used.

2.9.2 LEDC Assembler Code

The C source code is compiled into assembler code and then into machine code. The list file in Program 2.4 shows the assembler code alongside the source text. It can be seen that some statements are converted into a single instruction, for example:

```
PortB++; >>> INCF 06,F
```

Others need several instructions:

```
if (!input(PIN_D0)) >>>    BSF 03.5
                           BSF 08.0
                           BCF 03.5
                           BTFSS 08.0
```

Thus the assembled program is longer than the source code, because each C statement is converted into several assembler instructions. As a result, the program written in C will normally occupy more memory than the equivalent assembler program with the same function, so microcontrollers with larger memory are needed. Therefore, the more powerful

```
.................... void main()
....................  {
0004:  CLRF    04
0005:  MOVLW   1F
0006:  ANDWF   03,F
0007:  BSF     03.5
0008:  BSF     1F.0
0009:  BSF     1F.1
000A:  BSF     1F.2
000B:  BCF     1F.3
....................       set_tris_b(0);
000C:  MOVLW   00
000D:  MOVWF   06
....................       PortB=0;
000E:  BCF     03.5
000F:  CLRF    06
....................
....................       while(1)
....................       {
....................           if (!input(PIN_D0))
0010:  BSF     03.5
0011:  BSF     08.0
0012:  BCF     03.5
0013:  BTFSS   08.0
....................                 PortB=0;
0014:  CLRF    06
....................           if (!input(PIN_D1))
0015:  BSF     03.5
0016:  BSF     08.1
0017:  BCF     03.5
0018:  BTFSS   08.1
....................                 PortB++;
0019:  INCF    06,F
....................           }
001A:  GOTO    010
....................       }
....................
001B:  SLEEP
```

Program 2.4
LEDC list file.

series of PIC chips are usually used for C applications. These also have additional instructions, such as multiply, which makes the conversion more compact.

We are not going to look at the C language in any further detail here, but the advantages of C programming for microcontrollers should be clear. When assembly language has been mastered, the developer can decide if C would be a better choice for given applications. For applications needing complex mathematical calculations and data handling, C is a much better choice. Some C language products, such as CCS C, also provide functions which make programming the more complex serial interfaces covered later in this book much easier. For programs comprising simple bit I/O operations and fewer calculations, assembler is generally faster and more compact. C and assembler can be mixed in the same program, to gain the advantages of both.

C programming in CCS C is covered in more detail in 'Programming 8-Bit PIC Microcontrollers in C' by the author. Details may be found at www.picmicros.org.uk.

2.9.3 Real-Time Operating Systems

The next step up the programming ladder is to use a real-time operating system (RTOS). This consists of a set of utilities which provide the common operations required in more complex MCU-based control systems, concentrating on optimising timing-critical routines and multitasking. Used in conjunction with applications written in C, this is a more efficient way of creating real-time applications utilising complex interfaces. Specialist texts should be consulted to explore this further; see for example application note AN777 'Multi-Tasking on the PIC16F877 with the Salvo RTOS' from www.microchip.com.

Questions 2

1. Describe the effect in the PIC MCU of the assembler instructions
   (a) 'MOVLW 0FF' and (b) 'CALL delay'. (4)
2. Describe briefly the usage of assembler directives
   (a) CODE, (b) EQU and (c) #INCLUDE. (6)
3. Identify two instructions, one of which must be placed last in the PIC source code. (3)
   What happens if one of these is not used?
4. Identify two types of label used in assembly language programming. (2)
5. State the function of configuration bits PWRTEN, WDTEN and FOSCx. (3)
6. Explain how conditional program jumps are implemented in the PIC MCU. (5)
7. Explain briefly the difference between a subroutine and macro, and the main (4)
   advantage of each.
8. Briefly compare the operation of a subroutine and an interrupt, explaining the role (5)
   of the stack, return address, interrupt flag and the special significance of address
   004 in the P16XXX.
9. Identify the five main symbols which are used in a flowchart. (5)
10. Explain briefly the advantages of C programming. (3)

Total (40)

Assignments 2

2.1 LED2 Simulation
Use the MPLABX debugging tools to single step the program LED2 and observe the changes in the MCU registers. Operate the simulated inputs to enable the output count to Port B. Set a break point at the output instruction and run one loop at a time, checking that Port B is incremented. Use the stopwatch to measure the loop time. Comment out the delay routine call in the source

code, reassemble and check that the delay does not execute, and note the effect on the loop time. Reinstate the delay, change the delay count to 03 and note the effect on the loop time.

2.2 LED2 Modification

Study program LED2 in MPLABX. Modify the source code to light only the least significant LED and then rotate it through each bit so that the output port appears to scan at a visible rate. Add code to detect the high bit in the carry flag and reverse the direction of travel at each end so the scanning is continuous from end to end.

CHAPTER 3

PIC Design

Summary

- Microcontroller circuits can be simulated using mathematical models
- ISIS provides schematic capture, circuit simulation and source code debugging
- VSM provides virtual instruments for interactive testing
- ARES provides PCB layout and outputs production files
- MPLABX and ICD hardware provide firmware downloading and in-circuit testing

In the past, the electronics engineer needed to have a fairly comprehensive knowledge of both electronic component operation and circuit analysis before setting out to design new applications. The circuit would be designed on paper and a prototype built to test the design, using a hardware prototyping technique such as stripboard; further refinement of the design would often then be required. When the circuit was fully functional, a production version could be developed, with the printed circuit board (PCB) being laid out by hand. Further testing would then be needed on the production prototype to make sure that the layout was correct and that the variation in component values due to tolerances would not prevent the circuit from functioning correctly. Learning how electronics systems worked also required a good imagination! Unlike mechanical systems, it is not obvious how a circuit works from simple observation. Instruments (voltmeters, oscilloscopes, etc.) must be used to see what is happening, and these also need complex skills to use effectively.

We now have computer-based tools that make the job much easier and perhaps more enjoyable. An early ECAD (Electronic Computer-Aided Design) tool was a system of mathematical modeling used to predict circuit behavior. SPICE was developed at University of Berkley, California, to provide a consistent and commonly understood set of models for components, circuits and signals. This system uses nodal analysis to predict the signal flow between each point in an electronic network, based on the connections between the components. The results would originally be displayed or printed numerically.

The simplest component is the resistor, and the simplest mathematical model Ohm's law, $V = IR$, which relates the current and voltage in the resistor. For two resistors in series this becomes $V = I(R1 + R2)$. The power dissipated in the resistor is given by $P = IV$. For a.c. signals, RMS voltages are used so that the same model can be applied. Reactive components

Interfacing PIC Microcontrollers.
DOI: http://dx.doi.org/10.1016/B978-0-08-099363-8.00003-0

need the frequency of the signal ($\omega = 2\pi f$) to be included, so $V = IX$ is used, where X is the reactance. For a capacitor, the magnitude of the reactance is $1/\omega C$, for an inductor ωL, where C is capacitance and L is inductance. Based on these component models, the phase relationship between voltage and current in analogue circuits can be represented using complex algebra.

Digital circuits are in principle easier, since they are modelled using simple logical relationships, such as $A = B \cdot C$, where the dot represents the 'and' operation. The other main operators are '+' representing logical 'or', and '!' representing logical invert. Thus, a simple logic function may appear as $A = (B \cdot C + !D)$. The microcontroller operation is modeled using this type of logical function representing the logic gates that form most of the internal architecture.

The next step is to model mixed mode circuits, with analogue and digital components connected together. Finally, the microcontroller program must be included as input to the MCU digital model. Computer graphics have now developed to the point that the modeling can be done in conjunction with an on-screen schematic and a simulation generated interactively. Components placed in the drawing have their models attached, and the nodes are identified from the connections on the schematic made by the circuit designer. Inputs can be supplied from simulated signal sources, and virtual instruments and on-screen graphics used to display the virtual outputs obtained.

Interactive circuit simulation now makes the job of analysing and designing electronic circuits quicker, easier and therefore cheaper. The circuit can be drawn and tested on screen, and a PCB layout also generated from the schematic. The layout can then be produced as a prototype and passed to a specialist production system when finalised. Once in production, assembly and testing can also be automated.

Proteus VSM (Virtual System Modelling), from Labcenter Electronics, has been used to create the circuit diagrams and test the designs in this book. It provides a comprehensive range of microcontroller models, combined with interactive simulation by animated schematic. The schematic capture and simulator component is named ISIS, and a PCB layout can be created from the circuit design using the associated application ARES.

VSM is probably the most complete package available at the current time for designing and testing embedded applications, providing a full range of passive and active components, mixed mode simulation and interactive peripheral hardware. Details of Proteus VSM can be found at www.labcenter.com. Version 8.0 (released January 2013) was used in this book, which provides closer integration between the elements of the package and additional project management features compared with earlier versions.

A low-cost starter pack with a licence for the PIC 16F877A model (plus two others) is currently available which will allow the user to create and test his or her own designs. The process for creating the design LED2 is described below, using Proteus 8.

3.1 Application Design and Test

A circuit to demonstrate the operation of LED2 application is shown in Figure 3.1. The schematic was drawn using ISIS and exported as a bit map for insertion into a document. As can be seen, ISIS allows circuit diagrams to be readily presented to a professional standard.

The microcontroller is a PIC 16F877A, our reference device. A set of LEDs is connected to port D, with push buttons on RB1 and RB2. A CR clock circuit is shown connected to CLKIN, with a pre-set pot providing variable resistance which allows the clock frequency to be adjusted. The clock frequency in hardware is inversely proportional to the CR product. Note that for simulation purposes the external clock circuit does NOT control the operating frequency of the PIC; it is set in the properties dialogue of the MCU component (see below).

Similarly, the MCLR (Master Clear) input does not have to be connected high for the program to run in simulation mode, whereas this is essential in the real circuit. In the schematic it is connected to the programmer connector, since MCLR is used by the host

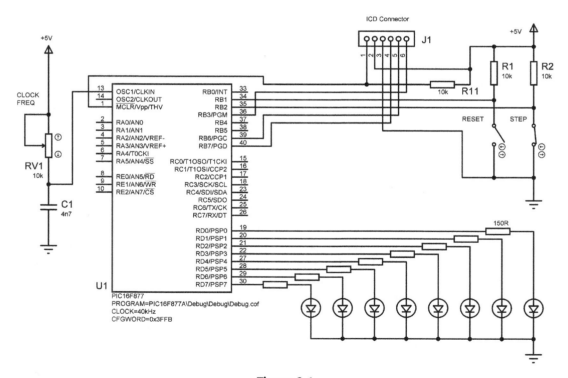

Figure 3.1
LED2 schematic.

to stop and start the MCU in hardware testing. If the MCU is replaced by the more recent 16F887, the I/O pinout and programming connections are the same, but an internal oscillator is available which eliminates the external clock components in the design.

The inputs are pulled up to 5V via 10k resistors and thus held at a logic high value when the button is off. The port does not need to be initialised for input, as this is also the default condition. On the other hand, the outputs do need to be initialised by the MCU program, by initialising Port D pins for digital I/O by loading the data direction register with zeros. The PIC outputs can typically provide up to 25mA, which is enough to light the LEDs without any additional current drivers; 150R resistors limit the current in the LEDs to about 20mA.

The ICD pins are connected to a programming unit for program downloading and disconnected when the application is running independently. The MCLR input is connected to +5V via a pull-up resistor to enable the chip in normal operation.

3.1.1 Schematic Capture

A new application can be created in VSM from the start screen using the new project wizard, or an existing project can be opened, such as the downloadable demo applications. If created from scratch, the project will be named LED2 and saved in a folder of the same name. In the following wizard dialogues the default schematic template may be selected, a firmware project created for the required controller (16F877A) and the assembler (MPASM) nominated as the project compiler.

A folder with the name of the selected MCU will also be created to store the firmware files. A default source code file is then displayed under the source code tab, which can be edited and renamed (right click) as required. To display the source code in the edit window, double-click on the filename in the project window. The schematic edit window is created under a separate tab.

It is generally most logical to create the schematic first and write the source code (as discussed in Chapter 2) afterwards. The ISIS schematic capture screen is shown in Figure 3.2. The main schematic edit window is accompanied by an overview window showing the whole drawing and an object select window which contains a list of selected components when in component mode. It also shows lists of other available devices for use in the edit window when other modes are selected.

The main editing window includes a sheet outline, which shows the edge of the drawing area, within which components must be placed. The component mode is normally enabled by default; components are then selected for placing on the schematic by hitting the P (pick devices) button and selecting the required category of components. The individual device type can then be chosen from a list. Figure 3.3 shows the selection dialogue for the MCU.

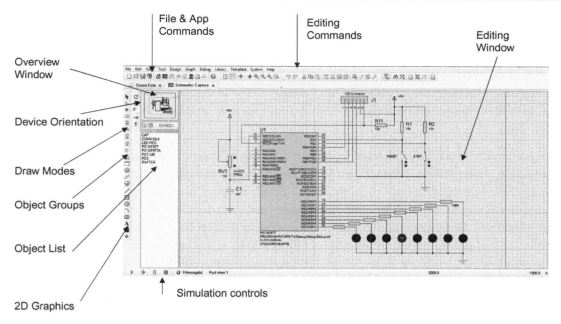

Figure 3.2
ISIS (v8) schematic capture.

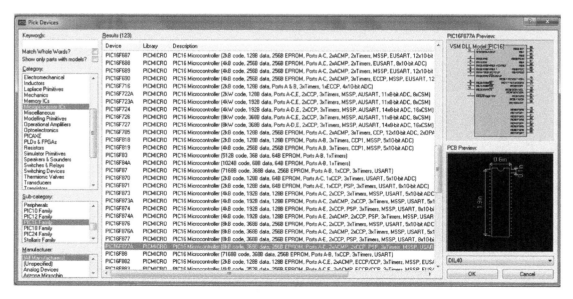

Figure 3.3
Picking a device in ISIS.

The components are categorised as microprocessors (includes microcontrollers), resistors, capacitors and so on, with sub-categories to narrow down the options. For now, generic active (and animated) components should be selected where possible (resistor, capacitor, switch, etc.). The selected components appear in the device list and, when highlighted, can be placed on the schematic with two left mouse clicks. In designs where a PCB will be laid out, suitable pinouts can be assigned to these components matching the physical component to be used.

The component pins are connected as required by clicking on a component pin and dragging a connection to another pin. Right click highlights a connection or component, and further right click deletes it. Right click, left click opens a connection or component properties dialogue. This allows the component value and labeling to be modified. The PIC chip property edit window allows the program file to be attached, the simulation clock frequency and configuration word entered (Figure 3.4). If we switch briefly to the source code window and build the default (not yet functional) source code, a 'debug.cof' file, which will contain the executable code, will be created and attached automatically. To create a new program, the default template may later be modified or replaced as preferred.

For the clock RC components shown in the schematic for LED2, the time constant is about $5\text{k} \times 4n7 \cong 25\mu\text{s}$, giving a frequency of 40kHz. This gives an instruction frequency of 10kHz and an instruction cycle time of 100μs; 40kHz should be entered into the MCU

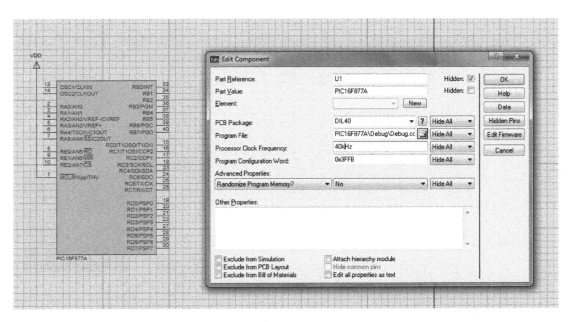

Figure 3.4
MCU properties dialogue.

properties as the clock frequency. The configuration word can be left at the default value 0x3FFB (RC clock, watchdog disabled) for this application. Note that most component properties can be selectively displayed or hidden.

To complete the LED2 schematic, power terminals must be added. Select the terminal button in the objects toolbar, and a list of terminals types is displayed in the device list. Power and ground terminals may then be added to the drawing. The power terminal voltage can be defined via its properties dialogue, or left as V_{DD} to recognise variation in a battery powered system; entering $+5V$ as the label actually defines the operating voltage. Note that the MCU does not have explicit power connections — but these must be generated if a PCB layout is created.

The project folder containing this complete design (LED2) can be downloaded from www.picmicros.org.uk by selecting the link within the section relating to this book. The schematic is stored as led2.pdsprj, and the firmware in a folder named after the selected MCU, 16F877A. The source code file led2.asm is stored in here, with a debug folder containing the executable debug.cof file, as well as the error message file, list file and object code.

3.1.2 Circuit Simulation

In the MPLAB development system, the application is tested with simulated inputs and numerical outputs, with the state of the output port displayed as a hex or binary number. Inputs are generated as asynchronous events by assigning on-screen buttons, or using a stimulus file to generate the same input sequence each time the simulation is run. It is a purely software simulator but with some advantages for the experienced developer in terms of more sophisticated project management tools. ISIS provides a more user-friendly development environment, particularly for the inexperienced designer, by providing interactive, on-screen, inputs and outputs, so that the circuit can be seen operating as it should in the real hardware.

The application firmware is edited in the source code window under the source code tab. At this stage we will assume the ready-made LED2 project has been opened from the initial VSM screen, so that the code does not have to be re-entered. In this case, the downloaded schematic appears under the Schematic Capture tab and the firmware under the Source Code tab. These can be arrange side by side by dragging and dropping one on a tab onto the desktop where it will be displayed in a separate window. The animated schematic and source code can then be observed simultaneously (Figure 3.5).

The simulation is run by clicking on the run button at the bottom of the screen. If the source code is changed, it is automatically saved and reassembled at this command, which provides one-click retesting. This saves a lot of time during the development process, especially for the inexperienced programmer.

Step controls

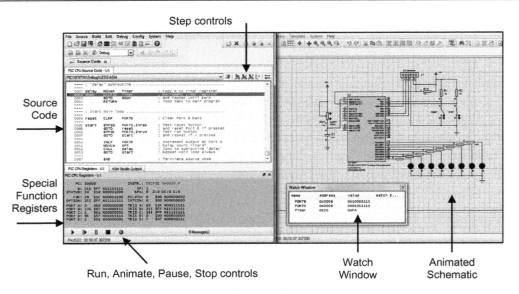

Source
Code

Special
Function
Registers

Run, Animate, Pause, Stop controls

Watch
Window

Animated
Schematic

Figure 3.5
Proteus VSM source code debugging.

The circuit will operate in real time (at full speed) if the simulation is not too complex. In this circuit, the output LEDs should show a visible binary count. As the delay between each increment is about 75ms, the whole count will take about 20s. The count is started by 'closing' the run button with the mouse. It should stop when released and start again at the same count. The reset button should clear the count to zero. While the simulation is running, the logic state of each line is indicted in red (1) or blue (0).

3.2 Software Debugging

The purpose of testing by simulation is to fault-find the software before downloading it to the real hardware. In ISIS, the hardware design can be tested (to some extent) at the same time. Changes to the hardware design simply require editing the component properties (e.g. to change a resistance), reconnecting or changing components.

Syntax errors (e.g. mis-spelling an instruction) and semantic errors (e.g. missing a label out) in the program should be identified at the initial program assembly stage. Simulation allows logical errors to be detected, i.e. incorrect operation of the program when executed. This assumes that the required operation is clearly defined in the program specification. Source code debugging means the source code execution sequence can be examined and readily changed to eliminate errors. Additional windows may be opened to monitor the MCU SFRs and GPRs, and program variables may also be tracked in a watch window.

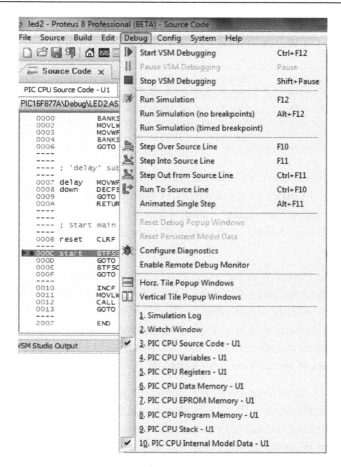

Figure 3.6
Source code debug menu.

3.2.1 Source Debug Window

In VSM8, the source code tab displays the program text for editing. When the program is run, then paused, the source code can be displayed in an execution window, with the current execution line highlighted. It is called up by selecting, when paused, PIC CPU Source Code in the Debug menu (Figure 3.6). Single stepping allows line by line scrutiny of the program, and break points to be set to run the program in stages. Program memory locations and hex code can also be displayed, if selected.

The buttons at the top of the source code window allow the program to be stepped or run between break points. The CPU (special function) registers can be displayed, and the CPU data memory window shows all the file registers, so that general purpose register contents can be monitored. The watch window displays only selected registers and remains visible in run mode. The debug control button (see Figure 3.5) functions are as follows:

Run. . .	*At full speed (source code window closes)*
Step Over. . .	*Step through instructions only in the current routine*
Step Into. . .	*Step through all instructions, including subroutines*
Step Out of. . .	*Run at full speed out of current subroutine*
Run to. . .	*Current cursor position*

If the overall operation is incorrect, these controls allow the program sequence to be inspected step by step, in order to see where it is going wrong. Subroutines may be executed in sequence using Step Into. . ., and, when correct, can be executed at full speed while stepping the calling routine. In this way the program can be debugged from the bottom up.

Break points allow the program to be run and stopped at a selected point. For example, if a break point is set at the beginning of a loop, it can be executed once to check the effect. A break point is set by simply left double-clicking in the left margin of the source code debug window and is displayed as a red dot. Additional options are available with a right click on the source debug window (e.g. clear all break points). Timing can be checked on the timer display adjacent to the simulator controls.

To make corrections to the source code, the debugger must be stopped and the program corrected in the edit window. It will be rebuilt automatically when re-run. Alternatively, the project can be rebuilt using the compiler control button 'Rebuild Project'. Note that sections of the source code can be commented out (; at the start of the line) to assist in debugging by disabling selected program sections. For example, the delay in LED2 is commented out to display the output on virtual instruments (see below).

3.2.2 Other Debug Windows

The CPU register window displays the special function registers, including the port data and direction registers, plus the working register and status flags. It also shows the stack pointer, which is not normally accessible in the real chip. This shows which of the eight return address locations is next available, i.e. how many levels of subroutine have been used up. The CPU data memory window shows all the file registers, so is a quick way to check on a general purpose register. For example, in LED2, the Timer register can be seen. When a register changes, it is highlighted, which helps to keep track of them.

A watch window (Figure 3.5) allows user selected registers to be monitored, in a variety of data formats. By right clicking on the window, the SFRs can be picked from a list by name, or GPRs added by address (number) and named. This allows only those registers which are of particular interest to be viewed. When undocked (click on the undock symbol top right of the debug window), the debug windows remain visible when the simulator is in run mode, which allows the source code and registers to be monitored while viewing the animated schematic. Windows can be re-docked at the right of the debug screen for a more convenient display.

3.3 System Testing

We have seen how to create the circuit in schematic form and to test the program. The simulation package also provides circuit monitoring devices and virtual instruments which can be used to measure circuit performance, as in the real hardware. These features are available via the mode buttons on the left of the schematic screen.

3.3.1 Probes and Meters

The Probe Mode button allows probes to be attached to the circuit connections to display the direct current and voltage. As an example, the current and voltage around one of the LEDs is shown in Figure 3.7(a). It may be necessary to open the System, Animation Options dialogue to display the voltage and current values. Probe values may also be recorded over time.

The Instruments button in the Gadgets toolbar provides a list of available instruments in the device window. It includes an oscilloscope, logic analyser, signal generator, voltmeters and ammeters. Virtual voltmeters and ammeters may be added to the circuit as an alternative to probes (Figure 3.7(b)). The voltmeter will measure voltage drop across a component, rather than with respect to ground. The range, internal resistance and other properties of a meter

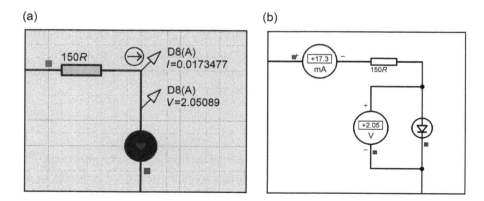

Figure 3.7
Circuit measurement: (a) voltage and current probes, and (b) virtual voltmeter and ammeter.

can be changed by right clicking on the instrument so that they can represent real circuit instruments.

3.3.2 Counter Timer

Figure 3.8 shows the LED2 application schematic with a scope, logic analyser and counter/ timer attached. A modified version of the program with the delay commented out to speed it up was used to give a more convenient output display.

The clock output (CLKOUT) from the PIC chip is displayed on the counter/timer instrument, reading 10kHz. This is the instruction clock, which is one-fourth of the clock oscillator frequency of 40kHz set in the MCU properties. The Timer Counter properties must be edited by right clicking on the instrument and selecting frequency measurement mode. Timing modes are available to measure pulse period as well as pulse counting.

CLKOUT in simulation has to be enabled in the MCU Properties, Advanced Properties dialogue. It might be useful to note at this stage the other features of the PIC which can be selectively enabled, e.g. the start-up timer delays. They are disabled by default so the program starts immediately when debugging, but in the real hardware the MCU should wait for the supply to settle.

3.3.3 Oscilloscope

A virtual oscilloscope allows analogue signals to be displayed in a similar way to a real oscilloscope. It can be selected from the Instruments list and dropped onto schematic as a minimised version. Four channels are available for connection to different points in the circuit, with 0V being implicit (internally connected).

A display version of the virtual scope appears when the simulation is run, and the controls are adjusted for a suitable signal display in the same way as the real instrument. The scope is primarily used to check overall signal characteristics and measure signal amplitude and period by setting the timebase and gain of each channel accordingly, with suitable triggering. Figure 3.9 shows the appearance of the virtual scope, with the four low bits of the output from LED2 application displayed.

3.3.4 Logic Analyser

The logic analyser allows multiple digital signals to be displayed simultaneously. They are captured by sampling a set of lines at regular intervals and storing the samples as binary bits. Unlike the oscilloscope, when the analyser is triggered, the data from *before* the trigger event, as well as after, can be displayed. Low-cost logic analysers now typically use a laptop as a front end display, while higher performance units typically have a dedicated

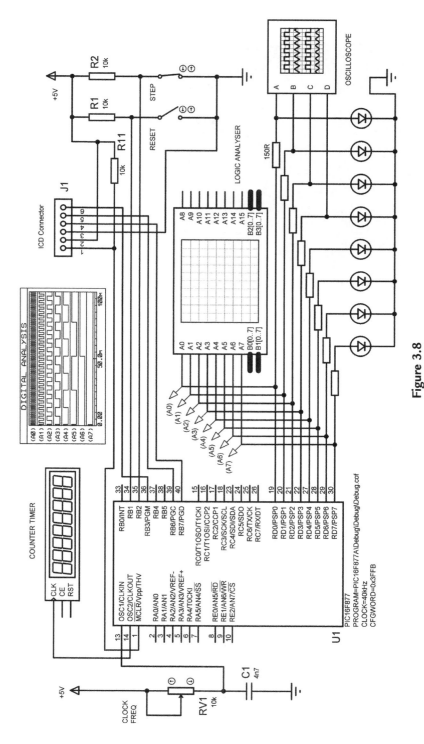

Figure 3.8

Test instruments on the schematic.

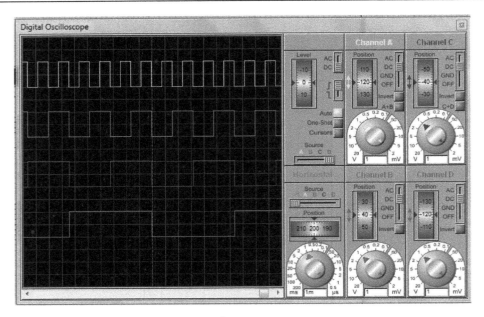

Figure 3.9
VSM oscilloscope display.

display and advanced data capture features operating at high speed for studying digital system signals in detail.

The logic analyser is particularly useful in testing conventional microprocessor systems, capturing the data signals flowing between the CPU, memory and I/O devices on the address and data busses. The actual signals can then be compared with those ideally expected to be generated by the application software. Precise signal timing and transient pulses are typical issues that may be causing system malfunctions in the physical hardware. A high signal sampling rate is important to capture all relevant signals and make accurate measurements. The data can be displayed as a timing diagram, as in a multichannel oscilloscope, or in numerical form. Complex triggering conditions may also be needed.

In comparison with a real instrument, the VSM virtual logic analyser is relatively simple to use. It has 16 discrete inputs (A0−A15) and 4 × 8 bit bus connections (B0−B3). In Figure 3.10, it is capturing the 8-bit output at Port B. The capture input must be operated while the program is running or paused. The data capture can be triggered on any channel by a high, low or edge condition, but note that there may be a delay before the data is displayed. Time intervals between events can be measured using multiple cursors which can be dropped onto the screen area. The capture and display timebases are adjusted separately for best effect.

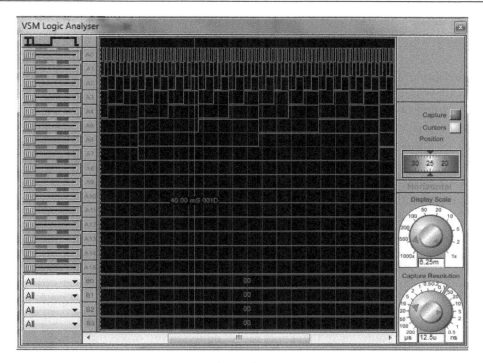

Figure 3.10
VSM logic analyser display.

3.3.5 Graphs

Another very useful feature of the Proteus simulation is the graph display, which allows a full screen view and a permanent record of the system performance in graphical form. Many of the graph mode options provide additional analogue signal analysis, including a more precise time-based display and frequency domain plots. In Figure 3.8, voltage probes have been attached to the same lines as the logic analyser so that the Port D outputs can be viewed at a larger scale and printed out if required. A miniaturised version of the graph is displayed on the schematic.

Graph mode is selected from the toolbar and the digital option highlighted in the list. A graph position holder can now be drawn in a convenient position on the schematic by dragging the pointer. Voltage probes are attached to the required digital lines, highlighted and dragged onto the graph area, where they are assigned to the next available graph line on the chart (they can be deleted by right clicking twice).

Now run the simulation, stop and hit the spacebar – the signals should appear in the graph window. If necessary, right, then left, click on the graph area to change the timescale settings in the graph properties dialogue. In this example, running LED2 without the delay, a timescale of 100ms is suitable. A single cursor is available to make timing measurements.

3.4 Hardware Design

When the correct operation of the design has been obtained in simulation mode, it needs to be converted to prototype hardware. A target board may be built on breadboard or stripboard, or a finished design produced using a PCB layout package such as ARES, the PCB layout part of the Proteus package. Hardware prototyping and implementation methods are discussed more fully in 'PIC Microcontrollers' by the author, but the process for creating a manufactured PCB is outlined below. It assumes ARES has been installed and is available in the main VSM toolbar.

3.4.1 Netlist

The components and the connections between them in the schematic are recorded in a netlist (Figure 3.11), which forms a complete description of the circuit, assuming that there is a SPICE model associated with each active component. This file can be studied by invoking the Netlist Compiler in the Tools menu, and viewing or saving it as a text file, LED2.SDF. It has two main parts, the component definitions and node list. The component characteristics are largely self-explanatory, but it is particularly useful to be able to see all the MCU options that are modelled.

The circuit nodes in the netlist can be identified by comparison with the circuit schematic. For example, the clock input CLKIN is defined as node 21:

```
#00021,3    Node number 21, 3 connections to:
C1,PS,2     Cap C1, passive terminal, pin 2
RV1,PS,2    Pot RV1, passive terminal, pin 2
U1,IP,13    Chip U1, input terminal, pin 13
```

The netlist is fed to the simulation engine along with any virtual signal inputs for network analysis, producing predicted outputs. Mixed mode analysis allows the logical operation of the digital components to be combined with the linear characteristics of the analogue components. The simulation, however, can never be 100% accurate, since it is calculated in discrete time steps with limited resolution and cannot incorporate the effects of the final layout on a PCB. For example, the capacitance between tracks depends on their length and separation, which are not yet defined. This will add a small capacitance at each node. Similarly, each track has a small inductance, but this is not usually so significant.

```
ISIS SCHEMATIC DESCRIPTION FORMAT 8.0
==========================================
Design:   led2.pdsprj
Doc. no.: <NONE>
Revision: <NONE>
Author:   <NONE>
Created:  13/12/2012
Modified: 13/12/2012

*PROPERTIES,0
*MODELDEFS,0
*PARTLIST,23

C1,CAP,4n7,EID=13,PACKAGE=CAP10,PINSWAP="1,2"

D1,LED-RED,LED-RED,BV=4V,EID=3,IMAX=10mA,PACKAGE=DIODE25,ROFF=100k,RS=3,STATE=7,VF=2V
D2,LED-RED,LED-RED,BV=4V,EID=4,IMAX=10mA,PACKAGE=DIODE25,ROFF=100k,RS=3,STATE=0,VF=2V
D3,LED-RED,LED-RED,BV=4V,EID=5,IMAX=10mA,PACKAGE=DIODE25,ROFF=100k,RS=3,STATE=7,VF=2V
D4,LED-RED,LED-RED,BV=4V,EID=6,IMAX=10mA,PACKAGE=DIODE25,ROFF=100k,RS=3,STATE=0,VF=2V
D5,LED-RED,LED-RED,BV=4V,EID=7,IMAX=10mA,PACKAGE=DIODE25,ROFF=100k,RS=3,STATE=7,VF=2V
D6,LED-RED,LED-RED,BV=4V,EID=8,IMAX=10mA,PACKAGE=DIODE25,ROFF=100k,RS=3,STATE=7,VF=2V
D7,LED-RED,LED-RED,BV=4V,EID=9,IMAX=10mA,PACKAGE=DIODE25,ROFF=100k,RS=3,STATE=7,VF=2V
D8,LED-RED,LED-RED,BV=4V,EID=A,IMAX=10mA,PACKAGE=DIODE25,ROFF=100k,RS=3,STATE=7,VF=2V

J1,CONN-SIL6,"ICD Connector",EID=16,PACKAGE=CONN-SIL6

R1,RES,10k,EID=1,PACKAGE=RES40,PINSWAP="1,2",PRIMTYPE=RESISTOR
R2,RES,10k,EID=2,PACKAGE=RES40,PINSWAP="1,2",PRIMTYPE=RESISTOR
R3,RES,220R,EID=B,PACKAGE=RES40,PINSWAP="1,2",PRIMTYPE=RESISTOR
R4,RES,220R,EID=C,PACKAGE=RES40,PINSWAP="1,2",PRIMTYPE=RESISTOR
R5,RES,220R,EID=D,PACKAGE=RES40,PINSWAP="1,2",PRIMTYPE=RESISTOR
R6,RES,220R,EID=E,PACKAGE=RES40,PINSWAP="1,2",PRIMTYPE=RESISTOR
R7,RES,220R,EID=F,PACKAGE=RES40,PINSWAP="1,2",PRIMTYPE=RESISTOR
R8,RES,220R,EID=10,PACKAGE=RES40,PINSWAP="1,2",PRIMTYPE=RESISTOR
R9,RES,220R,EID=11,PACKAGE=RES40,PINSWAP="1,2",PRIMTYPE=RESISTOR
R10,RES,150R,EID=12,PACKAGE=RES40,PINSWAP="1,2",PRIMTYPE=RESISTOR
R11,RES,10k,EID=17,PACKAGE=RES40,PINSWAP="1,2",PRIMTYPE=RESISTOR
RV1,POT-LIN,10k,EID=14,STATE=5

U1,PIC16F877,PIC16F877,ADC_ACQUISITION_TIME=20u,ADC_RCCLOCK_PERIOD=4u,ADC_SAMPLE_DELAY=100n,CFGWORD=0x3FFB,CLOCK=40
kHz,DBG_ADC_BREAK=0,DBG_ADC_WARNINGS=0,DBG_ADDRESSES=0,DBG_DUMP_CFGWORD=0,DBG_GENERATE_CLKOUT=1,DBG_I2C_OPERATIONS=
1,DBG_RANDOM_DMEM=0,DBG_RANDOM_PMEM=0,DBG_STACK=1,DBG_STARTUP_DELAY=0,DBG_UNIMPLEMENTED_MEMORY=1,DBG_UNIMPLEMENTED_
OPCODES=1,DBG_WAKEUP_DELAY=0,EID=15,EPR_WRITECODE_DELAY=10m,EPR_WRITEDATA_DELAY=10m,ITFMOD=PIC,MODDATA="256,255",MO
DDLL=PIC16,PACKAGE=DIL40,PORTTDHL=0,PORTTDLH=0,PROGRAM=PIC16F877A\Debug\Debug.cof,WDT_PERIOD=18m
```

```
*NETLIST,49          #00014,2          #00030,1          #00045,1
                     R5,PS,1           U1,IO,10          U1,IO,5
#00000,2             U1,IO,28
R1,PS,1                                #00031,1          #00046,1
U1,IO,34             #00015,2          U1,OP,14          U1,IO,15
                     R6,PS,1
#00002,2             U1,IO,27          #00032,1          #00047,1
R2,PS,1                                U1,IO,16          U1,IP,1
U1,IO,35             #00016,2
                     R7,PS,1           #00033,1          #00051,2
#00003,2             U1,IO,22          U1,IO,17          J1,PS,6
D1,PS,A                                                  R11,PS,1
R3,PS,2              #00017,2
                     R8,PS,1
#00005,2             U1,IO,21          #00034,1
D2,PS,A                                U1,IO,18
R4,PS,2              #00018,2
                     R9,PS,1           #00035,2
#00006,2             U1,IO,20          U1,IO,40          +5V,7
D3,PS,A                                J1,PS,3           +5V,PT
R5,PS,2              #00019,2                            RV1,PS,1
                     R10,PS,1          #00036,2          RV1,PS,3
#00007,2             U1,IO,19          U1,IO,39          R1,PS,2
D4,PS,A                                J1,PS,2           R2,PS,2
R6,PS,2              #00021,3                            R11,PS,2
                     C1,PS,2           #00037,1          J1,PS,5
#00008,2             RV1,PS,2          U1,IO,38
D5,PS,A              U1,IP,13                            VDD,3
R7,PS,2                                #00038,1          VDD,PT
                     #00023,1          U1,IO,37          U1,PP,11
#00009,2             U1,IO,2                             U1,PP,32
D6,PS,A                                #00039,2
R8,PS,2              #00024,1          U1,IO,36          VSS,3
                     U1,IO,3           J1,PS,1           VSS,PT
#00010,2                                                 U1,PP,12
D7,PS,A              #00025,1          #00040,1          U1,PP,31
R9,PS,2              U1,IO,4           U1,IO,33
                                                         GND,11
#00011,2             #00026,1          #00041,1          GND,PT
D8,PS,A              U1,IO,6           U1,IO,26          J1,PS,4
R10,PS,2                                                 C1,PS,1
                     #00027,1          #00042,1          D1,PS,K
#00012,2             U1,IO,7           U1,IO,25          D8,PS,K
R3,PS,1                                                  D7,PS,K
U1,IO,30             #00028,1          #00043,1          D6,PS,K
                     U1,IO,8           U1,IO,24          D5,PS,K
#00013,2                                                 D4,PS,K
R4,PS,1              #00029,1          #00044,1          D3,PS,K
U1,IO,29             U1,IO,9           U1,IO,23          D2,PS,K
```

Figure 3.11
LED2 circuit netlist.

3.4.2 PCB Layout

Each component needs a pinout corresponding to the physical device to be used in final construction. Most components will offer a default PCB Package that is selected in its properties dialogue. If there is a choice of physical packages, ARES will allow the user to select the most suitable from a library of standard pinouts. If not already available, a package can be created using the same tools as for the general layout. Alternatively, the component can be excluded from the PCB layout initially and placed manually on the layout later.

When ARES is invoked from the main toolbar, the components from the netlist are shown in the component window and can then be selected and placed manually, or all placed automatically. Temporary connections are displayed between the pins, producing a ratsnest display (Figure 3.12(a)).

It is often preferable to place the main components manually so that the external connectors, manual controls, displays and MCU are conveniently placed initially within a suitable board area. When all components have been placed, the auto-router can then be invoked to place the tracks. Some manual adjustment is usually required for optimum neatness and simplicity. A board edge can then be drawn and defined. In Figure 3.12(b), the layout for circuit LED2 is shown for a single-sided board. ARES also offers a 3D view of the board (Figure 3.12(c)) to check the overall arrangement before committing to hardware prototyping.

If the PCB is to be produced via a CNC machine or fabrication system, a Gerber (or alternative format) output file can be passed to the manufacturing system. This specifies the position and dimensions of all the component pin pads, tracks and other features of the final layout, plus machining information for drilling holes, milling the board edge and other features. The tracks and pads are laid out on a blank PCB and the circuit produced in copper (chemically or mechanically), and the board populated with the selected components. Small volumes of PCBs can be produced by specialist manufacturers at reasonable cost direct from the design file.

3.5 Hardware Testing

When the application hardware (prototype or PCB) has been produced, the MCU program must be downloaded into the chip. An overview of the available programming tools and techniques has already been provided in Chapter 1. In-circuit programming and debugging (ICPD) allows the program to be run under the control of a host PC, and debugged in hardware, with single stepping and break point control. This allows the interaction with the actual hardware to be checked and any final problems that did not appear in simulation to be resolved.

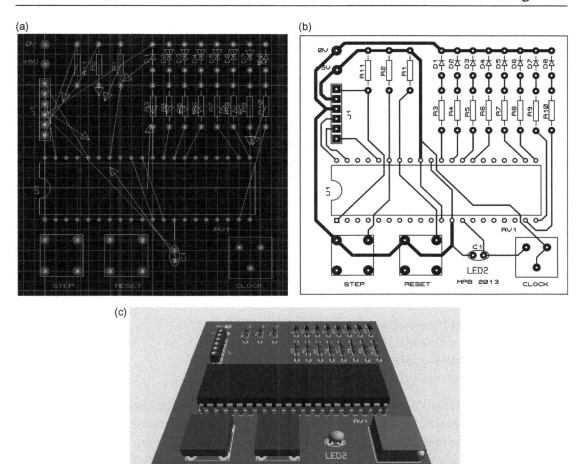

Figure 3.12
LED2 PCB layout: (a) ratsnest, (b) PCB layout and (c) 3D view.

3.5.1 ICPD Requirements

In order to facilitate ICPD, the simulation version of our test program LED2.ASM needs a slight modification − the insertion of an NOP instruction in the first program memory location. This allows the debugger to insert a jump to the debugging code which will be loaded into higher memory locations. Space must be allowed by the user (256 locations), so the maximum user program size when using ICD is $8192 - 256 = 7936$ instructions. In addition, the file registers 70h and 1EBh to 1EFh are required (see Table 1.3, the 16F877A file register map), and one stack level. If hardware debugging is to be used, remember also that RB6 and RB7 cannot be used for I/O. Other than the NOP at location 0000, these restrictions do not significantly affect the simple demo programs used in the rest of this book.

3.5.2 ICPD Testing

The 16F877A is programmed via RB6 and RB7, so it is preferable not to use these for other I/O connections. A pull-up resistor should be connected to !MCLR to enable normal running, while allowing the programming voltage (about 13V) to be applied without interfering with the target system 5V supply. Suitable connections are seen in the LED2 schematic, Figure 3.1. The 6-pin in-line connector is designed to connect the PICkit3 (or compatible) programmer.

If the current project is active in MPLAB and the programmer connected, the relevant downloading and debugging features will become available. It may be necessary to reselect the required debugging tool in the project properties. The final stage of testing in hardware can then be carried out, using the same source level debugging tools as used for simulation:

- *View source code execution*
- *View file register changes*
- *Run, stop, single step program*
- *Set break points and trace*

In this way, the interaction of the MCU with the target circuit can be checked, and the real input and output devices tested. When the operation of application has been confirmed as in accordance with the specification, the program is downloaded again with the ICD option turned off, so that it will run independently in the target hardware.

3.5.3 16F877A Electrical Characteristics

Appendix A contains a summary of the more useful electrical characteristics of the PIC 16F877A, based on the data sheet. It specifies maximum and typical values for such parameters as the power supply voltage and current, I/O pin drive capabilities, clock options and timer performance. These may need to be considered when designing the application hardware. For example, a mains power supply must be able to supply enough total current within the voltage limits specified, or a battery supply will have a limited life before reaching the minimum supply voltage. When the final testing is carried out, the effect of the application circuit characteristics, such as the residual capacitance of the tracks, also needs to be taken into account. The connecting tracks for an external crystal oscillator should be as short as possible for this reason, since the additional capacitance can affect its stability.

Questions 3

1. Explain briefly the significance of SPICE models in electronic simulation. (3)
2. Why does the PIC clock circuit not have to be included in simulation? (3)
3. Why is MCLR pulled high via a resistor in the LED2 circuit? (3)
4. Estimate the signal frequency at CLKOUT if the clock circuit has components $C = 10$nF and $R = 25$k. (3)
5. Explain briefly the advantages of simulation in VSM compared with MPLAB. (3)
6. State the difference between 'step into' and 'step over' in PIC debugging. (3)
7. Explain briefly why break points are useful in source code debugging. (3)
8. Compare briefly the functions of a logic analyser and oscilloscope. (3)
9. Explain briefly the function of the VSM netlist. (3)
10. Explain briefly the advantages of in-circuit debugging. (3)

Total (30)

Assignments 3

3.1 Development System Comparison

Compare in detail the functionality of the MPLAB and Proteus simulation environments, and identify the advantages of each.

Assignments 3.2 and 3.3 require access to Proteus VSM.

3.2 LED2 Simulation

Download the demo files from www.picmicros.org.uk and test the application LED2 in Proteus VSM. Confirm correct operation by operating the virtual inputs and observing the LED activity. Pause and open the source code debugging window. Single step the code, using step over to run through the delay. Set a break point at the label 'start' and measure the loop time, hence calculate the overall count cycle time. Check this by simulation.

Comment out the delay and use the virtual instruments in ISIS simulation mode to display the output as shown in Figure 3.8. Print out the graph on a suitable timescale showing all the outputs are active, and each is double the frequency of the last bit.

3.3 VSM Debugging

Load the LED2 project into the ISIS simulator environment. Introduce the following errors into LED2.ASM:

- *Omit (comment out) the PROCESSOR directive*
- Omit (comment out) the label equate for 'Timer'
- Replace 'CLRF' with 'CLR' (invalid mnemonic)
- Delete label 'Start' (label missing)
- Replace '0' with 'O' in the literal 0FF
- Omit (comment out) 'END' directive

Note the effect of each error type and the message produced by the assembler. What general type of error are they? Warning 'default destination being used' should be received in the list file. What does this mean? Eliminate it by changing the assembler error level to suppress messages and warnings.

With the program restored so that it assembles correctly:

- Replace 'BTFSS' with 'BTFSC'
- Omit (comment out) 'GOTO reset'

Note the effect of these errors. What general type of error are they? Describe the process used to detect each one. Restore the source code to its original condition.

PART 2

PIC Interfacing

CHAPTER 4

Input and Output

Summary
- A simple switch sometimes needs debouncing for correct operation
- Debouncing can be implemented in software or hardware
- The hardware timer and interrupts are often used in input processing
- LEDs can be used as simple indicators or numeric displays
- Keypads provide a simple array of switches for numeric input
- The alphanumeric LCD provides a versatile, low-power display device

This chapter outlines a range of input and output techniques for microcontroller applications. The typical MCU-based consumer product can be quite complex, containing a range of peripherals around the main controller. The mobile phone is a good example; in addition to the sophisticated digital communications sub-system which provides its main functions, the smart phone has a high-resolution LCD touch screen, camera and internet connectivity, and all the firmware necessary to support these functions.

The touch screen input needs a sensing matrix in front of the display, and a powerful processor to provide a fast response, as well as controlling the other functions of the phone. It is probably one of the most complex electronic products available, yet is now commonplace. This type of application demonstrates the highest end of microcontroller performance, compared with the simple examples considered here.

A detailed understanding of these technologies requires a high level of engineering knowledge. Some simpler, but equivalent, techniques will be described here. A prototype application board is shown in Figure 4.1, featuring simple switched inputs, a keypad, two-digit 7-segment display and other control outputs. It is designed to operate as a temperature controller, using externally connected temperature sensors, motors and heaters. A simple keypad and LED display system, then an alphanumeric display, will then be analysed.

4.1 Switch Inputs

If a PIC input is an open circuit, it is pulled up to V_{DD} (nominally $+5V$) internally, i.e. logic high. Switches are therefore generally connected active low, with a pull-up resistor,

Interfacing PIC Microcontrollers.
DOI: http://dx.doi.org/10.1016/B978-0-08-099363-8.00004-2

Figure 4.1
Prototype application board.

as seen in LED2 hardware (Figure 3.1). On some ports, additional weak pull-ups can be enabled to eliminate the need for external pull-up resistors. It is generally acceptable to leave unused inputs open circuit, configured as default inputs, as long as they are not exposed to external static charge, which can potentially damage the MOSFET inputs.

4.1.1 Switch Interface

The simplest input is a manual switch with sprung metallic contacts. A basic toggle switch operates by leverage and a spring to retain it in one of two positions. A slider is a simpler and a cheaper two-way switch. Banks of miniature toggle or slider switches are available in dual-in-line packages. Miniature push buttons (tactile switches) provide momentary inputs, and rotary switches allow one input to be manually selected from several. Sample devices are shown in Figure 4.2(a).

The switch symbol assumes toggle mode operation − the switch remains in the set position until changed. Push buttons are normally assumed to be closed only when held on, but latching operation can be implemented in software if required, to obtain a push-on, push-off operation.

Switches are generally classified according to the number of poles (sets of contacts operating simultaneously) and the number of switch positions (single or double throw). Thus, an SPDT switch is single pole (one set of contacts), double throw (two-way switching). The alternative is simply to specify x way, y pole operation. Most toggle or slider switches are DPDT (double pole, double throw).

The main problems associated with interfacing switches are the effects of mechanical contact operation. Apart from their limited lifespan due to wear, contact bounce or other

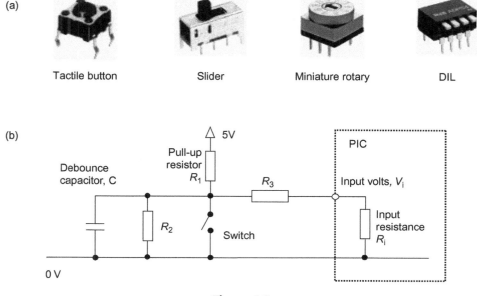

(a)

Tactile button · Slider · Miniature rotary · DIL

(b)

Figure 4.2
Input switches: (a) PCB switches and (b) switch interfacing.

intermittent behavior on changeover is a significant problem. Gold contact coating on the more expensive switches improves reliability and durability, but measures to counter switch bounce are normally required in digital systems, since multiple transitions on changeover switches are very likely to be detected at the input and cause a firmware malfunction. One countermeasure is to detect the switch opening rather than closing, but contact bounce must still be considered. Alternatively, switches working on capacitance or magnetic (hall effect) principles can be used.

Switches can operate with just one support component, a pull-up resistor, as seen in the LED2 hardware. In this case, the input is switched between 0V and the supply voltage. The pull-up resistor prevents a short circuit across the supply when the switch is closed, while allowing the input to rise to the supply value with the switch open. This is adequate for testing in simulation, but possibly not in the real hardware.

A more reliable switch circuit is shown in Figure 4.2(b), using a voltage divider and a debounce capacitor. The minimum value of a logic high input required is given in the electrical characteristics of the 16F877A (Appendix A) as $0.8V + 0.25V_{DD}$. With a supply of 5V, this works out to about 2V. The maximum logic low input voltage is $0.15V_{DD}$, or 0.75V with a 5V supply.

The switch is connected in parallel with a voltage divider, R_1 and R_2. The values of the resistors must be calculated to ensure the input high voltage is at least 2V with a 5V supply, and are high enough to dissipate minimum power, while being low compared with the input resistance. The input leakage current is quoted as $1\mu A$, giving an input resistance of $5M\Omega$ (R_i). For values of $R_1 = 10k$ and $R_2 = 40k$ (39k NPV, nearest preferred value), V_{ih} (switch open) will be 4V, and V_{il} will still be 0V. Power consumption will be about $25/50k = 0.5mW$ for each open switch. An input resistor is often placed in series with the input pin to insure against the effect of any transient or static overvoltages on the input.

4.1.2 Supply Voltage

If the target system is powered from a regulated 5V supply, the design of input circuits can be based on this value for V_{DD}. However, if it is battery powered, the possible variation in supply voltage as the battery discharges *must* be taken into account when designing interface circuits. PIC chips are typically designed to operate over the voltage range 2−6V. The input threshold voltage (switching level) reduces in line with supply voltage. The figures may be obtained from the electrical characteristic for minimum and maximum V_{IN} versus V_{DD} (Figures 18−20 in the 16F877A data sheet, typical values at 25°C):

Supply Voltage(V_{DD})	2.0	5.0	5.5
Threshold Voltage(V_{IN})	0.75	1.2	1.3

We also need to ensure that output devices are operated correctly at reduced supply voltage by checking their electrical specifications.

4.1.3 Hardware Switch Debouncing

Contact bounce in switches generally lasts a few milliseconds, so if the switch is sampled repeatedly by the firmware within this timeframe, it can appear that it has been operated several times. If the program timing or sequence ensures that the input is not sampled until the bouncing has finished anyway, debouncing is not necessary. However, it is a sensible precaution to incorporate some form of debouncing in most instances. The capacitor seen in Figure 4.2(b) provides hardware debouncing.

In Figure 4.3(a), the output voltage from a switch jumps back up to 5V due to the switch contacts bouncing open. If a suitable capacitor is connected across the switch, it is charged up to 5V when the switch is open. When the contacts close, it is quickly discharged by the short circuit. However, it can only recharge via the pull-up resistor, which takes more time. If the switch closes again before the logic 0, minimum threshold, is crossed (0.75V), the voltage is prevented from going back to logic 1 (Figure 4.3(b)).

The capacitor needs to be a high enough value to give a slow voltage rise in the charging phase, while not being so large as to cause a large discharge current through the switch

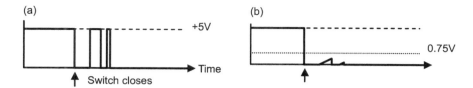

Figure 4.3

Switch hardware debounce (a) without debounce capacitor and (b) with debounce capacitor.

contacts when the contacts close or making the rise time too long when the switch is opened. With a 40k pull-up resistor, a 1nF capacitor would give a time constant of 40ms, which should be more than adequate.

4.1.4 Software Switch Debouncing

The alternative to the hardware debouncing is to introduce a delay in the program which allows time for the switch to settle down, but which is not noticeable to the user, say 50ms. The hardware developed for the LED2 demo, shown in Figure 3.1, will be used to illustrate this. A binary display count is to be incremented under manual control, via a switch input, one step at a time. If the switch input is not debounced, several counts are likely to be registered each time a switch is closed, as the contacts will reopen momentarily.

The debounce process shown in Program 4.1 (VSM project COUNT1) uses the same software delay routine previously used to provide delays between the each output step. The delay count is loaded before the CALL so the routine could be used for a different delay within the same program. The delay count has been adjusted in simulation using the stopwatch to give exactly 50ms if the clock is set to 40kHz.

In Proteus VSM, the simulated switch model has a default delay of 1 ms built in to the model to represent the bounce effect. This is only an approximation to real switch behaviour and can cause confusion if not taken into account when debugging by simulation. Folder COUNT1 in the demo fileset contains the project files for this application.

4.1.5 Timer Switch Debouncing

The main problem with the delay loop method is that MCU time is being wasted (at 5MHz instruction rate, 5000 instruction cycles could be completed in 1ms). In a high-performance application, this inefficient use of MCU time may be unacceptable, so the use of a hardware timer to perform the debouncing function will be preferred. This option allows the MCU to proceed with other tasks while carrying out a timing operation concurrently. The switch problem is also a good opportunity to examine the use of hardware timers and interrupts in general.

```
;;;;;;;;;;;;;;;;;;;;;;;;;;;;;;;;;;;;;;;;;;;;;;;;;;;;;;;;;;;
;
;       Source File:    COUNT1.ASM
;       Author:         MPB
;       Date:           10-01-13
;       Version:        2.0
;
;       Output counts number of switch input pulses
;       Demonstrates software delay switch debounce
;       ICD downloading version
;       RC Clock = 40kHz
;       Updated for Proteus VSM v8 & MPLABX
;
;;;;;;;;;;;;;;;;;;;;;;;;;;;;;;;;;;;;;;;;;;;;;;;;;;;;;;;;;;;

        PROCESSOR 16F877A      ; Define MCU type
;       __CONFIG 0x3733        ; Set config. fuses

; Register Label Equates.................................

PORTB   EQU             06     ; Port B Data Register
PORTD   EQU             08     ; Port D Data Register
TRISD   EQU             88     ; Port B Direction Register
Timer   EQU             20     ; GPR used as delay counter


;;;;;;;;;;;;;;;;;;;;;;;;;;;;;;;;;;;;;;;;;;;;;;;;;;;;;;;;;;;

        CODE    0              ; Start address
        NOP                    ; ICD location

; Initialise Port B (Port A defaults to inputs)............

        BANKSEL TRISD          ; Select bank 1
        MOVLW   b'00000000'    ; Port B Direction Code
        MOVWF   TRISD          ; Load the DDR code into F86
        BANKSEL PORTD          ; Select bank 0
        GOTO    reset          ; Jump to main loop

; 'delay' subroutine .....................................

delay   MOVWF   Timer          ; Copy W to timer register
down    DECFSZ  Timer          ; Decrement timer register

        GOTO    down           ; and repeat until zero
        RETURN                 ; Jump back to main program

; Start main loop ........................................

reset   CLRF    PORTD          ; Clear LEDs

start   BTFSS   PORTB,1        ; Test RESET button
        GOTO    reset          ; and clear LEDs
        BTFSC   PORTB,2        ; Test STEP button
        GOTO    start          ; and again if not pressed
        MOVLW   d'165'         ; Delay count 50ms
        CALL    delay          ; Delay after STEP pressed

wait    BTFSS   PORTB,2        ; Test STEP button again
        GOTO    wait           ; and wait if not released
        INCF    PORTD          ; Increment LEDs
        GOTO    start          ; Repeat always

        END                    ; Terminate source code
```

Program 4.1
Software switch debouncing.

If a hardware timer is started when the switch is first closed, the closure can be confirmed by retesting the input after a time delay to check if it is still closed. Alternatively, the switch input can be processed after the button is released, rather than when it is closed. The same virtual hardware LED2 (Figure 3.1) will be used to demonstrate this process with Program 4.2 (VSM project COUNT2).

TMR0 (Timer0) is located at file register address 01. It operates as an 8-bit binary up counter, driven from an external or internal clock source. The count increments with each input pulse, and a flag is set when it overflows from FF to 00. It can be preloaded with a value so that the timeout flag is set after the required interval. For example, if it is preloaded with the number 156_{10}, it will overflow after 100 counts (256_{10}). A block diagram of Timer0 is shown in Figure 4.4.

Timer0 has a prescaler available at its input, which divides the number of input pulses by a factor of 2, 4, 8, 16, 32, 64, 128 or 256, which increases the range of the count but reduces its accuracy. For timing purposes, the internal clock is usually selected, which is the same as the instruction clock seen at CLKOUT in RC mode ($f_{osc}/4$). Without the prescaler, the register is therefore incremented once per instruction cycle. At a clock rate of 40kHz, the counter input will be 10kHz and will therefore count in steps of 100μs.

In program COUNT2, Timer0 is initialised to make a full count of 256 instruction cycles. At 10kHz, this gives a delay of just over 25ms. This is more than enough for debouncing. The OPTION register is therefore set up for the timer to operate from the instruction clock with no prescaling. The INTCON register contains the timeout flag for Timer0 (bit 2), which is set when the counter rolls over from 11111111 to 00000000. The program waits for this flag before incrementing the output count and restarting the loop.

4.1.6 Switch Input Interrupts

In simple programs, or where program execution speed is not important, inputs can be checked regularly within the main loop (polled). However, this generally wastes too much processor time, so interrupts are often used to read the inputs. Say we want the output to operate at the highest possible frequency, but also allow adjustment via input switches; polling the inputs will slow it down. Assigning interrupts to the inputs (using RB4–RB7 in the 16F877A) is the answer.

Interrupts allow external devices or internal events to force a change in the execution sequence of the MCU. When an interrupt occurs in the PIC, the program jumps to address 004 and continues from there until it sees a Return From Interrupt (RETFIE) instruction. It then resumes at the original point, the return address having been stored automatically on the stack as part of the interrupt process.

```
;;;;;;;;;;;;;;;;;;;;;;;;;;;;;;;;;;;;;;;;;;;;;;;;;;;;;;;;;;;;;;
;
;       Source File:    COUNT2.ASM
;       Author:         MPB
;       Date:           10-01-13
;       Version:        2.0
;
;       Output counts number of switch input pulses
;       Uses hardware timer to debounce input switch
;       ICD downloading version
;       RC Clock = 40kHz
;       Updated for Proteus VSM v8 & MPLABX
;
;;;;;;;;;;;;;;;;;;;;;;;;;;;;;;;;;;;;;;;;;;;;;;;;;;;;;;;;;;;;;;

        PROCESSOR 16F877            ; Define MCU type
        __CONFIG 0x3733            ; Set config fuses

; Register Label Equates..................................

PORTB   EQU     06       ; Port B Data Register
PORTD   EQU     08       ; Port D Data Register
TRISD   EQU     88       ; Port D Direction Register

TMR0    EQU     01       ; Hardware Timer Register
INTCON  EQU     0B       ; Interrupt Control Register
OPTREG  EQU     81       ; Option Register


;;;;;;;;;;;;;;;;;;;;;;;;;;;;;;;;;;;;;;;;;;;;;;;;;;;;;;;;;;;;;;

        CODE    0                  ; Start of program memory
        NOP                        ; ICD location

; Initialise Port D (Port B defaults to inputs)............

        BANKSEL TRISD              ; Select bank 1
        MOVLW   b'00000000'        ; Port B Direction Code
        MOVWF   TRISD              ; Load the DDR code into F86

; Initialise Timer0 ......................................

        MOVLW   b'11011000'        ; TMR0 initialisation code
        MOVWF   OPTREG             ; Int clock, no prescale
        BANKSEL PORTD              ; Select bank 0

; Start main loop ........................................

reset   CLRF    PORTD              ; Clear Port B Data

start   BTFSS   PORTB,1            ; Test reset button
        GOTO    reset              ; and reset Port B if pressed
        BTFSC   PORTB,2            ; Test step button
        GOTO    start              ; and repeat if not pressed

        CLRF    TMR0               ; Reset timer
wait    BTFSS   INTCON,2           ; Check for time out
        GOTO    wait               ; Wait if not
        BCF     INTCON,2           ; Reset TMR0 interrupt flag

stepin  BTFSS   PORTB,2            ; Check step button
        GOTO    stepin             ; and wait until released
        INCF    PORTD              ; Increment output at Port B
        GOTO    start              ; Repeat main loop always

        END                        ; Terminate source code......
```

Program 4.2
Hardware timer debouncing.

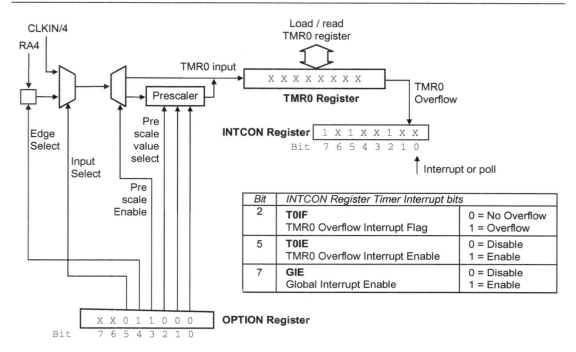

Figure 4.4
Timer0 operation.

Typically, a GOTO ISR (Interrupt Service Routine) is placed at the interrupt address 004, and the ISR placed higher up memory. The interrupt source must be identified by the associated flag, e.g. the Timer0 timeout flag. This has an associated interrupt enable bit which allows the MCU to respond to (or ignore) this particular source. A global interrupt enable bit allows all interrupts to be enabled or disabled together; they are disabled by default. If more than one interrupt source is enabled, the program must test, as part of the ISR, the individual interrupt flags to see which is active.

Program 4.3 (VSM project COUNT3) uses the RB0 interrupt, which allows a change on that pin to trigger an interrupt. The test program runs a simple output loop incrementing Port D. When a button connected to RB0 is pressed, an interrupt is called which stores the current output value, switches on all the outputs for about 3 s (using Timer0 with prescale factor of 128) to represent the interrupt in progress and then continues the output from the same value by recovering from storage.

When an interrupt is called, we normally want the interrupted task to be restarted as though the interrupt had never happened. Therefore, the contents of any relevant register must be saved and later restored, particularly the status register and any other status flags. Ideally, each task will be allocated its own set of working data registers, but SFRs used by both tasks must also be saved at the start of the ISI and restored at the end, before the return from interrupt. This is called context saving.

4.2 Display Outputs

The simplest display output is a light-emitting diode (LED). These are now available for a wide range of applications other than simply status indicators. The light output frequency (color) variation covers not only all visible wavelengths but also infrared (IR) and ultraviolet (UV) rays. IRLEDs are used in remote controls so that the receiver is not affected by ambient light. Laser LEDs that produce a single frequency coherent light output are used in communications as data transmitters in optical fibre systems.

LEDs can be modulated (switched on and off) at high frequency to produce wide bandwidth communications with multiple simultaneous data streams, hence the advantage of optic fibre over copper for internet access. High-power white light (the full spectrum of visible frequencies) LEDs are now cheap enough to use as a high-efficiency lighting source. A selection of LED-based components is shown in Figure 4.5.

4.2.1 LED Output Circuit

The basic LED output circuit is very simple (Figure 1.6). The only other component required is a current limiting resistor, which is calculated according to the supply voltage. A typical indicator LED requires a forward current of about 15mA to light up and produces a forward volt drop of about 2V (depending on the LED type). We can use a simple formula to estimate the resistor value required:

$$\mathtt{Resistor\ value = (Vs - 2) \ / \ 15 \times 10^{-3}}$$

So if the supply is 5V, the resistor value required is 200Ω. Low-power or high-efficiency LEDs can use a higher value, thus saving power. The PIC output can sink or source a maximum current of about 25mA, so LEDs can be connected directly to the outputs. The LED can just as easily be used to indicate an active a.c. supply, since it acts as a rectifier

```
;;;;;;;;;;;;;;;;;;;;;;;;;;;;;;;;;;;;;;;;;;;;;;;;;;;;;;;;;;;;
;
;       Source File:          COUNT3.ASM
;       Author:               MPB
;       Date:                 10-01-13
;
;       Output count is captured and stored on RB0 interrupt
;
;;;;;;;;;;;;;;;;;;;;;;;;;;;;;;;;;;;;;;;;;;;;;;;;;;;;;;;;;;;;

        PROCESSOR 16F877       ; Define MCU type
        __CONFIG 0x3733        ; Set config fuses

; Register Label Equates.................................

PORTB   EQU     06      ; Port B Data Register
PORTD   EQU     08      ; Port D Data Register
TRISD   EQU     88      ; Port D Direction Register

TIMER0  EQU     01      ; Hardware Timer Register
INTCON  EQU     0B      ; Interrupt Control Register
OPTREG  EQU     81      ; Option Register

Store   EQU     20      ; Temp store for count


;;;;;;;;;;;;;;;;;;;;;;;;;;;;;;;;;;;;;;;;;;;;;;;;;;;;;;;;;;;;

        CODE    0              ; Start of program memory
        NOP                    ; ICD location
        GOTO    start          ; Jump to program start

; Interrupt Service Routine ...........................

        ORG     004            ; ISR start location
        MOVF    PORTD,W        ; Get current count
        MOVWF   Store          ; Save it
        MOVLW   0FF            ; All on code
        MOVWF   PORTD          ; Switch all outputs on

        CLRF    TIMER0         ; Reset Timer0 count
        BCF     INTCON,2       ; Reset timeout flag
check   BTFSS   INTCON,2       ; Check timeout flag
        GOTO    check          ; Until timeout

        MOVF    Store,W        ; Get count
        MOVWF   PORTD          ; Restore output
        BCF     INTCON,1       ; Clear interrupt flag
        RETFIE                 ; Return to main loop

; Initialise Port D (Port B defaults to inputs)...........

start   MOVLW   b'00000000'    ; Port B Direction Code
        BANKSEL TRISD          ; Select bank 1
        MOVWF   TRISD          ; Load the DDR code into F86

; Initialise RB0 Interrupt ............................

        MOVLW   b'10010110'    ; Interrupt and timer setup
        MOVWF   OPTREG         ; Int clock, no prescale
        BANKSEL PORTD          ; Select bank 0
        MOVLW   b'10010000'    ; INTCON code
        MOVWF   INTCON         ; Enable RB0 interrupt

; Start main loop .....................................

        CLRF    PORTD          ; Switch off outputs
count   INCF    PORTD          ; Increment output at Port B
        GOTO    count          ; Loop fast

        END                    ; Terminate source code......
```

Program 4.3
Switch interrupt source code.

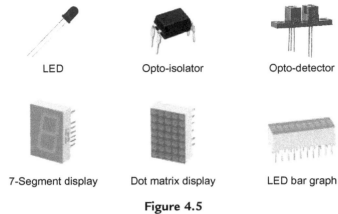

LED   Opto-isolator   Opto-detector

7-Segment display   Dot matrix display   LED bar graph

Figure 4.5
LED components.

diode. The current is calculated from the average of a half-wave rectified sinusoidal waveform, hence the required value of the current limiting resistor.

4.2.2 LED Opto-Isolator and Detector

Sometimes, an input signal needs to be electrically isolated from the microcontroller input to protect it from high voltages and electrical noise, which are often found in industrial environments. The supply voltage used in many industrial controllers is 24V d.c., so the opto-isolator can provide level shifting down to 5V, as well as safe operation.

The opto-isolator (or opto-coupler) incorporates an LED and phototransistor in one package. This component can be seen in Figure 8.4, which is used as an output isolator with a triac that controls the current to a $240V_{a.c.}$ load. A similar circuit is fitted internally at the inputs to PLCs (programmable controllers) that are used in manufacturing systems.

When switched on, via a suitable current limiting resistor, the LED in the opto-isolator illuminates the base of the phototransistor, causing it to conduct. The transistor must be saturated (fully on), producing a minimal forward volt drop across the collector—emitter junction. A load resistor in the collector of the transistor connected to the digital supply produces a logic output. A typical opto-isolator inverts the logic level.

The same components can be used to make an opto-detector. The LED and photo-detector are mounted either side by side for detecting a reflective object in front of the sensor or on either side of a slot so that the light beam is interrupted by a moving object. Often, a metal or plastic slotted disc or graduated strip is used to form a position or speed detector. Typical applications of this type include print head positioning in an inkjet printer and speed measurement of a motor shaft. Figure 4.6 shows the circuit for opto-isolation or

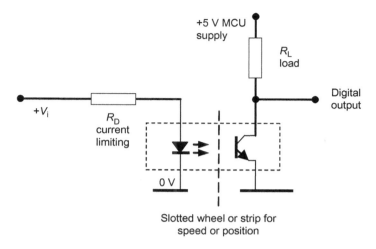

Figure 4.6
Opto-isolator or detector circuit.

photo-detection. The applications of opto-couplers and detectors are discussed further in later chapters.

4.2.3 7-Segment LED Display

The standard 7-segment LED display consists of illuminated segments arranged to show numerical symbols when switched on in the appropriate combination. Each segment is driven separately from an output port via a current-limiting resistor. Numbers 0−9 can be displayed, but for a full range of alphanumeric characters, displays with more segments or a dot matrix are available. A 7-segment LED display can be seen in the prototype hardware in Figure 4.1. It is an active high display with a common cathode and individual anodes requiring a logic 1 and sufficient current to switch it on. An active low type, requiring a logic 0 at each cathode, will have a common anode.

The 7-segment codes for 0−9, * and #, are shown in Table 4.1. The segments are labeled a−g and are assumed to operate active high (1 = ON). The binary code required must then be worked out for each character to be displayed, depending on the order in which the outputs are connected to the segments. In this case, bit 1 = a, through to bit 7 = g, with bit 0 unused. Hash is displayed as 'H' and star as three horizontal bars. As only 7 bits are needed, the LSB (least significant bit) is assumed to be 0 when converting to hexadecimal. In any case, it is preferable to put the binary code in the program. Codes for other types of display or connections can be worked out in the same way.

An alternative to the plain 7-segment display is a BCD module. This receives a Binary Coded Decimal (BCD) input and displays the corresponding number, using an internal

Table 4.1: 7-Segment Codes.

Key	Segment	Hex
	g f e d c b a -	LSB = 0
1	0 0 0 0 1 1 0 0	0C
2	1 0 1 1 0 1 1 0	B6
3	1 0 0 1 1 1 1 0	9E
4	1 1 0 0 1 1 0 0	CC
5	1 1 0 1 1 0 1 0	DA
6	1 1 1 1 1 0 1 0	FA
7	0 0 0 0 1 1 1 0	0E
8	1 1 1 1 1 1 1 0	FE
9	1 1 0 0 1 1 1 0	CE
#	1 1 1 0 1 1 0 0	EC
0	0 1 1 1 1 1 1 0	7E
*	1 0 0 1 0 0 1 0	92

decoder. In BCD $0 = 0000_2$, $1 = 0001_2$ and so on to $9 = 1001_2$. It therefore only needs four inputs (plus a common terminal) and displays binary numbers from 0 to 9 without encoding.

4.3 Keypad System

The keypad is simply an array of push buttons connected in rows and columns so that each can be tested for closure with the minimum number of connections (the physical component is shown in the prototype hardware in Figure 4.1). Either the row or column connections can be chosen as inputs, the other set becoming the outputs. These usually have pull-up resistors to ensure an input high to the MCU when inactive. The MCU generates an active low on each keypad input line in turn, and the outputs are tested to detect this low, which is only generated when a button is pressed. The keys are counted by the program, and the count used to select a display code when a key is detected.

There are 12 keys on a phone type pad (0–9, #, *), arranged in a 3 × 4 matrix. Another common type is the calculator keypad, with arithmetic operators and a total of 16 keys (see Figure 6.1). In the phone keypad, columns are labelled 1, 2, 3 and the rows A, B, C, D.

4.3.1 Keypad Interface

An application using simple I/O is shown in Figure 4.7 (VSM project KEYPAD2). It has a telephone style keypad and active high 7-segment LED display. The PIC outputs provide

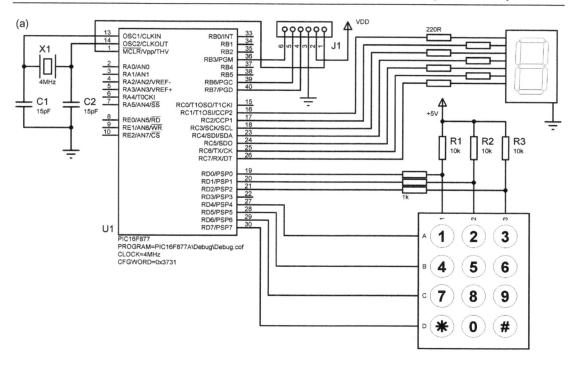

(a)

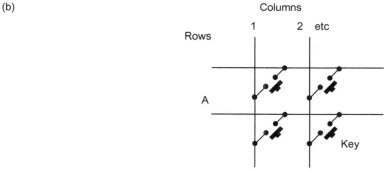

Figure 4.7
Keypad operation: (a) keypad schematic and (b) keypad connections.

enough current (max 25 mA) to drive one LED segment. The output codes to selectively light the display segments can be stored in a data table within the application program. Displays are considered in more detail in Section 4.4.

A 4-MHz crystal clock is used in this application. The components have included the schematic, although they will generally be omitted in the demo application circuits, since the clock is internally specified in simulation mode. The ICPD connections are also shown in this case but will similarly be omitted in future schematics (note that the SIL connector properties may need to be adjusted to exclude them from the simulation). All circuit components and connections must be included if a PCB or other prototype hardware implementation is planned.

4.3.2 Keypad Program

In the demo circuit, the seven keypad pins are connected to Port D. Bits 4–7 are initialised as outputs and bits 0–2 as inputs. The input pins are pulled high by external resistors, and the output rows are also initially set to 1. The keypad scanning process outputs a zero to each row and checks the inputs. If no key has been pressed, none of the inputs will be low. If a key has been pressed, it can be identified by the combination of the output and input low bits. For example, if the number 7 has been pressed, it is detected when row output 3 is low, and a low input is seen on column input 1.

A simple way to identify the active key is to increment a count of keys tested before each is checked so that when a button is detected, the scan of the keyboard is terminated with current key number in the counter. This works for the first three rows because the (non-zero) numbers on the keypad are arranged in order:

```
Row A = 1,2,3
Row B = 4,5,6
Row C = 7,8,9
Row D = *,0,#
```

In the last row, the star symbol is represented by a count of 10 (0Ah), zero by 11 (0Bh) and hash by 12 (0C). The keypad read operation (Program 4.4) steps through the buttons in this way, incrementing a key count, and quits the scanning routine when a button is detected, with the corresponding count of keys stored. If no button is pressed, it repeats. The program then displays the button number on a 7-segment display, with arbitrary symbols representing star and hash.

4.3.3 Keyboards

Most keyboards operate by row and column scanning. PC keyboards generally have their own processor which converts the key detected into a suitable character code, which is then transferred to the host controller in serial form, saving on I/O lines. Originally this was an RS232 data stream (see Chapter 9) with a D-type connector. More recently, a USB connection was the norm, then wireless. These communication standards are discussed in Chapter 8.

```
;;;;;;;;;;;;;;;;;;;;;;;;;;;;;;;;;;;;;;;;;;;;;;;;;;;;;;;;;;;;;
;
;         KEYPAD.ASM        MPB  Ver 1.0 28-8-05
;
;         Reads keypad and shows digit on display
;         Design file KEYPAD.DSN
;
;;;;;;;;;;;;;;;;;;;;;;;;;;;;;;;;;;;;;;;;;;;;;;;;;;;;;;;;;;;;;

              PROCESSOR 16F877

PCL           EQU       002            ; Program Counter
PORTC         EQU       007            ; 7-Segment display
PORTD         EQU       008            ; 3x4 keypad

TRISC         EQU       087            ; Data direction
TRISD         EQU       088            ; registers

Key           EQU       020            ; Count of keys

; Initialise ports.......................................

              BANKSEL   TRISC          ; Display
              CLRW                     ; all outputs
              MOVWF     TRISC          ;
              MOVLW     B'00000111'    ; Keypad
              MOVWF     TRISD          ; bidirectional

              BANKSEL   PORTC          ; Display off
              CLRF      PORTC          ; initially
              GOTO      main           ; jump to main

; Check a row of keys ...................................

row           INCF      Key            ; Count first key
              BTFSS     PORTD,0        ; Check key
              GOTO      found          ; and quit if on

              INCF      Key            ; and repeat
              BTFSS     PORTD,1        ; for second
              GOTO      found          ; key

              INCF      Key            ; and repeat
              BTFSS     PORTD,2        ; for third
              GOTO      found          ; key
              GOTO      next           ; go for next row

; Scan the keypad........................................

scan          CLRF      Key            ; Zero key count
              BSF       3,0            ; Set Carry Flag
              BCF       PORTD,4        ; Select first row
newrow        GOTO      row            ; check row

next          BSF       PORTD,3        ; Set fill bit
              RLF       PORTD          ; Select next row
              BTFSC     3,0            ; 0 into carry flag?
              GOTO      newrow         ; if not, next row
              GOTO      scan           ; if so, start again

found         RETURN                   ; quit with key count

; Display code table.....................................

table         MOVF      Key,W          ; Get key count
              ADDWF     PCL            ; and calculate jump
              NOP                      ; into table
              RETLW     B'00001100'    ; Code for '1'
              RETLW     B'10110110'    ; Code for '2'
              RETLW     B'10011110'    ; Code for '3'
              RETLW     B'11001100'    ; Code for '4'
              RETLW     B'11011010'    ; Code for '5'
              RETLW     B'11111010'    ; Code for '6'
              RETLW     B'00001110'    ; Code for '7'
              RETLW     B'11111110'    ; Code for '8'
              RETLW     B'11001110'    ; Code for '9'
              RETLW     B'10010010'    ; Code for '*'
              RETLW     B'01111110'    ; Code for '0'
              RETLW     B'11101100'    ; Code for '#'
```

Program 4.4

Keypad program.

```
; Output display code......................................

show       CALL      table              ; Get display code
           MOVWF     PORTC              ; and show it
           RETURN


;;;;;;;;;;;;;;;;;;;;;;;;;;;;;;;;;;;;;;;;;;;;;;;;;;;;;;;;;;;;;;

; Read keypad & display....

main       MOVLW     0FF                ; Set all outputs
           MOVWF     PORTD              ; to keypad high
           CALL      scan               ; Get key number
           CALL      show               ; and display it
           GOTO      main               ; and repeat

           END       ;;;;;;;;;;;;;;;;;;;;;;;;;;;;;;;;;;;;;;;;;;;;
```

Program 4.4
(Continued)

4.4 Liquid Crystal Display

The liquid crystal display (LCD) is now a common choice for graphical and alphanumeric displays. It is more versatile and consumes less power that an LED display, since it works electrostatically. The liquid crystal segments or pixels darken against the background due to an applied voltage across the liquid crystal layer, as in a capacitor. It can also be backlit, or operate purely by passive reflection, whereas an LED needs to be fairly bright to be viewed in daylight. The downside is that the drive requirements are usually a bit more complicated.

LCD displays range from small, 7-segment monochrome numerical types such as those used in digital multimeters (typically 3½ digits, maximum reading 1.999) to large, full colour, high-resolution touch screens which can display full video, as well as flat screen televisions. Here, we shall concentrate on the simpler type which displays alphabetical, numerical and symbolic characters from the standard ASCII character set. This type can also display low-resolution graphics, but we will discuss only alphanumeric operation here.

4.4.1 LM016L LCD Application

A 16F877A-based demonstration board, which can function as a digital voltmeter using the LM016L display described below, is shown in Figure 4.8. It also has the programmer/debugger module PICkit2 plugged in, ready to program and test.

The LM016L LCD displays 2 lines of 16 characters (16×2) using a standard interface. Each character is 5×8 pixels, making it 80×16 pixels overall. The display receives ASCII codes for each character at the data inputs (D0–D7). The data is presented to the display inputs by the MCU and is latched in by pulsing the E (Enable) input. The RW (Read/Write) line can be tied low (write mode), as the LCD is receiving data only.

Figure 4.8
LCD demo hardware with ICPD module.

The RS (Register Select) input allows commands to be sent to the display. RS = 0 selects command mode, RS = 1 data mode. The connections to the MCU are shown in Figure 4.9.

The display module contains its own microcontroller, the Hitachi HD44780, a standard chip for this type of interface. It must be initialised according to the data and display options required. In this example, the data is being sent in 4-bit mode. The 8-bit code for each ASCII character is sent in two halves; high nibble first, low nibble second. This saves on I/O pins and allows the LCD to be driven using only 6 lines of a single port, while making the software only slightly more complex.

The command set for the display controller, derived from the data sheet for the controller chip, is given in Table 4.2, with the RAM addresses for the display codes and the initialisation sequence required for 4-bit operation.

4.4.2 LCD Demo Program

In the demo program (Program 4.5, VSM project LCD2), a fixed message is displayed on line 1, showing all the numerical digits. The second line finishes with a character that counts up from 0 to 9 and repeats to demonstrate a variable display. It can be seen that the display must be initially set to default operating mode, before selecting the required mode (4-bit, 2 lines) and resetting. Note that the commands are differentiated by the number of leading zeros.

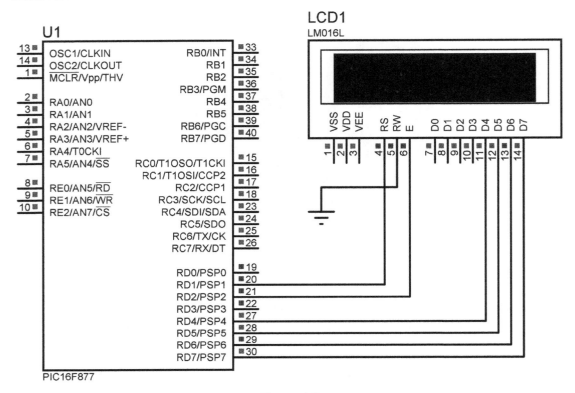

Figure 4.9
LCD display connections.

We will analyse the LCD program in detail as it contains a number of features which we will see again. In order to clarify the overall structure, a program outline is shown in Figure 4.10. It has three main processes:

1. *Output line 1 fixed message 'CONST:0123456789'*
2. *Output line 2 fixed message 'VARIABLE ='*
3. *Output variable count 0–9 at line 2, position 12*

The main program (last in the source code list) is very short, comprising the following:

1. *Initialise the MCU and LCD*
2. *Output fixed characters*
3. *Output count*

The code is divided into these functional blocks. Note that standard register labels are defined by including the standard file P16F877.INC, which contains a list of all labels for the SFRs and control bits, e.g. PORT D, STATUS, Z.

Table 4.2: LM016L LCD Operation.

(a) Commands

Instruction	Code	Description
Clear display	0000 0001	Clear display and reset address
Home cursor	0000 001x	Reset display location address
Entry mode	0000 01MS	Set cursor move and display shift
Display control	0000 1DCB	Display and cursor enable
Shift control	0001 PRxx	Moves cursor and shifts display
Function control	001L NFxx	Data mode, line number, font
CGRAM address	01gg gggg	Send character generator RAM address
DDRAM address	1ddd dddd	Send display data RAM address

X	don't care
M	cursor move direction 1 = right 0 = left
S	enable whole display shift = 1
D	whole display on = 1
C	cursor on = 1
B	blinking cursor on = 1
P	display shift = 1, cursor move = 0
R	shift right = 1, shift left = 0
L	8 bits = 1, 4 bits = 0
N	2 lines = 1, 1 line = 0
F	5×10 character = 1, 5×8 = 0
G	Character generator RAM address bit
D	Data RAM address bit

(b) Character Addresses (16 $\times$ 2 Display)

00	01	02	03	04	05	06	07	08	09	0A	0B	0C	0D	0E	0F
40	41	42	43	44	45	46	47	48	49	4A	4B	4C	4D	4E	4F

(c) LCD Initialisation Command Code Sequence

Hex	Binary	Type	Meaning
32	0011 0010	Function control	8-bit data, 1 line, 5×8 character
28	0010 1000	Function control	4-bit data, 2 lines, 5×8 character
0C	0000 1100	Display control	Enable display, cursor off, blink off
06	0000 0110	Entry mode	Cursor auto-increment right, shift off
01	0000 0001	Clear display	Clear all characters
80	1000 0000	DDRAM address	Reset display memory address to 00

To send the data and commands to the display, the output data is initially masked so that only the high nibble is sent. The low bits are cleared. However, since the low bits control the display (RS and E), these have to be set up after the data has been output in the port high bits. In particular, an RS flag bit is set up in a dummy register 'Select' to indicate whether the current output is command or data, and copied to RD1 after the data set-up.

```
;;;;;;;;;;;;;;;;;;;;;;;;;;;;;;;;;;;;;;;;;;;;;;;;;;;;;;;;;;;;
;
;         LCD2.ASM    MPB       11-01-13
;
;         Outputs fixed and variable characters
;         to 16x2 LCD in 4-bit mode
;         Updated for Proteus VSM v8
;
;;;;;;;;;;;;;;;;;;;;;;;;;;;;;;;;;;;;;;;;;;;;;;;;;;;;;;;;;;;;

          PROCESSOR 16F877A
;         Clock = XT 4MHz, standard fuse settings
          __CONFIG 0x3731

; LABEL EQUATES      ;;;;;;;;;;;;;;;;;;;;;;;;;;;;;;;;;;;;;;;;;;;

          INCLUDE "P16F877A.INC"      ; Standard register labels

Timer1    EQU       20               ; 1ms count register
TimerX    EQU       21               ; Xms count register
Var       EQU       22               ; Output variable
Point     EQU       23               ; Program table pointer
Select    EQU       24               ; Copy of RS bit
OutCod    EQU       25               ; Temp store for output code

RS        EQU       1                ; Register select output bit
E         EQU       2                ; Display enable

; Program code ;;;;;;;;;;;;;;;;;;;;;;;;;;;;;;;;;;;;;;;;;;;;;;;

          CODE      0                ; Place machine code
          NOP                        ; for ICD mode

          BANKSEL   TRISD            ; Select bank 1
          CLRW                       ; All outputs
          MOVWF     TRISD            ; Initialise display port
          BANKSEL   PORTD            ; Select bank 0
          CLRF      PORTD            ; Clear display outputs

          GOTO      Start            ; Jump to main program


; SUBROUTINES ;;;;;;;;;;;;;;;;;;;;;;;;;;;;;;;;;;;;;;;;;;;;;;;;

; 1ms delay with 1us cycle time (1000 cycles)..............

Onems     MOVLW     D'249'           ; Count for 1ms delay
          MOVWF     Timer1           ; Load count
Loop1     NOP                        ; Pad for 4 cycle loop
          DECFSZ    Timer1           ; Count
          GOTO      Loop1            ; until Z
          RETURN                     ; and finish


; Delay Xms, X received in W ...............................

Xms       MOVWF     TimerX           ; Count for X ms
LoopX     CALL      Onems            ; Delay 1ms
          DECFSZ    TimerX           ; Repeat X times
          GOTO      LoopX            ; until Z
          RETURN                     ; and finish


; Generate data/command clock siganl E ...................

PulseE    BSF       PORTD,E          ; Set E high
          CALL      Onems            ; Delay 1ms
          BCF       PORTD,E          ; Reset E low
          CALL      Onems            ; Delay 1ms
          RETURN                     ; done


; Send a command byte in two nibbles from RB4 - RB7 ........

Send      MOVWF     OutCod           ; Store output code
          ANDLW     0F0              ; Clear low nybble
          MOVWF     PORTD            ; Output high nybble
          BTFSC     Select,RS        ; Test RS bit
          BSF       PORTD,RS         ; and set for data
          CALL      PulseE           ; and clock display register
          CALL      Onems            ; wait 1ms for display
```

Program 4.5
LCD2 display source code.

```
            SWAPF     OutCod            ; Swap low and high nibbles
            MOVF      OutCod,W          ; Retrieve output code
            ANDLW     0F0               ; Clear low nibble
            MOVWF     PORTD             ; Output low nibble
            BTFSC     Select,RS         ; Test RS bit
            BSF       PORTD,RS          ; and set for data
            CALL      PulseE            ; and clock display register
            CALL      Onems             ; wait 1ms for display
            RETURN                      ; done
; Table of fixed characters to send --------------------------

Line1       ADDWF     PCL               ; Modify program counter
            RETLW     'C'               ; Pointer = 0
            RETLW     'O'               ; Pointer = 1
            RETLW     'N'               ; Pointer = 2
            RETLW     'S'               ; Pointer = 3
            RETLW     'T'               ; Pointer = 4
            RETLW     ':'               ; Pointer = 5
            RETLW     '0'               ; Pointer = 6
            RETLW     '1'               ; Pointer = 7
            RETLW     '2'               ; Pointer = 8
            RETLW     '3'               ; Pointer = 9
            RETLW     '4'               ; Pointer = 10
            RETLW     '5'               ; Pointer = 11
            RETLW     '6'               ; Pointer = 12
            RETLW     '7'               ; Pointer = 13
            RETLW     '8'               ; Pointer = 14
            RETLW     '9'               ; Pointer = 15

Line2       ADDWF     PCL               ; Modify program counter
            RETLW     'V'               ; Pointer = 0
            RETLW     'A'               ; Pointer = 1
            RETLW     'R'               ; Pointer = 2
            RETLW     'I'               ; Pointer = 3
            RETLW     'A'               ; Pointer = 4
            RETLW     'B'               ; Pointer = 5
            RETLW     'L'               ; Pointer = 6
            RETLW     'E'               ; Pointer = 7
            RETLW     ' '               ; Pointer = 8
            RETLW     '='               ; Pointer = 9
            RETLW     ' '               ; Pointer = 10


; Initialise the display ----------------------------------

Init        MOVLW     D'100'            ; Load count 100ms delay
            CALL      Xms               ; and wait for display
            MOVLW     0F0               ; Mask for select code
            MOVWF     Select            ; High nibble not masked

            MOVLW     0x30              ; Load initial nibble
            MOVWF     PORTD             ; and output it to display
            CALL      PulseE            ; Latch initial code
            MOVLW     D'5'              ; Set delay 5ms
            CALL      Xms               ; and wait
            CALL      PulseE            ; Latch initial code again
            CALL      Onems             ; Wait 1ms
            CALL      PulseE            ; Latch initial code again
            BCF       PORTD,4           ; Set 4-bit mode
            CALL      PulseE            ; Latch it

            MOVLW     0x28              ; Set 4-bit mode, 2 lines
            CALL      Send              ; and send code
            MOVLW     0x08              ; Switch off display
            CALL      Send              ; and send code
            MOVLW     0x01              ; Clear display
            CALL      Send              ; and send code
            MOVLW     0x06              ; Enable cursor auto inc
            CALL      Send              ; and send code
            MOVLW     0x80              ; Zero display address
            CALL      Send              ; and send code
            MOVLW     0x0C              ; Turn on display
            CALL      Send              ; and send code

            RETURN                      ; Done
```

Program 4.5
(Continued)

```
; Send the fixed message to the display -----------------------

OutMes   CLRF      Point           ; Reset table pointer
         BSF       Select,RS       ; Select data mode

Mess1    MOVF      Point,W         ; and load it
         CALL      Line1           ; Get ASCII code from table
         CALL      Send            ; and do it
         INCF      Point           ; point to next character
         MOVF      Point,W         ; and load the pointer
         SUBLW     D'16'           ; check for last table item
         BTFSS     STATUS,Z        ; and finish if 16 done
         GOTO      Mess1           ; Output character code

         MOVLW     0xC0            ; Move cursor to line 2
         BCF       Select,RS       ; Select command mode
         CALL      Send            ; and send code
         CLRF      Point           ; Reset table pointer
Mess2    MOVF      Point,W         ; and load it
         CALL      Line2           ; Get fixed character
         BSF       Select,RS       ; Select data mode
         CALL      Send            ; and send code
         INCF      Point           ; next character
         MOVF      Point,W         ; Reload pointer
         SUBLW     D'11'           ; and check for last
         BTFSS     STATUS,Z        ; Skip if last
         GOTO      Mess2           ; or send next
         RETURN                    ; done

; Output variable count to display (0-9) endlessly -------------

OutVar   CLRF      Var             ; Clear variable number
         MOVLW     0X30            ; Load offset to be added
         ADDWF     Var             ; to make ASCII code (30-39)

Next     MOVF      Var,W           ; Load the code
         BSF       Select,RS       ; Select data mode
         CALL      Send            ; and send code

         MOVLW     0xCB            ; code to move cursor back
         BCF       Select,RS       ; Select command mode
         CALL      Send            ; and send code
         MOVLW     D'250'          ; Load count to wait 250ms
         CALL      Xms             ; so numbers are visible

         INCF      Var             ; Next number
         MOVF      Var,W           ; Load number
         SUBLW     0x3A            ; Check for last (10=A)
         BTFSS     STATUS,Z        ; and skip if last
         GOTO      Next            ; or do next number
         GOTO      OutVar          ; Repeat from number Z

; MAIN PROGRAM ;;;;;;;;;;;;;;;;;;;;;;;;;;;;;;;;;;;;;;;;;;;;;;;;;;

Start    CALL      Init            ; Initialise the display
         CALL      OutMes          ; Display fixed characters
         GOTO      OutVar          ; Display an endless count

         END                       ; of source code ;;;;;;;;;;
```

Program 4.5
(Continued)

After each output, a 1ms delay is executed to allow the LCD controller time to process the input and display it. An exact timing loop (1ms) is achieved by padding the delay loop to 4 cycles with an NOP and executing it 249 times. Including the initial instructions and subroutine jumps, the delay is exactly $250 \times 4 = 1000\mu s$. This is then used by another loop (X ms) to obtain delays in whole milliseconds. It is also used to generate a 1ms pulse at E to latch the data and commands into the LCD controller input port.

```
Project LCD
Program to demonstrate fixed and
variable output of alphanumeric characters
 to 16x2 LCD (simulation only)

HARDWARE
        ISIS design file LCD.DSN
        MCU     16F877A
                Clock = XT 4MHz
        LCD     Data = RD4 – RD7
                RS = RD, E = RD2, RW = 0

FIRMWARE

        Initialise
                LCD output = Port D
                Wait 100ms for LCD to start
                LCD: 4-bit data, 2 lines, auto cursor
                Reset LCD

        Display message line 1
                Reset table pointer
                REPEAT
                     Get next code
                     Send ASCII code
                UNTIL 16 characters done

        Display message line 2
                Position cursor
                Reset table pointer
                REPEAT
                     Get next code
                     Send ASCII code
                UNTIL 11 characters done

        Display incrementing count
           REPEAT
              Set Count = 0
                 LOOP
                     Calculate ASCII
                     Send ASCII code
                     Increment Count
                     Reset cursor
                     Delay 250ms
                 UNTIL Count = 9
           ALWAYS

        Send ASCII code
           Mask low nibble
           Output high nibble
           Pulse E
           Wait 1ms
           Swap nibbles
           Output low nibble
           Pulse E
           Wait 1ms
           Return
```

Figure 4.10
LCD program outline.

4.4.3 ASCII Codes

This is a standard character set where the basic keyboard characters are represented by a 7-bit code. The codes are shown in Table 5.7, which describes all the basic data types. There is a standard code for all the characters found on a typical full keyboard: letters A−Z (upper and lower case), numbers (0−9) and punctuation characters. The basic set of 128 characters can be encoded with 7 bits each, leaving an eighth bit for error checking if required. When the alphanumeric LCD receives a data input, it will display the default ASCII character, unless an alternative character set has been selected. The upper case characters A−Z have the codes 41h−5Ah, lower case a−z have 61h−7Ah and numbers 0−9 have 30h−39h, respectively.

The PIC assembler generates an ASCII code in response to single quotes enclosing the character in the source code. In the LCD demo program, the display messages are generated by outputting the characters sequentially from a data table. The usual technique, as seen in the keypad demo program, of adding a table offset to the program counter is used to generate the output codes. The table pointer is checked each time to see the end of the table has been reached. A more elegant method is to use the assembler directive DATA to define the fixed messages as text strings (see MPASM Assembler User Guide).

By contrast, the variable output count from 0−9 is calculated in real time. To obtain the corresponding ASCII code, 30h must be added to the number, as the ASCII for 0 is 30h, for 1 is 31h and so on. A 250ms delay is executed between each output to make the count visible.

4.4.4 LCD Display Modes

The LCD display module typically has its own controller, RAM and display latch/drivers, as shown in the block diagram in Figure 4.11. The RAM has to be loaded with codes which will produce the required pattern; these are then sent to the row and column drivers, which contain shift registers which store the data for one scan of the display. The drivers have one output for each row or column in the display matrix, which intersect to control each pixel.

The controller data sheet needs to be consulted for details of the operating procedure. In character mode, the display is divided into cells (typically 6 × 8 pixels) which are filled with character data from the controller ROM, corresponding to a pre-coded character set. In graphics mode, the user must work out the data code for each location which will produce the required pattern. The display addresses generally start at the top left and run across the screen in rows. Each pixel is controlled by the corresponding

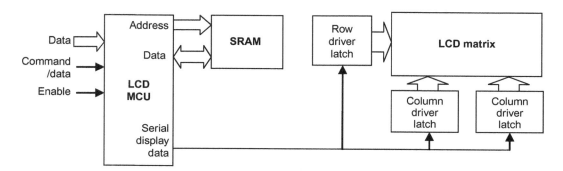

Figure 4.11
LCD display block diagram.

row and column driver output. The display RAM needs at least 1 bit per pixel; for example, if the display is 64×128, at least 1 kB is needed. More memory can store alternate screens.

Questions 4

1. Explain why a pull-up resistor is needed with a switch input. (3)
2. State three methods of switch debouncing. (3)
3. Briefly explain why hardware timers are useful in MCUs. (3)
4. Explain how a prescaler extends the timer period. (3)
5. Explain why the plain 7-segment LED display needs a code table. (3)
6. Explain why a BCD encoded display does not need a code table. (3)
7. Briefly explain the scanning process used to read a keypad input. (3)
8. Explain briefly how the LM016L LCD can be driven with only six outputs. (3)
9. Explain briefly the meaning of ASCII code. (2)
10. Draw a block diagram of a PIC system with a keypad and LCD display indicating the main components and signals in the system. (4)

Total (30)

Assignments 4

Access to Proteus VSM for 16F877A is required.

4.1 Keypad Test
Run the keypad project, KEYPAD2, in VSM. Ensure that the correct display is obtained when the keypad is operated. Represent the program using a flowchart and pseudocode. Explain the advantages of each. Explain why switch debouncing is not necessary in this particular application.

4.2 LCD Test
Run the LCD project, LCD2, in VSM. Ensure that the correct display is obtained; fixed messages should be followed by an incrementing count. Enable debug mode by hitting pause instead of

run, and select PIC CPU Source Code in the debug menu; also select PIC CPU Registers. Ensure that the debug windows are opened.

Single step in the source code window using the 'step into' button. Follow through the initialisation until the delay routine is entered. Why is the 'step out' button now useful? Start again using 'step over' — note the effect. Why is 'step over' useful? Single step through the output sequence, displaying the fixed characters one at a time, then the count.

Connect probes to the LCD data inputs and control signals, and generate a graph showing the codes entering the display. Demonstrate that the signals are consistent with the data being displayed.

CHAPTER 5

Data Processing

Summary
- Digital information is stored as numerical or character data
- A number system uses a base set of digits and column weighting
- Bases 2, 10 and 16 are most useful in microsystems
- The decimal equivalent is the sum of column weighted products
- Multiply and divide can be implemented using add and subtract
- The main numerical types are integers and floating point
- Negative numbers are represented by a sign bit or in 2s complement
- BCD is useful as it does not need converting for input and output
- ASCII is the standard character code, with text stored as strings
- Parallel memory can be used to expand data storage
- A simple calculator program demonstrates numerical processing

Most microcontroller programs need, at some point, to process data using arithmetic or logical operations, although large scale numerical storage and manipulation is not normally undertaken by the microcontroller, as it has, by definition, limited RAM.

The two main types of data are numbers and characters. There are only 26 letters in the alphabet and 10 numerical characters, and, even allowing for upper and lower case letters, only 62 codes are required to represent this minimal set. The basic ASCII character set uses a 7-bit code and includes most of the additional symbols found on a standard computer keyboard.

Numerical data is a bit more of a problem, as an 8-bit code can only represent numbers 0–255; various methods are needed to handle larger (and smaller) numbers, so that calculations can be performed with a useful degree of precision. The floating point number (as used in calculators) has decimal and exponent parts so that the whole range of useful values can be represented and processed.

5.1 Number Systems

Computers use binary numbers, but humans prefer to work in decimal. However, all number systems follow the same basic rules, relying principally on the concept of digits

Interfacing PIC Microcontrollers.
DOI: http://dx.doi.org/10.1016/B978-0-08-099363-8.00005-4
© 2014 Martin Bates and Elsevier Ltd. All rights reserved.

with column weighting for counting and arithmetic. The counting process can be described as follows:

1. *Select a range of digit symbols (e.g. base 10 symbols 0–9) and start at 0*
2. *Count up to the maximum value in the least significant column*
3. *When the maximum is reached in the column, increment the digit to the left*
4. *Reset the current column to 0*
5. *Repeat Steps 3 and 4, using more columns as necessary for larger numbers*

Hopefully, we all know already how to count in decimal (denary is the official name). The base of the number system is the number of digits used (10 in denary). Binary is base 2, octal is base 8, hex is base 16. Any base can be used in theory, but in practice some are more useful than others.

Historically, base 12 has been used extensively (hours, minutes, angles) and is useful because 12 can be divided by 2, 3, 4 and 6. But this is not a true number system because there are no discrete symbols for 10 and 11. Similarly, binary code decimal (BCD) is not a proper number system (see later for details) because its binary count stops at 9. Hexadecimal is a true number system because it uses discrete symbols for 10 (A) through to 15 (F). It is useful because it provides a compact way of representing binary, the native number system of all conventional digital computers and controllers.

Before digital hardware was available, computers were developed in which analog voltages represented denary values, with linear amplifier (op-amp) circuits providing the processing functions (see Chapter 7). These are capable of mathematical processes such as addition, subtraction, integration and differentiation, but the accuracy was limited by the signal and component quality. Nevertheless, analogue computers were successfully applied in, for example, ballistics and space travel to obtain useful results.

5.1.1 Denary

Zero was a very important innovation in mathematics, because all number systems depend on it. Another important idea is column weighting, and you can see the significance by simply analysing how the denary system works (Table 5.1). The number 7395 is analysed, with the column weighting following the progression, from right to left, 10^0, 10^1, 10^2 etc. to 10^n for the highest column. The total value of the number is then given by:

```
Total value = sum of(column weighting x digit value)
```

Thus, the column weighting is the base raised to incrementing powers, and the digit value is all the values from zero to base minus 1 (9). This may seem obvious, but it is important to state it clearly, as we will be implementing numerical processing which relies on this structure. Denary is the reference system, i.e. other systems are evaluated by reference to the base 10 system.

Table 5.1: Structure of a Denary Number.

Digit	7	3	9	5
Column weight	1000 (10^3)	100 (10^2)	10 (10^1)	1 (10^0)
Digit value	7×10^3	3×10^2	9×10^1	5×10^0
Total value		$7000 + 300 + 90 + 5 = \mathbf{7395}$		

5.1.2 Binary

Computers work in binary because it was found to be the most reliable and precise way to represent numbers electronically. The accuracy of the digital computer can be increased (in theory) to any required level by simply increasing the number of data bits used in the calculations. Processing in 32 bits provides a potential degree of precision of 1 part in 2^{32}, and even a modest 8 bits gives an accuracy of 1 part in 256, better than 0.5% error at full scale. Floating point binary extends the range of values represented at the expense of precision.

The analysis of the structure of a binary number (Table 5.2) shows that the decimal value can be calculated by multiplying the digit value (1 or 0) by the column weight, which is given by the power series of 2, the base value. Since the contribution of the columns containing 0 is 0, the decimal value can be worked out as follows:

```
Total value = sum of the column weight of non-zero bits
```

The maximum value that can be represented for a given number of bits is obtained when all the bits are 1. In an 8-bit number, the maximum value is $11111111_2 = 255_{10}$ (the subscript indicates the number base). This is calculated as $2^8 - 1$, i.e. two to the power of the base minus 1, and the number of different values available, including 0, is $2^8 = 256$. This is useful in defining memory capacity. The number of data locations (each usually storing 1 byte) addressed by N address bits is calculated as 2^N. Some useful reference values are:

```
2^8  = 256 bytes
2^10 = 1024 = 1 kb
2^16 = 65,536 = 64 kb
2^20 = 1,048,576 = 1 Mb
2^24 = 16,777,216 = 16 Mb
2^30 = 1,073,741,824 = 1 Gb
2^32 = 4,294,967,296 = 4 Gb
2^40 = 1,099,511,627,776 = 1 Tb
```

Note that for each additional bit, the binary range doubles.

Table 5.2: Structure of a Binary Number.

Digit	1	0	0	1
Column weight	$8\ (2^3)$	$4\ (2^2)$	$2\ (2^1)$	$1\ (2^0)$
Digit value	$1 \times 8 = 8$	$0 \times 4 = 0$	$0 \times 2 = 0$	$1 \times 1 = 1$
Total value		$8 + 0 + 0 + 1 = \mathbf{9}$		

Table 5.3: Structure of a Hex Number.

Digit	9	B	0	F
Column weight	$16^3 = 1000_{16} = 4096$	$16^2 = 100_{16} = 256$	$16^1 = 10_{16} = 16$	$16^0 = 1_{16} = 1$
Digit value	$9 \times 4096 = 36{,}864$	$B = 11 \times 256 = 2816$	$0 \times 16 = 0$	$F = 15 \times 1 = 15$
Total value		$36{,}864 + 2816 + 0 + 15 = \mathbf{36{,}695}$		

5.1.3 Hexadecimal

The same principle applies to the number system using base 16, hexadecimal (hex). The problem here is that extra numerical symbols are required, so symbols that are normally used as letters (A, B, C, D, E, F) are adopted as numbers, where $A_{16} = 10_{10}$ and so on up to $F_{16} = 15_{10}$. The subscript indicates the base of the number where there might be any ambiguity. The structure of the random hex number $9B0F_{16}$ is shown in Table 5.3.

Note the pattern in the progression of the hex column weight — the weighting is 0_{16}, 10_{16}, 100_{16}, 1000_{16}, etc. This applies to all number systems — the column weight is a progression of 0_n, 10_n, 100_n, 1000_n, etc., where n is the base. It can also be seen that the conversion from hex to denary is not simple, but the conversion from hex to binary is easier, which is why hex is useful.

5.1.4 Other Number Systems

Numbers can be represented using any base by following the rules outlined earlier. Octal (base 8) is sometimes used in industrial controllers but will not be considered any further here. Numbers with a base greater than 16 would need additional symbols; in theory, one could carry on using letters up to Z (base 36!). This could be useful as 36 is divisible by 18, 12, 6, 4, 3 and 2. 144 has also historically been used as a standard quantity because of its divisibility.

5.2 Numerical Conversion

Conversion between numerical types is often required in microprocessor systems. We have seen in Chapter 4 that input from a keypad may be acquired in BCD, processed in binary and output in ASCII. Machine code is normally displayed in hexadecimal, but we need to know how to convert back to binary. Conversion between decimal, binary and hex will be explained, and the usage of BCD, ASCII and floating point numbers outlined.

5.2.1 Binary to Decimal

As seen earlier, the value of a binary number is found by multiplying each digit by its decimal column weight and adding. The weighting of the digits in binary is from the least significant bit upwards: 1, 2, 4, 8, 16... or 2^0, 2^1, 2^2, 2^3..., i.e. the base of the number system is raised to the power 1, 2, 3... The conversion process for a sample 8-bit binary number is therefore:

$$\begin{aligned}
\mathbf{1001\ 0110_2} \quad &= (128 \times 1) + (64 \times 0) + (32 \times 0) + (16 \times 1) + (8 \times 0) + (4 \times 1) + (2 \times 1) + (1 \times 0) \\
&= 128 + 16 + 4 + 2 = 150_{10}
\end{aligned}$$

We can see that the process can be simplified to just adding the column weight for the bits that are not 0.

5.2.2 Decimal to Binary

The process is reversed for conversion from decimal to binary. The binary number is divided by two, the remainder recorded as a digit, and the result divided by two again, until the result is zero. For the same number 150_{10}:

```
150/2  =  75   rem 0   (Least Significant Bit, LSB)
 75/2  =  37   rem 1   Bit 6
 37/2  =  18   rem 1   Bit 5
 18/2  =   9   rem 0   Bit 4
  9/2  =   4   rem 1   Bit 3
  4/2  =   2   rem 0   Bit 2
  2/2  =   1   rem 0   Bit 1
  1/2  =   0   rem 1   (Most Significant Bit, MSB)
```

We then see that the binary result is obtained by transcribing the column of remainder bits from the bottom upwards (MSB to LSB).

5.2.3 Binary to Hex

Binary to hex conversion is simple—that is why hex is used. Each group of 4 bits is converted to the corresponding hex digit, starting with the least significant four, and padding with leading zeros if necessary:

```
1001   1111   0011   1101    = 9F3D₁₆
9      F      3      D
```

The reverse process is just as trivial, where each hex digit is converted to a group of 4 bits, in order. The result can be checked by converting both to decimal. First binary to decimal:

```
Bit  15  14  13  12  11  10  9  8  7  6  5  4  3  2  1  0
      1   0   0   1   1   1   1  1  0  0  1  1  1  1  0  1
```

$$= 2^{15} + 2^{12} + 2^{11} + 2^{10} + 2^9 + 2^8 + 2^5 + 2^4 + 2^3 + 2^2 + 2^0$$
$$= 32{,}768 + 4096 + 2048 + 1024 + 512 + 256 + 32 + 16 + 8 + 4 + 1$$
$$= 40{,}765_{10}$$

Now hex to decimal:

$$9F3D_{16} = (9 \times 16^3) + (15 \times 16^2) + (3 \times 16^1) + (13 \times 16^0)$$
$$= 36{,}864 + 3840 + 48 + 13$$
$$= 40{,}765_{10}$$

5.3 Binary Arithmetic

Some form of calculation is needed in most programs, even if it is a simple subtraction to determine whether an input is greater or less than a required level. At the other extreme, a computer-aided design program will carry out thousands of high precision operations per second when drawing a 3D graphic. Games programs are also among the most demanding of processor power, because 3D graphics must be generated at maximum speed. Here, we will cover just the basics so that control and communication processes that use simple arithmetic operations can be attempted later. Some of the operations outlined later will be demonstrated in a calculator demo program (see Section 5.6).

5.3.1 Addition

The simplest calculation is adding two numbers whose result is 255 or less, the maximum value for an 8-bit location. In PIC assembler, ADDWF (Add W to F) will give the right result with no further adjustment required. An example of binary addition is shown below; the conversion to decimal of each binary number is also shown to confirm that the result is correct.

Add (result <256)

```
              0111 0100   = 64 + 32 + 16 + 4   = 116
          +   0011 0101   = 32 + 16 + 4 + 1    =  53
              1010 1001   = 128 + 32 + 8 + 1   = 169
Carry bits    111   1
```

The carry bits within the 8-bit result are handled within the processor ALU. The PIC ALU records a carry from bit 3 to bit 4 in the DC bit of the status register (low to high nibble) since this is useful for BCD calculations.

Add (result > 255)

```
                  0111 0100   = 116
              +   1001 0000   = 144
Carry out  1      0000 0100   = 260₁₀
Carry bits        111
```

When the result of the 8-bit addition is greater than 255, the carry out of the MSB is recorded in the Carry (C) bit of the status register. This must be included in the result, and any further processing, for correct results. It must be added to the next most significant byte in the result of a multi-byte addition. A sample calculation of this type is shown below in binary and hex.

Add (multiple bytes)

```
              0111 0101 0101 0111   = 7557
              0001 1000 1100 1011   = 18CB
          +   1000 1110 0010 0010   = 8E22₁₆
Carry bits    111     11 1  11 111         11
```

5.3.2 Subtraction

Subtraction is straightforward if one number is subtracted from a larger one. In decimal, a digit borrowed from one column has a weight of 10 in the next lower weighted column. Similarly, in binary, the borrow has a value of two in the next lower column.

Subtract (positive result)

```
Borrowed digits    11           11
              1100 1011  =   203
            − 0110 0010  = − 98
              0110 1001  =   105
```

If a borrow is required into the MSB, the carry flag is used. Therefore, the carry flag must be set before a subtract operation so that a '1' is available to borrow. If the carry flag is found to be clear after the subtraction, a negative result is indicated, i.e. a larger number has been subtracted from a smaller.

Subtract (negative result)

```
Borrow into MSB 1 1100 1011  = 256 + 203
              − 1110 0010  =     − 226
                1110 1001  =       233
```

In this example, the borrow bit represents the least significant bit of the next byte, which has a value of 256_{10}. In multi-byte subtraction, the carry flag is used to transfer the borrow from one byte to the next. If the borrow is taken, the next highest byte must be decremented to 'take' the borrow from it. As can be seen in the decimal result, the 8-bit answer taken in isolation is incorrect. For this reason, subtraction is usually implemented using the 2s complement format, which is described later.

5.3.3 Multiplication

There are two basic methods for multiplication of binary numbers, successive addition and shift and add.

5.3.3.1 Successive Addition

A simple algorithm for multiplication is successive addition. For example:

$$3 \times 4 = 4 + 4 + 4$$

That is, add four, three times. The general process is outlined below:

```
Load Num1 & Num2 registers
Clear Result register
Load Count register with Num1
Loop Add Num2 to Result
       Decrement Count
Until Count = 0
Read Result
```

One number is loaded into a Count register, and the other added to a Result register, which has been initialised to zero. Count is decremented and the addition is repeated until the count is zero. The result of the multiplication is then in the Result register. Carry handling is required as described earlier if the result overflows from one register to the next.

5.3.3.2 Shift and Add

An alternative method is shift and add, which is more efficient for larger numbers. It is based on conventional long multiplication, but when implemented in binary, the process can be simplified because the multiplier contains only 1s and 0s:

```
    1101        =  13  (multiplicand)
x   0110        =  06  (multiplier)
    0000
   11010
  110100
 0000000
 01001110       =  78
```

Where the multiplier is a 0, the result must be 0, so that operation can be skipped, and the non-zero sub-totals obtained by shifting, and then adding to a running total, as follows:

```
Clear a Result register
Get Bit0 in multiplier
Loop IF multiplier bit is 1
         add multiplicand to Result
    Shift multiplicand left
    Get next multiplier bit
Until last bit done
```

The multiplier bit can be tested by rotating it into the carry bit. This process is implemented internally in hardware in the higher performance MCUs. A multiply operation is then available in the instruction set.

5.3.4 Division

Divide is the inverse of multiply, so can be implemented using successive subtraction. The divisor is subtracted from the dividend, and a counter incremented. This process is repeated until the result goes negative; this is detected by the carry flag being cleared, so it must be set before the process starts. The remainder is then corrected by adding the divisor back on to the negative dividend, leaving a positive remainder in the dividend register, and decrementing the result in the counter to compensate for going one step too far.

```
Load Dividend & Divisor register
Set Carry flag
Loop Subtract Divisor from Dividend
        Increment Result
Until Carry flag clear
Add Divisor back onto Dividend
Decrement Result
```

As for multiplication, a divide instruction may be provided in hardware in higher performance MCUs.

5.4 Numerical Types

There are several types of numerical variable used in computer and microcontroller systems (Table 5.4). The range of numbers and their precision is determined by the number of bits used and how they are allocated. The simplest are unsigned integers (whole numbers), usually having 8, 16 or 32 bits, representing positive numbers only, up to a limited maximum. Signed variables use the most significant bit to represent the sign, positive or negative. However, the sign bit must be processed separately from the numerical value, according to the usual rules of arithmetic, to obtain the correct result. This problem is overcome using 2s (twos) complement numbers, where the sign is represented by the carry flag and included in the calculations.

To represent the whole range of large, small and fractional numbers with a fixed number of bits requires a different approach. In the scientific calculator, it is achieved using a decimal number (mantissa) with an exponent multiplier that positions the decimal point. Floating point numbers

Table 5.4: Numerical Variable Types.

Numerical Format	Range	Precision
8-bit unsigned integer	0 to 255	1%
16-bit unsigned integer	0 to 65,535	0.002%
32-bit unsigned integer	0 to 4,294,967,296	10^{-9}
8-bit signed integer	± 127	2%
16-bit signed integer	$\pm 32,767$	0.01%
32-bit signed integer	$\pm 2,147,483,647$	10^{-9}
8-bit 2s complement	0 to -255	1%
16-bit 2s complement	0 to $-65,535$	0.002%
32-bit 2s complement	0 to $-4,294,967,296$	10^{-9}
16-bit floating point	$\pm 10^5$	0.2%
32-bit floating point	$\pm 10^{38}$	10^{-6}
64-bit floating point	$\pm 10^{308}$	10^{-15}

in processor systems typically use 16, 32 or 64 bits with groups of bits assigned to represent the mantissa and exponent. This numerical format is detailed in Section 5.4.5.

The precision of each type is indicated in the table as the change between adjacent values in the mid-range. For example, for the 8-bit unsigned integer, this is 1/128, or approximately 1%. When the value is displayed, the number of digits available determines the precision. For example, a typical digital meter has a maximum display value of 1.999, which corresponds to a precision of just under 1 in 2000, or 0.05%, at full scale, and 0.1% at mid range (1.000).

5.4.1 Positive Integers

An integer is a whole number without fractional part. An 8-bit location can store integers from 0 to 255 in binary. This is an obvious limitation in programs that may need to calculate results up to, say, four significant decimal digits (0–9999). 16-bit integers can take us up to 65,535 ($2^{16} - 1$) as a maximum value, which covers this range comfortably, while 32-bit integers have a maximum value of 4,294,967,295. As long as the result is in the positive range, the results from arithmetic operations will be correct. However, negative and fractional numbers cannot be handled.

5.4.2 Negative Integers

Negative numbers can be represented by an integer value with the most significant bit assigned to represent the sign, where 0 = positive and 1 = negative. This reduces the range by half, and the sign bit must be manipulated explicitly to maintain its correct value after a calculation. Fortunately, a more coherent method for implementing integer arithmetic is available, using 2s complement numbers.

When a register is incremented beyond its range, it rolls over from all 1s to all 0s, and repeats the count from 0. Conversely, when decremented past 0, it rolls under to the maximum value. In an 8-bit register, the value goes from 00h to FFh. The negative going count from zero is shown below in binary in the left column, with its hex equivalent in the next column. These values can represent negative numbers, as shown in the last column:

```
0000 0000₂   ==   00₁₆   ==   000₁₀
------------------------------------
1111 1111    ==   FF     ==   − 001
1111 1110    ==   FE     ==   − 002
1111 1101    ==   FD     ==   − 003
1111 1100    ==   FC     ==   − 004
1111 1011    ==   FB     ==   − 005
---- ----         --          ---
etc
---- ----         --          ---
```

```
1000 0010    = =    82    = =    − 126
1000 0001    = =    81    = =    − 127
1000 0000    = =    80    = =    − 128
0111 1111    = =    7F    = =    − 129
---- ----           --           ---
etc
---- ----           --           ---
0000 0010    = =    02    = =    − 254
0000 0001    = =    01    = =    − 255
```

Taking a random value, the binary number 0000 0010 could mean either $+2$ or -254. However, if we use the carry bit (C) in the ALU to represent the sign, (1)00000010 means $+2$ and (0)00000010 means -254. In this way, negative numbers down to -255 can be represented in the working register of the PIC MCU, where the carry flag stores the sign bit, with one (1) signifying a positive number and zero (0) negative (this is the reverse of a signed binary convention).

5.4.3 2s Complement Arithmetic

This principle can be used in signed arithmetic using the carry flag. If the carry flag is set before the operation, it will be cleared afterwards to indicate a negative result. An example is shown below in hex and binary:

```
Calculate: 2−7 = −5
```

```
+ 02    = = (1)0000 0010
− 07    = = −  0000 0111
──────
− 05    = = (0)1111 1011 (FB)
```

The operation (e.g. SUBWF) causes a borrow into the MSB of the first number, clearing the carry flag. 7 is then subtracted from 102_{16} (02_{16} plus 100_{16} borrowed) to give the result FB_{16}. Because the carry flag is clear, this result is interpreted as the 2s complement of 5, i.e. -5.

If a positive number is added to a 2s complement negative number, giving a positive result, the correct positive integer is obtained and the carry flag set, due to the carry out from the destination register, signifying a positive result. For example:

```
Calculate: −4 + 7 = +3
```

```
− 04    = = (0)1111 1100
+ 07    = = +  0000 0111
──────
+ 03    = = (1)0000 0011
```

5.4.4 2s Complement Conversion

If necessary, a corresponding positive number can be derived from the 2s complement negative number by applying the process, **invert all bits and add 1**. For example:

$$- 000_{10} = = 00 = = 0000\ 0000 \rightarrow 1111\ 1111 \rightarrow 0000\ 0000 = = +000_{10}$$
$$- 003_{10} = = FD = = 1111\ 1101 \rightarrow 0000\ 0010 \rightarrow 0000\ 0011 = = +003_{10}$$
$$- 127_{10} = = 81 = = 1000\ 0001 \rightarrow 0111\ 1110 \rightarrow 0111\ 1111 = = +127_{10}$$
$$- 129_{10} = = 7F = = 0111\ 1111 \rightarrow 1000\ 0000 \rightarrow 1000\ 0001 = = +129_{10}$$
$$- 255_{10} = = 01 = = 0000\ 0001 \rightarrow 1111\ 1110 \rightarrow 1111\ 1111 = = +255_{10}$$

Note that when converting 0, the carry is discarded. Conversion to the 2s complement negative form from the positive equivalent number is achieved by the inverse process, i.e. **subtract 1 and invert all bits**. For example:

$$1 = = 0000\ 0001 \rightarrow 0000\ 0000 \rightarrow 1111\ 1111 = = FF = = -001$$
$$63 = = 0011\ 1111 \rightarrow 0011\ 1110 \rightarrow 1100\ 0001 = = C1 = = -063$$
$$128 = = 1000\ 0000 \rightarrow 0111\ 1111 \rightarrow 1000\ 0000 = = 80 = = -128$$
$$255 = = 1111\ 1111 \rightarrow 1111\ 1110 \rightarrow 0000\ 0001 = = 01 = = -255$$

The 2s complement form allows the usual arithmetic operations to be applied to negative binary numbers and the correct result obtained. An 8-bit operation will produce results precise to about 0.4%. Multi-byte integers can be processed using 2s complement arithmetic, provided that appropriate carry and borrow handling is incorporated. The 16-bit integers can then represent the range $+65,535$ to $-65,535$, giving results precise to about 0.0015%.

5.4.5 Floating Point Numbers

The plain integer format has a limited range and cannot represent fractional decimal numbers easily. An alternative format is needed to represent positive and negative numbers with a greater range, from very small to very large. On a calculator, scientific notation is used. An example of a large number stored in this format is 2.3615×10^{73}, while 6.9248×10^{-23} is a small one. The decimal part is called the mantissa and the range multiplier is the exponent.

These values are generally input and output as base 10 numbers, but internally are represented by an equivalent binary floating point value. An example of a large negative number could be $-1.011001010 \times 2^{10011}$, and a small positive one $1.0101001101 \times 2^{-01001}$. The common standard is IEEE 754 floating point format, which specifies 32-bit single precision and 64-bit double precision numbers. There is also a 16-bit low-precision option.

The allocation of bits in the 32-bit form are shown in Table 5.5(a), with representative bit values. The sign is the MSB (Bit 31), the exponent the next eight bits (30−23), and the

Table 5.5a: Floating Point Number.

Structure of a 32-Bit Floating Point Number											
Sign				**Exponent**					**Exponent Values**		
Bit#	31	30	29	28	27	26	25	24	23	0000 0000 = −126	
Value	1	1	0	0	1	0	0	1	0	0111 1111 = 0	
										1111 1111 = +127	

Mantissa											
Bit#	22	21	20	19	18	17	16	15	14	13	12	11
Weight	2^{-1}	2^{-2}	2^{-3}	2^{-4}	2^{-5}	2^{-6}	2^{-7}	2^{-8}	2^{-9}	2^{-10}	2^{-11}	2^{-12}
Value	1	0	1	1	0	1	0	0	0	0	0	0
Bit#	10	9	8	7	6	5	4	3	2	1	0	
Weight	2^{-13}	2^{-14}	2^{-15}	2^{-16}	2^{-17}	2^{-18}	2^{-19}	2^{-20}	2^{-21}	2^{-22}	2^{-23}	
Value	0	0	0	0	0	0	0	0	0	0	0	

significand (significant digits) the remaining 23 bits (22−0). The required range of positive and negative exponents from −126 to +127 are represented by an 8-bit unsigned integer, with 01111111 equivalent to 0. The mantissa always starts with a bit 1, so this does not need to be encoded. The bits have the fractional bit weightings of 0.5, 0.25, 0.125. . ., so the significand covers the range from exactly 1 to approximately 2. It is calculated by assigning weightings of 2^{-1}, 2^{-2}, 2^{-3}, . . ., 2^{-22}, 2^{-23} to the bits.

The representative 32-bit floating point number is analysed below in order to explain how it represents a decimal value. A limited number of the significant bits are used to simplify the calculation.

Bits:	*Sign(31), Exponent(30−24), Significand(23−0)*
Example:	**1 10010010 10110100000000000000000**
Sign:	*Bit 31 = 1 → negative number*
Exponent:	*Bits 30−23 = 10010010*
	Denary = 128 + 16 + 2 = 146
	Exponent index = 146−127 = +19
	Exponent multiplier = 2^{+19} = 524,288
Significant Fraction:	$2^{-1} + 2^{-3} + 2^{-4} + 2^{-6}$ *= 1/2 + 1/8 + 1/16 + 1/64*
	= 0.5 + 0.125 + 0.0625 + 0.015625 = 0.703125
Significand:	*Fraction + 1 = 1.703125*
Decimal:	*Significand × Exponent Multiplier = −1.703125 × 524,288 = −892,928*
Result:	**−8.92928 × 10⁵**

The exponent multiplier is calculated by converting the binary exponent value to decimal, subtracting 127 to normalise the range, and raising 2 to this power. The mantissa is found by adding the weightings of the fraction bits to form a number in the range 0 to 1. One is

Table 5.5b: Floating Point Number.

Bit Assignment in Floating Point Numbers				
Size	Sign	Exponent	Significand	Precision
16-bit	1	5	10	Half
32-bit	1	8	23	Single
64-bit	1	11	52	Double

then added to give a mantissa range of 1.0000 to 1.9999. The result can then be calculated as a decimal and converted to scientific notation.

The maximum absolute value represented in 32-bit floating point format is given approximately by the maximum exponent value, 2^{+127}, or about 10^{38}. The resolution is determined by the smallest fraction, 2^{-23}, or 10^{-7}. This illustrates the advantages of the floating point format very clearly — the range and resolution are both extremely high.

The main floating point formats are compared in Table 5.5(b). A high-resolution option uses 64 bits, and a low-precision format is also specified in the standard using 16bits. This has 5 exponent bits (0 offset at 15_{10}) and 10 significant bits. This means the range is about 2^{+17}, or about 10^5, and the resolution is 2^{-10}, or about 10^{-3} (0.1%). This could generate results precise enough for a standard 3½ digit display (0.000−1.999), with an error of 2 LSD (least significant digits) at full scale.

An alternative format, 32-bit format, used in Microchip C compilers, assigns bit 23 as the sign bit, leaving the complete high byte representing the exponent, which is probably easier to process. Otherwise, the numerical representation is the same. When programming in higher level languages, the available floating point formats will be predefined and internal or library functions provided to handle floating point calculations.

5.5 BCD and ASCII

BCD (Binary Coded Decimal) is not a proper number system, but it is nevertheless very useful. In BCD, the numbers 0−9 are represented by their binary equivalent and stored as 4-bit numbers. These are easily converted into ASCII code (see below) for sending to a display or communications port, or into pure binary for processing.

5.5.1 BCD Calculations

Calculators have traditionally used BCD arithmetic, since it does not need to be converted between the input, process and output. Integers, fractions and decimals may be processed one digit at a time using the rules of conventional arithmetic. It therefore has significant attractions for the assembly programmer, despite being less elegant than floating point arithmetic. The PIC ALU also has some helpful 4-bit features, such as the digit carry flag and nibble swapping, which facilitate BCD arithmetic.

Table 5.6 shows some examples of the basic operations. Simple binary addition and subtraction are used (ADDWF, SUBWF, etc.). After addition, the result is tested, to check if the result is greater than 9 (not shown). This means setting the carry bit, loading a spare register with 9 and subtracting the result from it. If the carry flag is cleared, the result was greater than 9. The low digit must then be calculated by subtracting 10 from the result (make sure it remains unchanged), and the next higher digit set to 1, or incremented. This process is extended in the multiply process, which is implemented by successive addition.

Figure 5.1 shows an algorithm for the addition of BCD digits. This type of process can be devised and implemented for all BCD arithmetic operations. The results can be input from a

Table 5.6: Sample BCD Calculations.

	Add (No Carry)	Add with Carry	Multiply by Addition
Decimal	$2 + 3 = 5$	$7 + 6 = 13$	$3 \times 9 = 27$
BCD	$\begin{aligned}&0010\\+&0011\\=&0101\end{aligned}$	$\begin{aligned}&0111 = 7\\+&0110 = 6\\=&1101 = 13\\-&1010 = 10\\=&0011 = 3\end{aligned}$	$\begin{aligned}&1001 = 9\\+&1001 = 9\\=&10010 = 18\\-&1010 = 10\\=&1000 = 8\\+&1001 = 9\\=&10001 = 17\\-&1010 = 10\\=&0111 = 7\end{aligned}$

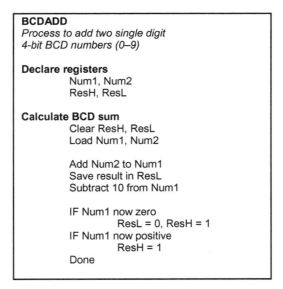

```
BCDADD
Process to add two single digit
4-bit BCD numbers (0–9)

Declare registers
        Num1, Num2
        ResH, ResL

Calculate BCD sum
        Clear ResH, ResL
        Load Num1, Num2

        Add Num2 to Num1
        Save result in ResL
        Subtract 10 from Num1

        IF Num1 now zero
                ResL = 0, ResH = 1
        IF Num1 now positive
                ResH = 1
        Done
```

Figure 5.1
Outline of BCD add with carry.

keyboard as BCD, and output as ASCII without conversion to binary. However, if the arithmetic required is more complex, conversion to signed binary or floating point format may be necessary.

5.5.2 BCD to Binary Conversion

The input from a numeric keypad is likely to be in BCD, i.e. binary numbers 0–9 representing each key. When multi-digit numbers are input, the keys are pressed in the sequence from the highest significant digit to the lowest. This sequence may need to be converted into the corresponding binary after the sequence is complete, usually indicated by pressing an enter, or other function key. The decimal input number may have several digits, from 1 to the maximum allowed by the binary format to which it is converted.

Let us assume the system handles 16-bit positive integers only; the range will be $0-65,535_{10}$. We will therefore limit the input to four digits, with a maximum of 9999_{10}. The key inputs will be stored in temporary registers, then converted to the equivalent 16-bit binary when an enter key is pressed. The process will have to detect if four, or fewer, digits have been entered. It must then add the lowest digit (last key) to a previously cleared register pair (2×8 bits), multiply the next digit by 10, add the result to the running total, multiply the next by 100, add it to the total, and multiply the highest digit (first key) by 1000, and add it to the total.

The process is illustrated in Figure 5.2 for four-digit input, but it can be extended as far as the integer size allows. A set of registers is assigned and cleared, and a keypad scanning routine reads in keys as 4-bit BCD codes stored in the low nibble of the BCD registers. The codes are shifted to the next higher digit register after each input stroke, to allow the next input digit to be stored in BCD0. A maximum of four digits are stored in BCD4–BCD1. If the enter code is detected as input before four keys have been entered, the loop quits with the digits in the correct registers, with the leading digits left at 0.

If a 12-button telephone style keypad is used for input, the star key (*) could be used as 'enter' to terminate the number, and hash (#) to restart the input sequence (clear). These could be assigned codes A_{16} and B_{16}, and checked in the BCD0 input register, with the keys 0–9 assigned the corresponding BCD code and shifted into the digit registers. The digits are then multiplied by their digit weighting and added to a running binary total in a pair of 8-bit registers. The binary multiplication by 10, 100 and 1000 can be implemented by an adding loop, which is simpler, or shifting and adding, which is more efficient. Note that the calculated sub-totals must be added in low byte, high byte order and the carry flag handled to obtain the correct 16-bit total.

Binary to BCD conversion for output may be implemented as the inverse process: divide by 1000, 100 and 10 and store the results as BCD digits; the last remainder is the units digit. These processes are implemented in the calculator program in Section 5.6.

```
BCDTOBIN
Converts 4 digits BCD keypad input into 16-bit binary
Inputs: Up to 4 BCD codes 0 – 9
Output: 16-bit binary code

Declare Registers
          BCD0, BCD1, BCD2, BCD3, BCD4
          BINHI, BINLO
          Keycount
          Clear all registers

Read in BCD digits from keypad
          REPEAT
                    Read key into BCD0
                    Shift all BCD digits left
                    Increment Keycount
          UNTIL Return OR Keycount = 4

Calculate binary
          Add BCD1 to BINLO
          Multiply BCD2 by 10
          Add BCD2 to BINLO
          Multiply BCD3 by 100
          Add BCD3 to BINHI+BINLO
          Multiply BCD4 by 1000
          Add BCD4 to BINHI+BINLO

          Done
```

Figure 5.2
Outline of BCD to binary conversion.

5.5.3 Characters and Strings

The standard coding method for text characters is ASCII (American Standard Code for Information Interchange). This provides a binary code for most of the characters found on a standard computer keyboard using a 7-bit code. The eighth bit can be used for simple error checking using parity (see Chapter 8) or for an extended character set. The 7-bit ASCII codes are shown in Table 5.7. The low and high bits for each character must be merged, e.g. upper case 'A' is 0100 0001 (41 in hex or 65 in decimal). ASCII code was developed primarily for use in serial communications interfaces and generated by so-called dumb terminals on mainframe computers. It is commonly output by keyboard interfaces and recognised by many text display devices, including the standard alphanumeric displays already seen in Chapter 4.

A sequence of characters is frequently needed to form a complete message. Therefore, ASCII codes are often stored in successive memory locations. A simple method of sequential access to output a fixed message is illustrated in the display demo Program 4.5, where the program counter is used as the table pointer. To create the text table, the

Table 5.7: ASCII Character Codes.

Low Bits	High Bits					
	0010	**0011**	**0100**	**0101**	**0110**	**0111**
0000	Space	0	@	P	`	p
0001	!	1	A	Q	a	q
0010	"	2	B	R	b	r
0011	#	3	C	S	c	s
0100	$	4	D	T	d	t
0101	%	5	E	U	e	u
0110	&	6	F	V	f	v
0111	'	7	G	W	g	w
1000	(	8	H	X	h	x
1001	)	9	I	Y	i	y
1010	*	:	J	Z	j	z
1011	+	;	K	[	k	{
1100	,	<	L	\	l	|
1101	—	=	M	]	m	}
1110	.	>	N	^	n	~
1111	—	?	O	_	o	

characters are placed in single quotes, and the assembler converts them to the corresponding ASCII codes.

Type conversion of numerical characters is straightforward, because the ASCII code for '0' is 30h, the code for '1' is 31h and so on until up to 39h for '9'. Therefore, to convert BCD or binary numbers up to 9 to ASCII, just add 30h. To convert ASCII to BCD, subtract 30h. These conversions are used after reading the numerical keys and to display BCD data on an LCD display in the calculator program that follows.

5.6 *Calculator Application*

A basic calculator application (VSM project CALC2) will be used to illustrate some of the techniques of arithmetic processing discussed earlier. The hardware, consisting of a 16F877A MCU, arithmetic keypad and 16 × 2 LCD display, is shown in Figure 5.3. The keypad has 16 keys: ten numeric buttons, four arithmetic operations, equals and clear. The results are displayed on the first line of the LCD display, which receives the characters as ASCII codes in 4-bit mode (see Chapter 4). To keep it simple, the program is limited to single-digit input and double-digit results. This allows the algorithms for the arithmetic operations to be more easily understood; the same principles can then be extended to multi-digit calculations.

To perform a calculation, the user presses a number key, then an operation key, then another number, then equals. The calculation and result are displayed. For the divide operation, the

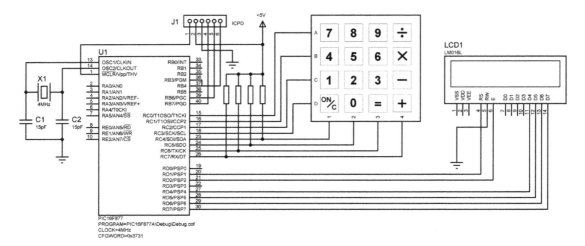

Figure 5.3
Calculator schematic.

result is displayed as an integer and remainder. The clear key will then erase the current display, and a new calculation can be entered. If an invalid key sequence is entered, the program must be restarted.

5.6.1 Calculator Hardware

In order to leave Port B available for in-circuit programming and debugging (ICD), the peripheral devices are connected to Ports C and D. The 16-button keypad is scanned by row and column as previously described (Section 4.3). The row outputs are programmed to default high. The column inputs also default high due to pull-up resistors. Each is then taken low in turn by outputting a logic 0 at RC0–RC3. Each input is tested for 0 when each row is active, and the numerical key value or symbol acquired and displayed if pressed.

In this program, the ASCII code is generated individually for each key, giving a rather lengthy scanning process, but one that is simple to understand. All keys are also output immediately to the display. The program checks the keypad twice, once to acquire the key and then to check if the key has been released. This ensures that the next keypad scan does not start until the previous one has been properly completed.

The LCD operates in 4-bit mode, as described in Section 4.4. The ASCII codes are sent in high nibble, low nibble order, and each nibble is latched into the display by pulsing input E. The R/W (read/not write) line is tied low, as reading from the display is not required. RS (register select) is toggled high for data input and low for commands.

5.6.2 Calculator Program

As the program is relatively lengthy, an outline of the algorithm is shown in Figure 5.4, and the corresponding source code is listed in Program 5.1.

The standard P16F877A register label file is included in the source code, and the initialisation of the ports carried out. A separate file for the LCD driver routines has been created (LCD.INC) to keep the source code size down and to provide a reusable file for future programs. This is included at the top of the subroutine section. It contains the LCD initialisation sequence (inid) and code transmission block (send), as seen in the LCD demo program in Chapter 4.

The main program sequence calls the keypad scanning routine to detect a key, in which the key code is stored, and then delays for switch debouncing and release. The input key is displayed, and the program then jumps to a routine to handle each input in a sequence of five buttons (Num1, Operation, Num2, Equals and Clear). The calculation routine then uses the operation input code to select the required process: add, subtract, multiply or divide. The binary result of the calculation is passed to a routine to convert it to BCD, then ASCII, and send it to the display. The result of the divide, being a single-digit result and remainder, is sent direct to the display. The clear operation sends a command to the display to clear the last set of characters.

5.7 Data Storage

The typical microcontroller has limited data memory. The 16F877A and 16F887 have only 368 bytes of RAM in the form of GPRs in the file register block. If more data storage is needed, for example, to save input data samples over a longer period of time, there are three choices:

1. *Use a standard PC and operating system with special interfaces*
2. *Design a conventional microprocessor-based system*
3. *Expand the memory in a microcontroller*

The first and second are relatively expensive options, both in terms of hardware cost and development time. It depends on the system specification as to whether this is justified. In this book, we will only consider the less-expensive microcontroller-based options, of which there are, again, three:

1. *Choose an MCU with more RAM*
2. *Fit a smaller one with external parallel memory*
3. *Fit a smaller one with external serial memory*

The first option will be considered later in Chapter 10. Serial memory will be described later in the section on the I^2C interface (Chapter 8, Section 3). Here, we will consider parallel memory, an established technology used for many years in conventional

```
CALC
Single digit calculator produces two digit results.
Hardware:    x12 keypad, 2x16 LCD, P16F887 MCU
+++++++++++++++++++++++++++++++++++++++++++++++++++++++++++++++++++++++
MAIN PROGRAM

        Initialise
                PortC = keypad
                        RC0 – RC3 = output rows
                        RC4 – RC7 = input columns
                PortD = LCD
                        RD1, RD2 = control bits
                        RD4– RD7 = data bits
                Initialise display

        Scan Keypad
                REPEAT
                        Keypad input
                        Delay 50ms for debounce
                        Keypad input
                        Check key released

                        IF first key, load Num1,
                                        Display character
                                        restart loop
                        IF second key,  load sign
                                        Display character
                                        restart loop
                        IF third key, load Num2
                                        Display character
                                        restart loop
                        IF fourth key
                                        Calculate result
                        IF fifth key
                                        Clear display
                ALWAYS

SUBROUTINES +++++++++++++++++++++++++++++++++++++++++++++++++++++

                Included LCD driver routines
                        Initialise display
                        Display character

        Keypad Input
                Check row
                                IF key pressed, load ASCII code
                        ELSE load zero code

        Calculate result
                IF key = '+', Add
                IF key = '-', Subtract
                IF key = 'x', Multiply
                IF key = '/', Divide

                Add             Add Num1 + Num2
                                Two digits

                Subtract        Subtract Num1 – Num2
                                IF result negative, add minus sign
                                Display character

                Multiply
                                REPEAT
                                        Add Num1 to Result
                                        Decrement Num2
                                UNTIL Num2 = 0
                                Two digits

                Divide
                                REPEAT
                                        Subtract Num2 from Num1
                                        Increment Result
                                UNTIL Num1 negative

                                Restore Remainder
                                Load Result
                        Display character
                                Load Remainder
                                Display character

        Two digits
                Divide result by 10
                Load MSD
                Display character
                Load LSD
                Display character
```

Figure 5.4
Outline of calculator program.

```
;;;;;;;;;;;;;;;;;;;;;;;;;;;;;;;;;;;;;;;;;;;;;;;;;;;;;;;;;;;
;
;          CALC2.ASM MPB        12-01-13
;
;          Simple calculator
;          Single digit input, two digit results
;          Integer handling only
;       Updated for VSM v8
;
;;;;;;;;;;;;;;;;;;;;;;;;;;;;;;;;;;;;;;;;;;;;;;;;;;;;;;;;;;;

            PROCESSOR 16F877
            __CONFIG 0x3731          ; Clock = XT 4MHz

; LABEL EQUATES     ;;;;;;;;;;;;;;;;;;;;;;;;;;;;;;;;;;;;;

            INCLUDE "P16F877A.INC"

Char      EQU      30        ; Display character code
Num1      EQU      31        ; First number input
Num2      EQU      32        ; Second number input
Result    EQU      33        ; Calculated result
Oper      EQU      34        ; Operation code store
Temp      EQU      35        ; Temporary register for subtract
Kcount    EQU      36        ; Count of keys hit
Kcode     EQU      37        ; ASCII code for key
Msd       EQU      38        ; Most significant digit of result
Lsd       EQU      39        ; Least significant digit of result
Kval      EQU      40        ; Key numerical value

RS        EQU      1         ; Register select output bit
E         EQU      2         ; Display data strobe

; Program begins ;;;;;;;;;;;;;;;;;;;;;;;;;;;;;;;;;;;;;;;;;

            CODE     0                  ; Default start address
            NOP                         ; required for ICD mode

            BANKSEL  TRISC              ; Select bank 1
            MOVLW    B'11110000'        ; Keypad direction code
            MOVWF    TRISC              ;
            CLRF     TRISD              ; Display port is output

            BANKSEL  PORTC              ; Select bank 0
            MOVLW    0FF                ;
            MOVWF    PORTC              ; Set keypad outputs high
            CLRF     PORTD              ; Clear display outputs
            GOTO     start              ; Jump to main program


; MAIN LOOP ;;;;;;;;;;;;;;;;;;;;;;;;;;;;;;;;;;;;;;;;;;;;;;;

start     CALL     inid               ; Initialise the display
          MOVLW    0x80               ; position to home cursor
          BCF      Select,RS          ; Select command mode
          CALL     send               ; and send code

          CLRW     Char               ; ASCII = 0
          CLRW     Kval               ; Key value = 0
          CLRW     DFlag              ; Digit flags = 0

scan      CALL     keyin              ; Scan keypad
          MOVF     Char,1             ; test character code
          BTFSS    STATUS,Z           ; key pressed?
          GOTO     keyon              ; yes - wait for release
          GOTO     scan               ; no - scan again

keyon     MOVF     Char,W             ; Copy..
          MOVWF    Kcode              ; ..ASCIIcode
          MOVLW    D'50'              ; delay for..
          CALL     xms                ; ..50ms debounce

wait      CALL     keyin              ; scan keypad again
          MOVF     Char,1             ; test character code
          BTFSS    STATUS,Z           ; key pressed?
          GOTO     wait               ; no - rescan
          CALL     disout             ; yes - show symbol
```

Program 5.1

Calculator program source code.

```
          INCF      Kcount             ; inc count..
          MOVF      Kcount,W           ; ..of keys pressed
          ADDWF     PCL                ; jump into table
          NOP
          GOTO      first              ; process first key
          GOTO      scan               ; get operation key
          GOTO      second             ; process second symbol
          GOTO      calc               ; calculate result
          GOTO      clear              ; clear display

first     MOVF      Kval,W             ; store..
          MOVWF     Num1               ; first num
          GOTO      scan               ; and get op key

second    MOVF      Kval,W             ; store..
          MOVWF     Num2               ; second number
          GOTO      scan               ; and get equals key


; SUBROUTINES ;;;;;;;;;;;;;;;;;;;;;;;;;;;;;;;;;;;;;;;;;;;;;;

; Include LCD driver routine

          INCLUDE   "LCD.INC"


; Scan keypad ...........................................

keyin     MOVLW     00F                ; deselect..
          MOVWF     PORTC              ; ..all rows
          BCF       PORTC,0            ; select row A
          CALL      onems              ; wait output stable

          BTFSC     PORTC,4            ; button 7?
          GOTO      b8                 ; no
          MOVLW     '7'                ; yes
          MOVWF     Char               ; load key code
          MOVLW     07                 ; and
          MOVWF     Kval               ; key value
          RETURN

b8        BTFSC     PORTC,5            ; button 8?
          GOTO      b9                 ; no
          MOVLW     '8'                ; yes
          MOVWF     Char
          MOVLW     08
          MOVWF     Kval
          RETURN

b9        BTFSC     PORTC,6            ; button 9?
          GOTO      bd                 ; no
          MOVLW     '9'                ; yes
          MOVWF     Char
          MOVLW     09
          MOVWF     Kval
          RETURN

bd        BTFSC     PORTC,7            ; button /?
          GOTO      rowb               ; no
          MOVLW     '/'                ; yes
          MOVWF     Char               ; store key code
          MOVWF     Oper               ; store operator symbol
          RETURN

rowb      BSF       PORTC,0            ; select row B
          BCF       PORTC,1
          CALL      onems

          BTFSC     PORTC,4            ; button 4?
          GOTO      b5                 ; no
          MOVLW     '4'                ; yes
          MOVWF     Char
          MOVLW     04
          MOVWF     Kval
          RETURN

b5        BTFSC     PORTC,5            ; button 5?
          GOTO      b6                 ; no
          MOVLW     '5'                ; yes
```

Program 5.1
(Continued)

```
                MOVWF      Char
                MOVLW      05
                MOVWF      Kval
                RETURN

b6              BTFSC      PORTC,6           ; button 6?
                GOTO       bm                ; no
                MOVLW      '6'               ; yes
                MOVWF      Char
                MOVLW      06
                MOVWF      Kval
                RETURN

bm              BTFSC      PORTC,7           ; button x?
                GOTO       rowc              ; no
                MOVLW      'x'               ; yes
                MOVWF      Char
                MOVWF      Oper
                RETURN

rowc            BSF        PORTC,1           ; select row C
                BCF        PORTC,2
                CALL       onems

                BTFSC      PORTC,4           ; button 1?
                GOTO       b2                ; no
                MOVLW      '1'               ; yes
                MOVWF      Char
                MOVLW      01
                MOVWF      Kval
                RETURN

b2              BTFSC      PORTC,5           ; button 2?
                GOTO       b3                ; no
                MOVLW      '2'               ; yes
                MOVWF      Char
                MOVLW      02
                MOVWF      Kval
                RETURN

b3              BTFSC      PORTC,6           ; button 3?
                GOTO       bs                ; no
                MOVLW      '3'               ; yes
                MOVWF      Char
                MOVLW      03
                MOVWF      Kval
                RETURN

bs              BTFSC      PORTC,7           ; button -?
                GOTO       rowd              ; no
                MOVLW      '-'               ; yes
                MOVWF      Char
                MOVWF      Oper
                RETURN

rowd            BSF        PORTC,2           ; select row D
                BCF        PORTC,3
                CALL       onems

                BTFSC      PORTC,4           ; button C?
                GOTO       b0                ; no
                MOVLW      'c'               ; yes
                MOVWF      Char
                MOVWF      Oper
                RETURN

b0              BTFSC      PORTC,5           ; button 0?
                GOTO       be                ; no
                MOVLW      '0'               ; yes
                MOVWF      Char
                MOVLW      00
                MOVWF      Kval
                RETURN

be              BTFSC      PORTC,6           ; button =?
                GOTO       bp                ; no
                MOVLW      '='               ; yes
                MOVWF      Char
                RETURN
```

Program 5.1
(Continued)

```
bp        BTFSC     PORTC,7         ; button +?
          GOTO      done            ; no
          MOVLW     '+'             ; yes
          MOVWF     Char
          MOVWF     Oper
          RETURN

done      BSF       PORTC,3         ; clear last row
          CLRF      Char            ; character code = 0
          RETURN


; Write display ........................................

disout    MOVF      Kcode,W         ; Load the code
          BSF       Select,RS       ; Select data mode
          CALL      send            ; and send code
          RETURN

; Process operations ....................................

calc      MOVF      Oper,W          ; check for add
          MOVWF     Temp            ; load input op code
          MOVLW     '+'             ; load plus code
          SUBWF     Temp            ; compare
          BTFSC     STATUS,Z        ; and check if same
          GOTO      add             ; yes, jump to op

          MOVF      Oper,W          ; check for subtract
          MOVWF     Temp
          MOVLW     '-'
          SUBWF     Temp
          BTFSC     STATUS,Z
          GOTO      sub

          MOVF      Oper,W          ; check for multiply
          MOVWF     Temp
          MOVLW     'x'
          SUBWF     Temp
          BTFSC     STATUS,Z
          GOTO      mul

          MOVF      Oper,W          ; check for divide
          MOVWF     Temp
          MOVLW     '/'
          SUBWF     Temp
          BTFSC     STATUS,Z
          GOTO      div
          GOTO      scan            ; rescan if key invalid

; Calculate results from 2 input numbers ..................

add       MOVF      Num1,W          ; get first number
          ADDWF     Num2,W          ; add second
          MOVWF     Result          ; and store result
          GOTO      outres          ; display result


sub       BSF       STATUS,C        ; Negative detect flag
          MOVF      Num2,W          ; get first number
          SUBWF     Num1,W          ; subtract second
          MOVWF     Result          ; and store result
          BTFSS     STATUS,C        ; answer negative?
GOTO      minus                     ; yes, minus result
          GOTO      outres          ; display result

minus     MOVLW     '-'             ; load minus sign
          BSF       Select,RS       ; Select data mode
          CALL      send            ; and send symbol

          COMF      Result          ; invert all bits
          INCF      Result          ; add 1
          GOTO      outres          ; display result


mul       MOVF      Num1,W          ; get first number
          CLRF      Result          ; total to Z
```

Program 5.1
(Continued)

```
add1     ADDWF    Result          ; add to total
         DECFSZ   Num2            ; num2 times and
         GOTO     add1            ; repeat if not done
         GOTO     outres          ; done, display result


div      CLRF     Result          ; total to Z
         MOVF     Num2,W          ; get divisor
         BCF      STATUS,C        ; set C flag
sub1     INCF     Result          ; count loop start
         SUBWF    Num1            ; subtract
         BTFSS    STATUS,Z        ; exact answer?
         GOTO     neg             ; no
         GOTO     outres          ; yes, display answer
neg      BTFSC    STATUS,C        ; gone negative?
         GOTO     sub1            ; no - repeat
         DECF     Result          ; correct the result
         MOVF     Num2,W          ; get divisor
         ADDWF    Num1            ; calc remainder

         MOVF     Result,W        ; load result
         ADDLW    030             ; convert to ASCII
         BSF      Select,RS       ; Select data mode
         CALL     send            ; and send result

         MOVLW    'r'             ; indicate remainder
         CALL     send
         MOVF     Num1,W
         ADDLW    030             ; convert to ASCII
         CALL     send
         GOTO     scan


; Convert binary to BCD ................................

outres   MOVF     Result,W        ; load result
         MOVWF    Lsd             ; into low digit store
         CLRF     Msd             ; high digit = 0
         BSF      STATUS,C        ; set C flag
         MOVLW    D'10'           ; load 10

again    SUBWF    Lsd             ; sub 10 from result
         INCF     Msd             ; inc high digit
         BTFSC    STATUS,C        ; check if negative
         GOTO     again           ; no, keep going
         ADDWF    Lsd             ; yes, add 10 back
         DECF     Msd             ; inc high digit


; display 2 digit BCD result ...........................

         MOVF     Msd,W           ; load high digit result
         BTFSC    STATUS,Z        ; check if Z
         GOTO     lowd            ; yes, dont display Msd

         ADDLW    030             ; convert to ASCII
         BSF      Select,RS       ; Select data mode
         CALL     send            ; and send Msd

lowd     MOVF     Lsd,W           ; load low digit result
         ADDLW    030             ; convert to ASCII
         BSF      Select,RS       ; Select data mode
         CALL     send            ; and send Msd

         GOTO     scan            ; scan for clear key


; Restart ..............................................

clear    MOVLW    01              ; code to clear display
         BCF      Select,RS       ; Select data mode
         CALL     send            ; and send code
         CLRF     Kcount          ; reset count of keys
         GOTO     scan            ; and rescan keypad


         END      ;;;;;;;;;;;;;;;;;;;;;;;;;;;;;;;;;;;;;;;;;
```

Program 5.1
(Continued)

microprocessor systems. The data in parallel memory is accessed a whole byte at a time, while serial memory is accessed via a single data line, so parallel memory is inherently faster but needs more I/O pins.

5.7.1 Memory System Hardware

A conventional microprocessor system contains separate CPU and memory chips. A similar arrangement can be used if we need extra memory in a PIC system and there is no shortage of I/O pins. A system schematic is shown in Figure 5.5 based on the PIC 16F877A. A pair of traditional 32k RAM chips is used to expand the memory to 64kbytes. The clock and programming connections are not included in the schematic, as they are not needed for simulation, but must be added in any hardware implementation.

Each RAM chip has eight data I/O pins (D0−D7) and fifteen address pins (A0−A14), so each location contains 8 bits, and there are $2^{15} = 32,768$ locations. To select the chip for access, the Chip Enable (!CE) pin must be taken low. To write a location, an address code is supplied, data presented at D0−D7, and the Write Enable (!WE) is pulsed low. To read data, the Output Enable (!OE) is set active (low) in addition to the chip enable, and the data from the address can then be read back.

In the demo VSM project (PARMEM2), Port C is used as a data bus, and Port D as an address bus. In order to reduce the number of I/O pins needed for external memory addressing, address latches (U3 and U5) are used to store the high byte of the 15-bit address (D7 unused).

The address is output in two stages: the high address byte is latched, selecting a memory block within the chip (A8−A14), and the low address byte is then output direct to the memory chip low address bits (A0−A7) to select the location within that block. This divides this memory into 128 pages of 256 bytes. The high address bytes are temporarily stored in 74LS273 latches (8-bit registers) operated by a master reset (RB3) and separate clocks (RB0 and RB1). The 7-bit high address is presented at the inputs, and the clock pulsed high to load the latch.

The address decoder chip has three inputs C, B and A that receive a binary select code from the processor (A = LSB). The corresponding output is taken low − e.g. in binary 6 is input (110), output Y6 is selected (low) − while all the others stay high. This decoder can generate 8-chip select signals and, if attached to the high address lines of a processor, enable the memory chips in different ranges of addresses. In our system here, only the least significant input and two outputs (Y0, Y1) are used. However, the additional address decoder outputs could be used to control extra memory chips attached to the same set of address and data lines.

The two memory chips in the test system are selected alternately via the address decoder, by toggling RB2. This allows different memory schemes to be implemented in firmware, where

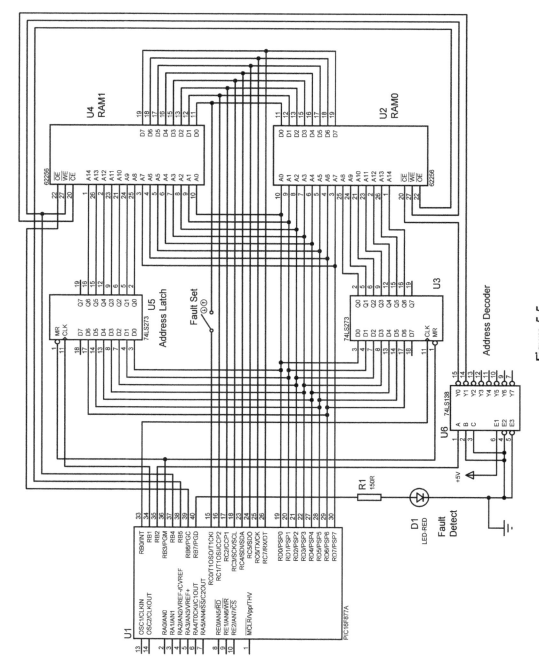

Figure 5.5

Parallel memory system.

the chips can be used one at a time or together to store 16-bit data. The memory can thus be organised as 64k × 8 bytes or 32k × 16-bit words. In the test program, all addresses are accessed in turn by incrementing the low address from 00 to FF for each high address (memory page select) and using the same address to write and read corresponding addresses in both chips at the same time.

This type of bus system operation where the outputs of the memory chips are both connected to the same data lines (Port C) depends on the presence of tri-state buffers at the output of the RAM chips, controlled by the enable inputs. These can be switched to allow data input (!CE and !WE = low), data output (!CE and !OE = low) or disabled (!CE and !OE = high). In the disabled state, the outputs of the RAM are effectively disconnected from the data bus. Only one RAM chip should be enabled at a time, otherwise there will be contention on the bus, with different data bytes attempting to use the bus at the same time.

5.7.2 Memory Test Program

The test program (Program 5.2) writes a traditional checkerboard pattern to the memory chips, placing the codes 01010101 (55h) and 10101010 (AAh) in successive locations. Adjacent memory cells are therefore all set to opposite voltage values, and any interaction between them, e.g. due to charge leakage, is more likely to show up. The memory is written and read, the data retrieved and compared with the correct value. If the write and read values do not agree, an error LED is lit. A switch has been placed in the data line D0 so that the error detection system can be tested. When the switch is open, data 0 will be written to all D0 bits (open circuit data input), so all the least significant bits of the test data 55h will be incorrect, with the value 54h read back.

In VSM simulation, the contents of the memory chips can be displayed by selecting them in the debug menu. It is also helpful to select 'Reset Persistent Data Model', so that the memory is cleared between simulation runs. The test data can be changed in the source code to check that new data has been stored. If a break point is set at the beginning of the read/write loop, the data can be viewed as it is being written into each pair of locations.

5.7.3 Extended Memory System

If this system were extended using six more RAM chips, there could be a total of 32k × 8 bytes = 256k (Figure 5.6(a)). A 3-bit input is required into the address decoder to extend the chip selection system. The high address (page select) would still be 7 bits, and the location select, 8 bits, giving a total address width of 18 bits.

Port E provides the chip selection inputs to the address decoder, Port D the location select (with high address latches) and Port C the data lines. To save I/O, RD7 might be used to select

```
;;;;;;;;;;;;;;;;;;;;;;;;;;;;;;;;;;;;;;;;;;;;;;;;;;;;;;;;;;;;;;;
;          PARMEM2.ASM    MPB    Ver:2.0          18-02-13
;..............................................................
;
;         Parallel memory system
;         PIC 16F877A operates with expansion memory
;         RAM = 2 x 62256 32kb
;          Updated for VSM v8
;;;;;;;;;;;;;;;;;;;;;;;;;;;;;;;;;;;;;;;;;;;;;;;;;;;;;;;;;;;;;;;

          PROCESSOR 16F877A      ; define MPU
          __CONFIG 0x3731        ; XT clock
          INCLUDE "P16F877A.INC" ; Standard register labels

ConReg  EQU     06              ; Port B = Control Register
DatReg  EQU     07              ; Port C = Data Register
AddReg  EQU     08              ; Port D = Address Register
HiAdd   EQU     20              ; High address store

CLK0    EQU     0               ; RAM0 address buffer clock
CLK1    EQU     1               ; RAM1 address buffer clock
SelRAM  EQU     2               ; RAM select bit
ResHi   EQU     3               ; High address reset bit
WritEn  EQU     4               ; Write enable bit
OutEn0  EQU     5               ; Output enable bit RAM0
OutEn1  EQU     6               ; Output enable bit RAM1
LED     EQU     7               ; Memory error indicator

; Initialise ;;;;;;;;;;;;;;;;;;;;;;;;;;;;;;;;;;;;;;;;;;;;;;;;;;;

          CODE    0             ; Place machine code
          NOP                   ; Required for ICD mode

          BANKSEL TRISB         ; Select bank 1
          CLRF    TRISB         ; Control output bits
          CLRF    TRISC         ; Data bus initially output
          CLRF    TRISD         ; Address bus output

          BANKSEL AddReg        ; Select bank 0
          CLRF    DatReg        ; Clear outputs initially
          CLRF    AddReg        ; Clear outputs initially
          BCF     ConReg,CLK0   ; RAM0 address buffer clock
          BCF     ConReg,CLK1   ; RAM1 address buffer clock
          BCF     ConReg,SelRAM ; Select RAM0 initially
          BCF     ConReg,ResHi  ; Reset high address latches
          BSF     ConReg,OutEn0 ; Disable output enable RAM0
          BSF     ConReg,OutEn1 ; Disable output enable RAM1
          BSF     ConReg,WritEn ; Disable write enable bit
          BCF     ConReg,LED    ; Switch of error indicator

; MAIN LOOP ;;;;;;;;;;;;;;;;;;;;;;;;;;;;;;;;;;;;;;;;;;;;;;;;;;;;
start   CALL    write           ; test write to memory
        CALL    read            ; test read from memory
        SLEEP                   ; shut down

; Write checkerboard pattern to both RAMs ;;;;;;;;;;;;;;;;;;;;;;
```

Program 5.2
Parallel memory test program.

```
write  BSF     ConReg,ResHi    ; Enable address latches
nexwrt MOVLW   055             ; checkerboard test data
       MOVWF   DatReg          ; output on data bus
       CALL    store           ; and write to RAM
       MOVLW   0AA             ; checkerboard test data
       MOVWF   DatReg          ; output on data bus
       CALL    store           ; and write to RAM
       BTFSS   ConReg,ResHi    ; all done?
       RETURN                  ; yes - quit
       GOTO    nexwrt          ; no - next byte pair

; Check data stored ;;;;;;;;;;;;;;;;;;;;;;;;;;;;;;;;;;;;;;;;;;;;;;;;;

read   NOP                     ; required for label
       BANKSEL TRISC           ; select bank 1
       MOVLW   0FF             ; all inputs..
       MOVWF   TRISC           ; ..at Port C

       BANKSEL ConReg          ; select default bank 0
       BSF     ConReg,ResHi    ; Enable address latches
       BCF     ConReg,SelRAM   ; select RAM0
       BCF     ConReg,OutEn0   ; set RAM0 for output
       CALL    nexred          ; check data in RAM0
       BSF     ConReg,SelRAM   ; select RAM1
       BCF     ConReg,OutEn1   ; set RAM1 for output
       CALL    nexred          ; check data in RAM1
       RETURN                  ; all done

; Load test data and check data ..............................

nexred MOVLW   055             ; load even data byte
       CALL    test            ; check data
       MOVLW   0AA             ; load odd data byte
       CALL    test            ; check data
       BTFSS   ConReg,ResHi    ; all done?
       RETURN                  ; yes - quit
       GOTO    nexred          ; no - next byte pair

; Write data to RAM .........................................

store  BCF     ConReg,SelRAM   ; Select RAM0
       BCF     ConReg,WritEn   ; negative pulse ..
       BSF     ConReg,WritEn
       BSF     ConReg,SelRAM   ; Select RAM1
       BCF     ConReg,WritEn   ; negative pulse ..
       BSF     ConReg,WritEn   ; ..write bit
       INCF    AddReg          ; next address
       BTFSC   STATUS,Z        ; last address?
       CALL    inchi           ; yes - increment high address
       RETURN                  ; no - next byte

; Test memory data .........................................

test   MOVF    DatReg,F        ; read data
       SUBWF   DatReg,W        ; compare data
       BTFSS   STATUS,Z        ; same?
       BSF     ConReg,LED      ; no - switch on LED
         INCF  AddReg          ; yes - next address
```

Program 5.2
(Continued)

```
              BTFSC   STATUS,Z        ; last address in block?
              CALL    inchi           ; yes - increment high address
              RETURN                  ; no - continue

; Select next block of RAM ...................................

inchi   INCF    HiAdd               ; next block
        BTFSC   STATUS,Z            ; all done?
        GOTO    alldon              ; yes
        MOVF    HiAdd,W             ; no - load high address
        MOVWF   AddReg              ; output it
        BSF     ConReg,CLK0         ; clock it into latches
        BSF     ConReg,CLK1
        BCF     ConReg,CLK0
        BCF     ConReg,CLK1
        CLRF    AddReg              ; reset low address to zero
        RETURN                      ; block done
alldon  BCF     ConReg,ResHi        ; reset address latches
        RETURN                      ; all blocks done

        END     ;;;;;;;;;;;;;;;;;;;;;;;;;;;;;;;;;;;;;;;;;;;;;;;;;;;;
```

Program 5.2
(Continued)

between the odd and the even chips via the address decoder clock input (master reset tied high). The memory would then operate as 32k blocks with the addresses shown in the memory map (Figure 5.6(b)). It contains a total of 256k locations, divided into 8 blocks of 32k (one chip), each containing 128 pages of 256 bytes. Alternatively, if RD0 were used as the high address clock, a 16-bit system would be created with high and low bytes stored in odd and even pairs of chips. A 32-bit memory system could be created in a similar manner.

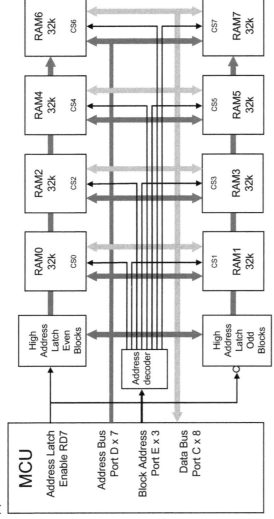

(a)

(b)

Block 32k	Start Address	End Address	Port E	RD7
0	00000	07FFF	000	0
1	08000	0FFFF	001	1
2	10000	17FFF	010	0
3	18000	1FFFF	011	1
4	20000	27FFF	100	0
5	28000	2F FFF	101	1
6	30000	37FFF	110	0
7	38000	3FFFF	111	1

Figure 5.6

256k Extended memory system: (a) block diagram and (b) memory map.

Questions 5

1. Show how to convert 10010011_2 to decimal. (3)
2. Show how to convert 1234_{10} to binary. (3)
3. Show how to convert $3FB0_{16}$ to decimal. (3)
4. Show how to multiply $1001_2 \times 0101_2$ in binary. (3)
5. Show how to calculate the binary 2s complement form of -99_{10}. (6)
6. By reference to the 32-bit IEEE 754 standard, deduce how to calculate the decimal (6) fraction represented by the 16-bit floating point number 0 00100 1001000000. Its format is: MSB = sign, 5 exponent bits (zero offset 15_{10}), 10 significant bits. Give a result to 3 significant figures.
7. Show how to convert the number 56_{10} from BCD to binary. (3)
8. State the ASCII codes for 'A', 'z' and '#' in hexadecimal. (3)
   Total (30)

Assignments 5

5.1 8-Bit Multiplication

a. Write a subroutine for the PIC 16F877A to multiply two 8-bit unsigned binary integers using simple adding loop using suitably labelled registers.
b. Write a subroutine for the PIC 16F877A to multiply two 8-bit unsigned binary numbers by shifting and adding using suitably labelled registers.
c. Calculate the total time taken (in instruction cycles) by each method and show which is more efficient (i.e. takes the shortest time).
d. Check your conclusion by simulation in MPLAB.

5.2 Floating Point Numbers

a. Investigate how to add, subtract, multiply and divide 16-bit floating point numbers, assuming the format is from MSB to LSB: sign bit, five exponent bits and ten significant bits.
b. Describe each process using a pair of sample numbers, converting them and the results to decimal scientific form and checking the result using a calculator.
c. Suggest advantages of using floating point numbers over integer formats in implementing 8-bit PIC MCU applications.

5.3 Memory Expansion

Sketch a schematic to show how to attach $4 \times 64k$ RAM (RAM0−RAM3) chips to the PIC 16F877A to provide 256k of external memory, organised as a continuous block of 8-bit data locations addressed from Ports C, D and E, with the data written and read via Port B. Do not use address latches, address the RAM chips direct, using Port E to control an address decoder. Draw a memory map indicating the address ranges occupied by each chip, and the utilisation of Ports C and D for addressing the memory.

CHAPTER 6

Analogue Interfacing

Summary
- The mid-range PIC ADC input provides a 10-bit analogue conversion
- An external reference voltage sets the conversion range
- 8-Bit results can be used for smaller range or lower resolution
- Op-amp-based input signal conditioning circuits are often required
- Non-inverting, inverting, summing, and difference configurations are available
- Gain and offset adjustment allow the interface to be calibrated
- Transient and frequency response must be specified
- The most suitable type of op-amp must be selected for an interface design
- The mid-range PIC usually has comparator inputs
- Amplitude and frequency of a.c. signals can be measured
- Parallel and serial DACs provide analogue output

Many control applications require the measurement of analogue variables, such as voltage, temperature, pressure and speed, using suitable sensors. This chapter will suggest a range of interfacing circuits for analogue sensors using IC amplifiers (op-amps) for signal conditioning and application software for converting these inputs into digital form via the PIC analogue to digital converter (ADC).

6.1 Analogue Input

Most PIC MCUs incorporate analogue inputs that are internally connected to an ADC that produces a binary representation of the input voltage. This is generally a 10-bit conversion, which is accurate to 1 part in 1024 (2^{10}). This is better than 0.1% at maximum output and precise enough for most purposes. Sometimes, an 8-bit result is sufficient, which gives an accuracy of 1 part in 256 (<0.5%) and is simpler to process.

The ADC produces a binary output that increases in steps with a rising analogue input. The input voltage is sampled at regular intervals by the ADC, and the analogue input converted into a corresponding binary code (Figure 6.1). The maximum sampling frequency may be significant when acquiring signals such as an audio input. The minimum sampling time in the PIC 16F877A is about 10μs or a maximum frequency of 100kHz. This is fast

Interfacing PIC Microcontrollers.
DOI: http://dx.doi.org/10.1016/B978-0-08-099363-8.00006-6

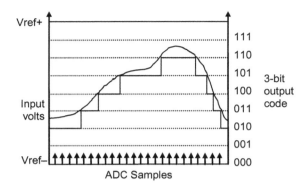

Figure 6.1
Stepwise ADC operation.

enough to acquire audio signals without distortion, since the minimum sampling rate required is twice the maximum signal frequency (20kHz). A standard audio sampling rate is 44kHz.

When setting up the ADC, reference voltages must be specified to set the minimum and maximum voltages for the conversion. These reference voltages are input from external circuits in the 16F877A, but internal reference voltages are available in more recent PIC chips. Often, the minimum reference voltage is set to 0V, which simplifies the conversion. In Figure 6.1, for clarity, only a 3-bit code is generated, which will allow 8 discrete input levels to be detected. In practice, 8-bit sampling produces 256 levels and 10-bit to 1024 levels. This determines the maximum resolution of the conversion.

The ADC uses a successive approximation conversion method. This uses a comparator to compare the input with the current sample and increase or decrease the binary code until all bits are converted. The diagram above shows that an offset of half a bit can be present, depending on the switching levels used, which may need correction for maximum accuracy.

In this chapter, programs to perform both 8-bit and 10-bit conversion will be provided. The ADC is controlled from special function registers ADCON0 and ADCON1 and can generate a peripheral interrupt if required. The output from the converter is automatically stored in SFRs ADRESH (analogue to digital conversion result, high byte) and ADRESL (low byte).

By default, the PIC ADC uses the internal supply voltage (V_{DD}, 5V nominal) to set the input range. However, a more convenient conversion factor is achieved by connecting an external 2.56V reference voltage (V_{REF+}). The ADC is then set up to convert an input voltage in the range of 0–2.55V to 10-bit binary. If only the high 8 bits of the result are used, a resolution of 10mV (2.56/256V) per bit is obtained.

6.1.1 8-Bit Test Circuit

A test application (VSM project ADC8BIT2) to demonstrate 8-bit conversion and display is shown in Figure 6.2. The 16F877A MCU has eight analogue inputs available, RA0, RA1, RA2, RA3, RA5, RE0, RE1 and RE2. These have alternate labels AN0–AN7 for this function. RA2 and RA3 are the reference voltage inputs, setting the minimum and maximum values for the measured voltage range. All these inputs default to analogue operation but will be explicitly initialised anyway.

The test voltage input at RA0 (analogue input AN0) is derived from a pot across the 5V supply. A reference voltage is provided at RA3 (AN3) which sets the maximum voltage to be converted, and thus the conversion factor required in the software. The minimum value defaults to 0V. The 2.7V zener diode provides a constant reference voltage; it is supplied via a current limiting resistor, so that the zener operates at the current specified for optimum voltage stability. This is then divided down across the reference voltage pot RV1 and a 10k fixed resistor. The range across the pot is about 2.7–2.4V and is adjusted for 2.56V. The LCD connected to Port D operates in 4-bit mode to display the voltage, as described in Chapter 4.

6.1.2 ADC Operation

A block diagram of the ADC module is shown in Figure 6.3. The inputs are connected to a function selector block which sets up each pin for analogue or digital operation according to the 4-bit control code loaded into the A/D port configuration control bits, PCFG0–PCFG3 in ADCON1. The code used, 0011, sets Port E as digital I/O, and Port A as analogue inputs with AN3 as the positive reference input.

The analogue inputs are fed to a multiplexer that allows one of the eight inputs to be selected at any one time. This is controlled by the three analogue channel select bits, CHS0–CHS2 in ADCON0. In this case, channel 0 is selected (000), RA0 input. If more than one channel is to be sampled, these select bits can be changed between ADC conversions. The conversion is triggered by setting the GO/DONE bit, which is later cleared automatically to indicate that the conversion is complete. An ADC interrupt is available, which will often be used, as the ADC conversion takes at least 10µs.

6.1.3 ADC Clock

The successive approximation ADC needs a clock to drive the synchronous logic that generates the binary output. The input voltage is fed to a comparator, and if the voltage is higher than 50% of the range, the MSB of the result is set high. The voltage is then checked against the mid-point of the remaining range, and the next bit set high or low accordingly, and so on for 10 bits. This takes a significant amount of time: the minimum conversion time

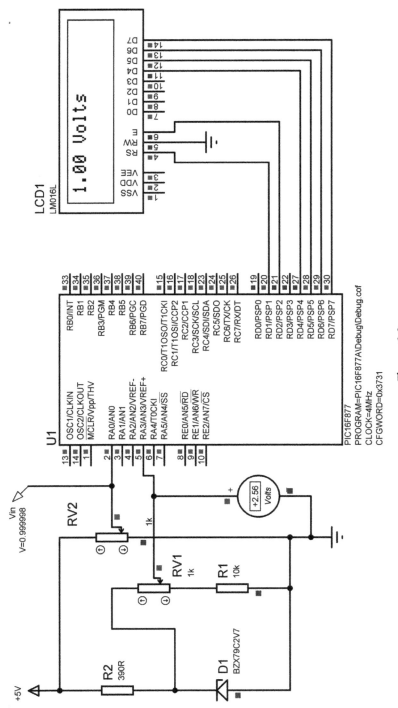

Figure 6.2

8-Bit ADC circuit.

(a)

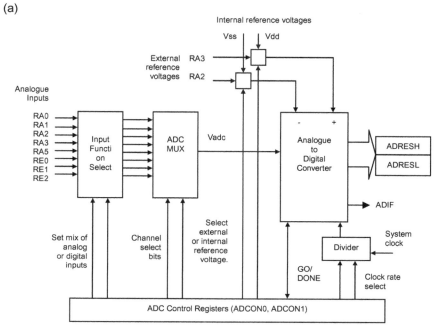

(b)

Register	Setting	Relevant bits	Function
ADRESH	**XXXX XXXX**	All	ADC result high byte
ADRESL	**XXXX XXXX**	All	ADC result low byte
ADCON0	0100 0X01	ADCS1,0 GO/DONE, ADON	Conversion frequency select ADC start, ADC enable
ADCON1	0000 **0011**	ADFM, PCFG3-0	Result justify, ADC input mode control
INTCON	**11**00 0000	GIE,PEIE	Peripheral interrupt enable
PIE1	0**1**00 0000	ADIE	ADC interrupt enable
PIR1	0**1**00 0000	ADIF	ADC interrupt flag

(c)

Justify selection		ADRESH	ADRESL
ADFM = 1	Right justified	xxxx xx98	7654 3210
ADFM = 0	Left justified	9876 5432	10xx xxxx

Figure 6.3
ADC configuration: (a) ADC block diagram, (b) ADC control registers set-up
for VINTEST and (c) 10-bit result justification (bit0−bit9).

is 1.6µs per bit, making 16µs for a 10-bit conversion. The MCU clock is divided down to provide an ADC clock period that produces this minimum conversion time.

The ADC register configuration for the 8-bit demo application is shown in Figure 6.3(b). The speed of the conversion is selected by bits ADSC1 and ADSC0. The simulated test circuit is clocked at 4 MHz. This gives a clock period of 0.25µs. We need a conversion time of at least 1.6µs, so if we select the divide by 8 option, the ADC clock period will then be $8 \times 0.25 = 2$µs, which is just longer than the minimum required. The select bits are therefore set to 01. Note that in hardware, a crystal clock circuit needs to be added, or the internal clock used in the 16F887, to run this program.

The input of the ADC has a sample and hold circuit to ensure that the voltage sampled is constant during the conversion process. This contains an *RC* low-pass filter with a time constant of about 20μs. Therefore, if the input voltage changes rapidly, or at high frequency, the sample and hold input may affect the response time. If sampling speed is not critical, a settling time delay of at least 20μs should be included in the conversion sequence.

6.1.4 Results Registers

When the conversion is complete, the result is placed in the result register pair, ADRESH and ADRESL, the GO/DONE bit cleared by the ADC controller, and the ADIF interrupt flag is set. Since the result is only 10 bits, the positioning in the 16-bit result register pair can be chosen so that the high 8 bits are in ADRESH (left justified), or the low 8 bits are in ADRESL (right justified) (Figure 6.3(c)). Obviously, to retain 10-bit resolution, both parts must be processed; right justification will probably be more convenient in this case.

If only 8-bit resolution is required, the process can be simplified. If the result is right justified, the low 8 bits in ADRESL will record the low bits of the conversion, meaning that only voltages up to 25% of the full range will be acquired, but at full resolution. If the result is left justified, the high byte will be obtained, which will represent the full voltage range, but at reduced resolution.

In our test circuit, the reference voltage is 2.56V. The justify bit ADFM = 0, selecting left justify. Only ADRESH then needs to be processed, giving results for the full range at 8-bit resolution, which is about 1% at mid-range. The result will be shown on the LCD as three digits, 0.00−2.55. The test input pot provides 0−5V but only 0−2.55V will be displayed. Over range inputs are displayed as 2.55V.

6.1.5 8-Bit ADC Program

The test program is outlined in Figure 6.4, and the source code listed in Program 6.1. The output port and ADC control registers are initialised in the first block, which include files providing the display initialisation and driver routines. The main loop contains subroutine calls to read the ADC input, convert from binary to BCD and display it. The routine to read the ADC sets the GO/DONE bit and then polls it until it is cleared at the end of the conversion. The 8-bit result from ADRESH is converted to three BCD digits using the subtraction algorithm described previously. Full scale input is 255, which is displayed as 2.55V.

Project: ADC8BIT

Function:	*Convert the analogue input to 8 bits and display*
Hardware:	*P16F877A (4MHz), Vref+ = 2.56, 16x2 LCD*

Initialise   Port A = Analogue inputs (default)
                Port C = LCD outputs
                ADC = Select f/8, RA0 input, left justify result, enable
                LCD = default setup (include LCD driver routines)

Main
        REPEAT
                Get ADC 8-bit input
                Convert to BCD
                Display on LCD
        ALWAYS

Subroutines

        Get ADC 8-bit input
                Start ADC and wait for done
                Store result

        Convert to BCD
                Calculate hundreds digit
                Calculate tens digit
                Remainder = ones digit

        Display on LCD
                Home cursor
                Convert BCD to ASCII
                Send hundreds, point, tens, ones
                Send 'Volts'

Figure 6.4
8-Bit ADC input program outline.

6.1.6 10-Bit ADC Input

Figure 6.5 shows a circuit which demonstrates 10-bit, full resolution, analogue to digital conversion (VSM project ADC10BIT2). The maximum binary result will be 1023 ($2^{10} - 1$). The zener circuit now provides a reference voltage of 4.096V, giving a range of 0−4.095V, at 4mV per bit. The result is displayed as a four-digit fixed point decimal. In order to provide a finer adjustment of the reference voltage, the zener voltage is divided down using fixed value resistors, and the final voltage tweaked by adjusting the current to the zener.

The data acquisition process is similar to the 8-bit system described earlier, but the binary to BCD conversion process is rather more complicated. The result is required in

```
;;;;;;;;;;;;;;;;;;;;;;;;;;;;;;;;;;;;;;;;;;;;;;;;;;;;;;;;;;
;
;        Project:              ADC8BIT2
;        Devised by:          MPB
;        Date:                14-01-13
;        Status:              Updated for VSM v8
;
;;;;;;;;;;;;;;;;;;;;;;;;;;;;;;;;;;;;;;;;;;;;;;;;;;;;;;;;;;
;
;        Demonstrates simple analogue input
;        using an external reference voltage of 2.56V
;        The 8-bit result is converted to BCD for display
;        as a voltage using the standard LCD routines.
;
;;;;;;;;;;;;;;;;;;;;;;;;;;;;;;;;;;;;;;;;;;;;;;;;;;;;;;;;;;

        PROCESSOR 16F877A
;       Clock = XT 4MHz, standard fuse settings
        __CONFIG 0x3731

;       LABEL EQUATES        ;;;;;;;;;;;;;;;;;;;;;;;;;;;;;;;;;;;;

        #INCLUDE "P16F877A.INC"      ; standard labels

; GPR 70 - 75 allocated to included LCD display routine

count   EQU     30          ; Counter for ADC setup delay
ADbin   EQU     31          ; Binary input value
huns    EQU     32          ; Hundreds digit in decimal value
tens    EQU     33          ; Tens digit in decimal value
ones    EQU     34          ; Ones digit in decimal value

; PROGRAM BEGINS ;;;;;;;;;;;;;;;;;;;;;;;;;;;;;;;;;;;;;;;;;

        CODE    0                   ; Default start address
        NOP                         ; required for ICD mode

; Port & display setup....................................

        BANKSEL TRISC               ; Select bank 1
        CLRF    TRISD               ; Display port is output
        MOVLW   B'00000011'         ; Analogue input setup code
        MOVWF   ADCON1              ; Left justify result,
                                    ; Port A = analogue inputs

        BANKSEL PORTC               ; Select bank 0
        CLRF    PORTD               ; Clear display outputs
        MOVLW   B'01000001'         ; Analogue input setup code
        MOVWF   ADCON0              ; f/8, RA0, done, enable

        CALL    inid                ; Initialise the display

; MAIN LOOP ;;;;;;;;;;;;;;;;;;;;;;;;;;;;;;;;;;;;;;;;;;;;;;

start   CALL    getADC              ; read input
        CALL    condec              ; convert to decimal
        CALL    putLCD              ; display input
        GOTO    start               ; jump to main loop

; SUBROUTINES ;;;;;;;;;;;;;;;;;;;;;;;;;;;;;;;;;;;;;;;;;;;;;

; Read ADC input and store ...............................

getADC  BSF     ADCON0,GO ; start ADC..
wait    BTFSC   ADCON0,GO ; ..and wait for finish
        GOTO    wait
        MOVF    ADRESH,W  ; store result high byte
        RETURN

; Convert input to decimal ...............................

condec  MOVWF   ADbin               ; get ADC result
        CLRF    huns                ; zero hundreds digit
        CLRF    tens                ; zero tens digit
        CLRF    ones                ; zero ones digit
```

Program 6.1
Display of 8-bit analogue input.

```
; Calclulate hundreds.......................................
        BSF      STATUS,C  ; set carry for subtract
        MOVLW    D'100'             ; load 100
sub1    SUBWF    ADbin              ; and subtract from result
        INCF     huns               ; count number of loops
        BTFSC    STATUS,C  ; and check if done
        GOTO     sub1               ; no, carry on

        ADDWF    ADbin              ; yes, add 100 back on
        DECF     huns               ; and correct loop count

; Calculate tens digit.................................
        BSF      STATUS,C  ; repeat process for tens
        MOVLW    D'10'              ; load 10
sub2    SUBWF    ADbin              ; and subtract from result
        INCF     tens               ; count number of loops
        BTFSC    STATUS,C  ; and check if done
        GOTO     sub2               ; no, carry on

        ADDWF    ADbin              ; yes, add 100 back on
        DECF     tens               ; and correct loop count
        MOVF     ADbin,W            ; load remainder
        MOVWF    ones               ; and store as ones digit

        RETURN                      ; done

; Output to display.......................................
putLCD  BCF      Select,RS ; set display command mode
        MOVLW    080                ; code to home cursor
        CALL     send               ; output it to display
        BSF      Select,RS ; and restore data mode

; Convert digits to ASCII and display.....................
        MOVLW    030                ; load ASCII offset
        ADDWF    huns               ; convert hundreds to ASCII
        ADDWF    tens               ; convert tens to ASCII
        ADDWF    ones               ; convert ones to ASCII

        MOVF     huns,W             ; load hundreds code
        CALL     send               ; and send to display
        MOVLW    '.'                ; load point code
        CALL     send               ; and output
        MOVF     tens,W             ; load tens code
        CALL     send               ; and output
        MOVF     ones,W             ; load ones code
        CALL     send               ; and output
        MOVLW    ' '                ; load space code
        CALL     send               ; and output
        MOVLW    'V'                ; load volts code
        CALL     send               ; and output
        MOVLW    'o'                ; load volts code
        CALL     send               ; and output
        MOVLW    'l'                ; load volts code
        CALL     send               ; and output
        MOVLW    't'                ; load volts code
        CALL     send               ; and output
        MOVLW    's'                ; load volts code
        CALL     send               ; and output

        RETURN                      ; done

; INCLUDED ROUTINES ;;;;;;;;;;;;;;;;;;;;;;;;;;;;;;;;;;;;;;

; Include LCD driver routines
;
        #INCLUDE 'LCD.INC'
;       Contains routines:
;       inid:    Initialises display
;       onems:   1 ms delay
;       xms:     X ms delay
;                Receives X in W
;       send:    Sends a character to display
;                Receives: Control code in W (Select,RS=0)
;                          ASCII character code in W (RS=1)

        END      ;;;;;;;;;;;;;;;;;;;;;;;;;;;;;;;;;;;;;;;;;
```

Program 6.1
(Continued)

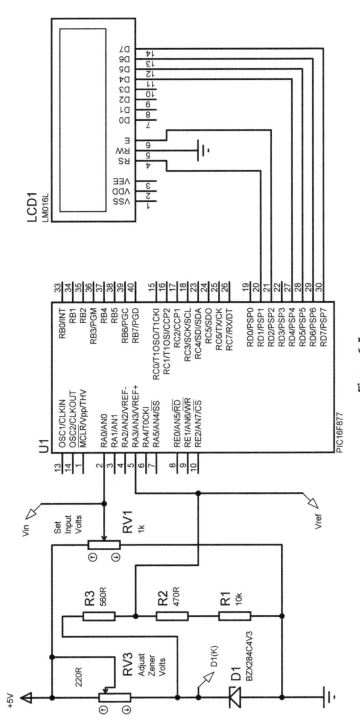

Figure 6.5

10-Bit analogue input test circuit.

CON4 ROUTINE

Convert 10-bit binary to four-digit BCD

Load 10-bit, right justified binary (0–1023)
Multiply by 4 (0–4092) by shift left
Clear BCD registers

REPEAT
          Subtract $E8_{16}$ from low byte
          Subtract 3_{16} from high byte
          Increment thousands digit
UNTIL remainder < $03E8_{16}$ (1000)

REPEAT
          Subtract 64_{16} from low byte
          Borrow from high byte
          Increment hundreds digit
UNTIL remainder < 64_{16} (100)

REPEAT
          Subtract 10 from low byte
          Increment tens digit
UNTIL remainder < 10

Remainder = ones digits

RETURN

Figure 6.6
10-Bit BCD conversion routine.

the range of 0–4095, so the original result (0–1023) is shifted left twice to multiply it by 4. One thousand (03E8h) is then loop subtracted from the result to calculate the thousands digit. Correct borrow handling between the high and the low byte is particularly important. The process stops when the remainder is less than 1000. The hundreds digit is calculated in a similar way, but the tens calculation is a little easier as the maximum remainder from the previous stage is 99, so the high byte borrow handling is not necessary. This process is outlined in Figure 6.6, and the source code shown in Program 6.2.

6.2 Op-Amp Interfaces

The output of a sensor may be only a few millivolts when the ADC typically needs 2–4V input. Op-amps are usually employed to provide the necessary signal voltage gain. We will consider direct voltages only initially.

```
;;;;;;;;;;;;;;;;;;;;;;;;;;;;;;;;;;;;;;;;;;;;;;;;;;;;;;;;;;;
;
;           Project:             ADC10BIT2
;           Devised by:          MPB
;           Date:                14-01-13
;           Status:              Updated for VSM v8
;
;;;;;;;;;;;;;;;;;;;;;;;;;;;;;;;;;;;;;;;;;;;;;;;;;;;;;;;;;;;
;
;           Demonstrates 10-bit voltage measurement
;           using an external reference voltage of 4.096V,
;           giving 4mV per bit, and a resolution of 0.1%.
;           The result is converted to BCD for display
;           as a voltage using the standard LCD routines.
;
;;;;;;;;;;;;;;;;;;;;;;;;;;;;;;;;;;;;;;;;;;;;;;;;;;;;;;;;;;;

           PROCESSOR 16F877
;          Clock = XT 4MHz, standard fuse settings
           __CONFIG 0x3731

;          LABEL EQUATES        ;;;;;;;;;;;;;;;;;;;;;;;;;;;;;;;;;;;;

           INCLUDE "P16F877A.INC"      ; standard register labels


;-----------------------------------------------------------
; User register labels
;-----------------------------------------------------------
; GPR 20 - 2F allocated to included LCD display routine

count    EQU      30         ; Counter for ADC setup delay
ADhi     EQU      31         ; Binary input high byte
ADlo     EQU      32         ; Binary input low byte
thos     EQU      33         ; Thousands digit in decimal
huns     EQU      34         ; Hundreds digit in decimal value
tens     EQU      35         ; Tens digit in decimal value
ones     EQU      36         ; Ones digit in decimal value

;-----------------------------------------------------------
; PROGRAM BEGINS
;-----------------------------------------------------------

           CODE     0                ; Default start address
           NOP                       ; required for ICD mode

;-----------------------------------------------------------
; Port & display setup

           BANKSEL  TRISD            ; Select bank 1
           CLRF     TRISD            ; Display port is output
           MOVLW    B'10000011'      ; Analogue input setup code
           MOVWF    ADCON1           ; Right justify result,
                                     ; Port A = analogue inputs
                                     ; with external reference

           BANKSEL PORTD             ; Select bank 0
           CLRF     PORTD            ; Clear display outputs
           MOVLW    B'01000001'      ; Analogue input setup code
           MOVWF    ADCON0           ; f/8, RA0, done, enable

           CALL     inid             ; Initialise the display

;-----------------------------------------------------------
; MAIN LOOP
;-----------------------------------------------------------

start    CALL     getADC           ; read input
         CALL     con4             ; convert to decimal
         CALL     putLCD           ; display input
         GOTO     start            ; jump to main loop

;-----------------------------------------------------------
```

Program 6.2
10-Bit analogue voltage input display.

```
; SUBROUTINES
;---------------------------------------------------------
; Read ADC input and store
;---------------------------------------------------------
getADC    MOVLW    007               ; load counter
          MOVWF    count
down      DECFSZ   count             ; and delay 20us
          GOTO     down

          BSF      ADCON0,GO ; start ADC..
wait      BTFSC    ADCON0,GO ; ..and wait for finish
          GOTO     wait
          RETURN


;---------------------------------------------------------
; Convert 10-bit input to decimal
;---------------------------------------------------------

con4      MOVF     ADRESH,W  ; get ADC result
          MOVWF    ADhi             ; high bits
          BANKSEL  ADRESL           ; in bank 1
          MOVF     ADRESL,W  ; get ADC result
          BANKSEL  ADRESH           ; default bank 0
          MOVWF    ADlo             ; low byte

; Multiply by 4 for result 0 - 4096 by shifting left........

          BCF      STATUS,C  ; rotate 0 into LSB and
          RLF      ADlo             ; shift low byte left
          BTFSS    STATUS,C  ; carry out?
          GOTO     rot1             ; no, leave carry clear
          BSF      STATUS,C  ; rotate 1 into LSB and
rot1      RLF      ADhi             ; shift high byte left

          BCF      STATUS,C  ; rotate 0 into LSB
          RLF      ADlo             ; rotate low byte left again
          BTFSS    STATUS,C  ; carry out?
          GOTO     rot2             ; no, leave carry clear
          BSF      STATUS,C  ; rotate 1 into LSB and
rot2      RLF      ADhi             ; shift high byte left


; Clear BCD registers.....................................

clrbcd    CLRF     thos             ; zero thousands digit
          CLRF     huns             ; zero hundreds digit
          CLRF     tens             ; zero tens digit
          CLRF     ones             ; zero ones digit


; Calclulate thousands low byte ...........................

tholo     MOVF     ADhi,F           ; check high byte
          BTFSC    STATUS,Z  ; high byte zero?
          GOTO     hunlo            ; yes, next digit

          BSF      STATUS,C  ; set carry for subtract
          MOVLW    0E8              ; load low byte of 1000
          SUBWF    ADlo             ; and subtract low byte
          BTFSC    STATUS,C  ; borrow from high bits?
          GOTO     thohi            ; no, do high byte
          DECF     ADhi             ; yes, subtract borrow

; Calculate thousands high byte............................

thohi     BSF      STATUS,C  ; set carry for subtract
          MOVLW    003              ; load high byte of 1000
          SUBWF    ADhi             ; subtract from high byte
          BTFSC    STATUS,C  ; result negative?
          GOTO     incth            ; no, inc digit and repeat

          ADDWF    ADhi             ; yes, restore high byte

; Restore remainder when done .............................
```

Program 6.2

(Continued)

```
            BCF       STATUS,C ; clear carry for add
            MOVLW     0E8               ; load low byte of 1000
            ADDWF     ADlo              ; add to low byte
            BTFSC     STATUS,C ; carry out?
            INCF      ADhi              ; yes, inc high byte
            GOTO      hunlo             ; and do next digit

; Increment thousands digit and repeat.....................

incth       INCF      thos              ; inc digit
            GOTO      tholo             ; and repeat

; Calclulate hundreds .....................................

hunlo       MOVLW     064               ; load 100
            BSF       STATUS,C ; set carry for subtract
            SUBWF     ADlo              ; and subtract low byte
            BTFSC     STATUS,C ; result negative?
            GOTO      inch              ; no, inc hundreds & repeat

            MOVF      ADhi,F            ; yes, test high byte
            BTFSC     STATUS,Z ; zero?
            GOTO      remh              ; yes, done
            DECF      ADhi              ; no, subtract borrow
inch        INCF      huns              ; inc hundreds digit
            GOTO      hunlo             ; and repeat

remh        ADDWF     ADlo              ; restore onto low byte


; Calculate tens digit.....................................

subt        MOVLW     D'10'             ; load 10
            BSF       STATUS,C ; set carry for subtract
            SUBWF     ADlo              ; and subtract from result
            BTFSS     STATUS,C ; and check if done
            GOTO      remt              ; yes, restore remainder
            INCF      tens              ; no, count number of loops
            GOTO      subt              ; and repeat


; Restore remainder........................................

remt        ADDWF     ADlo              ; yes, add 10 back on
            MOVF      ADlo,W            ; load remainder
            MOVWF     ones              ; and store as ones digit

            RETURN                      ; done



;---------------------------------------------------------
; Output to display
;---------------------------------------------------------

putLCD      BCF       Select,RS ; set display command mode
            MOVLW     080               ; code to home cursor
            CALL      send              ; output it to display
            BSF       Select,RS ; and restore data mode

; Convert digits to ASCII and display.....................

            MOVLW     030               ; load ASCII offset
            ADDWF     thos              ; convert thousands to ASCII
            ADDWF     huns              ; convert hundreds to ASCII
            ADDWF     tens              ; convert tens to ASCII
            ADDWF     ones              ; convert ones to ASCII

            MOVF      thos,W            ; load thousands code
            CALL      send              ; and send to display
            MOVLW     '.'               ; load point code
            CALL      send              ; and output
            MOVF      huns,W            ; load hundreds code
            CALL      send              ; and send to display
            MOVF      tens,W            ; load tens code
            CALL      send              ; and output
            MOVF      ones,W            ; load ones code
            CALL      send              ; and output
```

Program 6.2
(Continued)

```
            MOVLW      ' '            ; load space code
            CALL       send           ; and output
            MOVLW      'V'            ; load volts code
            CALL       send           ; and output
            MOVLW      'o'            ; load volts code
            CALL       send           ; and output
            MOVLW      'l'            ; load volts code
            CALL       send           ; and output
            MOVLW      't'            ; load volts code
            CALL       send           ; and output
            MOVLW      's'            ; load volts code
            CALL       send           ; and output

            RETURN                    ; done

;----------------------------------------------------------
; INCLUDED ROUTINES
;----------------------------------------------------------
; Include LCD driver routine
;
            INCLUDE    "LCD.INC"
;
;           Contains routines:
;           init:      Initialises display
;           onems:     1 ms delay
;           xms:       X ms delay
;                      Receives X in W
;           send:      sends a character to display
;                      Receives: Control code in W (Select,RS=0)
;                                ASCII character code in W (RS=1)
;
;
;----------------------------------------------------------
            END                       ; of source code
;----------------------------------------------------------
```

Program 6.2
(Continued)

6.2.1 Ideal Amplifier

The op-amp is a high gain integrated circuit (IC) amplifier with inverting and non-inverting inputs, whose output voltage is determined by the input differential voltage. An equivalent is shown in Figure 6.7. Since the differential gain (A) is very high, typically $>100,000$, the operating input differential voltage (V_d) is very small for output voltages within the supply range (5V in most of the demo circuits). If the output voltage is 1V, the input will be less than $10\mu V$. Similarly, the input resistance is large, typically at least $1M\Omega$, and the output resistance small, normally just a few ohms. As a result, the gain and bandwidth (frequency response) are broadly controlled by the external components and are assumed to be independent of the amplifier itself.

If we assume the op-amp has ideal characteristics:

1. *Differential gain, $A = \infty$ (for voltage applied between $+$ and $-$ terminals)*
2. *Differential voltage, $V_d = 0$ (terminals $+$ and $-$ are at the same voltage)*
3. *Input resistance, $R_{in} = \infty$ (zero input current at $+$ and $-$ terminals)*
4. *Output impedance, $R_o = 0$ (infinite current can be sunk or sourced at the output)*
5. *Bandwidth $= \infty$ (all frequencies are amplified equally)*

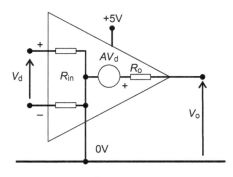

Figure 6.7
Op-amp equivalent circuit.

These rules allow amplifier circuit analysis and design to be greatly simplified and give results which are accurate enough for most applications, where most error is due to tolerances in the external components.

6.2.2 Basic Configurations

A basic set of configurations can be analysed on this basis, as shown in Figure 6.8. When used as a linear amplifier, the feedback in an op-amp circuit must be negative. Essentially, this means the feedback signal path must be connected to the minus input terminal.

The currents and voltages in the external resistors are then combined into a simple linear equation which predicts the overall circuit function that applies to all d.c. and a.c. signals. Bandwidth and other practical limitations will be considered later.

Figure 6.9 shows these basic amplifier circuits connected to the PIC MCU (VSM project AMPS2). They are connected to RA0 by a multi-way switch, so that the output of each may be displayed, using the 8-bit conversion and display program (Program 6.1) described earlier. This allowed the general models developed for each configuration to be tested by simulation to confirm correct function and examine op-amp performance.

The LM324 was used in this test circuit to illustrate performance limitations in op-amps. It has a bipolar input with relatively low impedance input and no offset adjustment. Improved performance can be obtained with alternative op-amps (see Section 6.5.4).

The inputs were adjusted to give the same output of 1V. The results are given in Table 6.1. The output error found is within the range of 2–12 mV. An error of 10mV in 1V corresponds to an accuracy of ±1%. The amplifiers are therefore adequate if 2% resistors are used.

The output error is mainly due to input offset in the amplifier – this can be demonstrated by varying the output, which results in a largely constant output error. It is significant

(a)

(b)

(c)

(d)

(e)

(f)

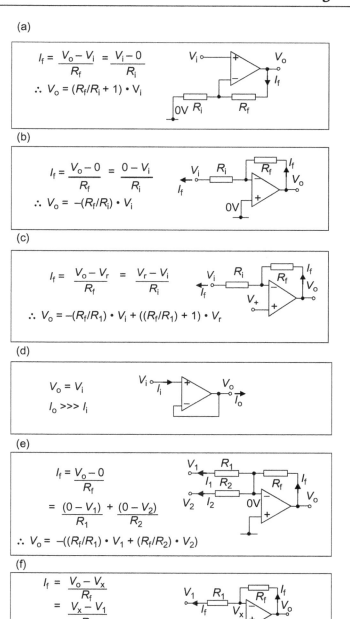

Figure 6.8
Basic amplifier configurations: (a) non-inverting amplifier, (b) inverting amplifier,
(c) inverting amplifier with offset, (d) unity gain buffer, (e) summing amplifier
and (f) difference amplifier.

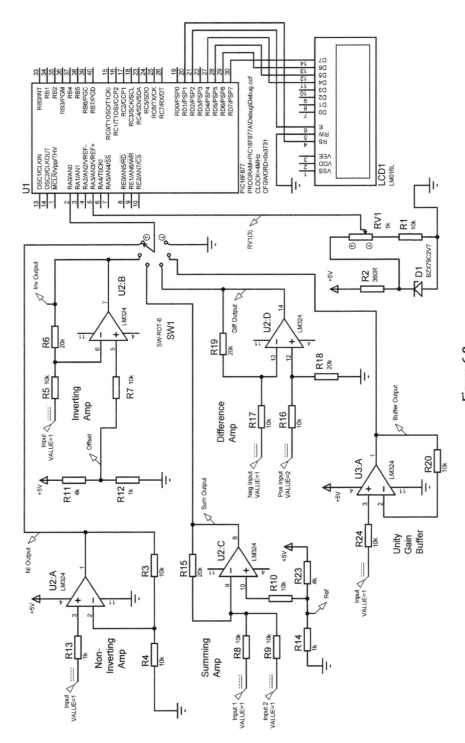

Figure 6.9

Basic amplifier interface test circuits.

Table 6.1: Amplifier Interface Simulation Results.

(Volts)	Gain	Input 1	Input 2	Offset	Output	Output Error (mV)
Non-inverting	+2	0.5	–	–	1.0035	4
Inverting	–2	1	–	1	1.0067	7
Summing	+2	1	1	1	1.0117	12
Difference	+2	1	1.5	–	1.0061	6
Unity gain	+1	1	1	–	1.0021	2

that the accuracy depends on the amplifier configuration and input common mode voltage. The worst case is the summing amplifier with a 1V offset. This problem can be alleviated by using a higher performance, low offset amplifier, now available at minimal extra cost. For maximum accuracy, high-precision, high-stability resistors must also be used.

6.2.3 Non-Inverting Amplifier

The basic configuration for the non-inverting amplifier is shown in Figure 6.8(a). The input is applied to the positive terminal, and feedback and gain controlled by the resistor network R_f and R_1. If we assume that the voltage between the terminals is zero (rule 2), the voltage at the negative terminal must be the same as the voltage at the positive terminal. We can then write down an equation for the feedback network using Ohm's law applied to each resistor, assuming the current flow is from the output through the resistors to ground. This is possible because it is assumed that none of the current is lost at the input terminal, as it has infinite input resistance (rule 3). A simple rearrangement of the equation allows us to predict the output voltage in terms of the resistor values. The gain can then be obtained by further rearrangement of the circuit values:

$$V_o/V_i = R_f/R_1 + 1$$

The main advantage of this configuration is that the input impedance is very high (in theory, infinite). The loading on the signal source is therefore negligible. The disadvantage is that the input is operating with an offset voltage, which reduces its accuracy, particularly with a single supply. In addition, the high input impedance makes it more susceptible to noise.

In the non-inverting amplifier shown in the test circuit (Figure 6.9), the feedback resistors are both 10k, giving a gain of 10k/10k + 1 = 2. The output should ideally range from 0 to 5.0V. In practice, the 324 simulation model provides a minimum output of 0.03V and a maximum of about 4.0V, representing the limits of the actual device. A constant output offset error of 4mV is demonstrated in the simulation results. If this is a potential problem, an op-amp with inherently low offset, or with offset adjustment, can be used. Alternatively, an external offset adjustment can be included in the circuit design, feeding a small additional offset current into the summing node (negative terminal).

6.2.4 Inverting Amplifier

The analysis is even simpler for the inverting amplifier (Figure 6.8(b)), since the input terminals are at 0V for the ideal analysis. The equation for the feedback current can be rearranged to give the gain:

$$V_o/V_i = -R_f/R_1$$

The negative sign indicates that the output is inverted, i.e. it goes negative when the input is going positive, and vice versa. Unfortunately, the input impedance is inherently low, being equal to the value of R_1. A significant input current is required to or from the signal source for this configuration to work correctly. However, with symmetrical supplies (± 5V) it can operate with zero offset, which reduces errors.

In the demo circuit, the inverting amplifier is operating with an offset of 1V, so that the output can remain positive with positive inputs. The positive op-amp terminal is connected to a reference voltage of 1.000V produced by a voltage divider across the supply. It is fed to the input terminal via a 10k, which helps to equalise the input offset currents at the + and − terminals. The gain (G) is 20k/10k = 2, and the output polarity inverted. Ideal analysis (Figure 6.8(c)) shows that the output voltage is given by:

$$V_o = (G+1) \cdot V_r - G \cdot V_i$$
$$\text{If } V_r = 1 \text{ and } G = 2 \qquad \text{then } V_o = 3 - 2V_i$$

In the test simulation, the input and offset are both 1V, resulting in an output of 1.007, or a 7mV offset. This is to be expected when there is no offset zero adjustment and the reference input is held at +1V.

6.2.5 Unity Gain Buffer

This is a special case of the non-inverting amplifier, where the feedback is 100%, i.e. zero feedback resistance, giving a gain of 1 (Figure 6.8(d)). The output voltage is then the same as the input voltages. So what is the point of the circuit? It is to provide current gain, to buffer a signal with a high source resistance, or inadequate available current. The input current is small (large input resistance at the + terminal), but the output current can be large, giving a high current gain. In practice, with standard op-amps, the output current would typically be limited to about 20mA, but high power output IC amps are available, or a further current driver stage can be added at the output using a power transistor (see Chapter 7). In simulation, it can be seen that the output offset error is minimal for this configuration.

6.2.6 Summing Amplifier

This is a development of the inverting amplifier, with additional inputs. Only two are shown in Figure 6.8(e) but more are possible. The output is determined by the sum of the input voltages, taking into account the input resistor weightings:

$$-V_o = G_1 \cdot V_1 + G_2 \cdot V_2 + G_3 \cdot V_3 + \dots$$

where $G_1 = R_f/R_1$, $G_2 = R_f/R_2$. ...

For a summing amplifier with offset, as seen in the demo circuits, it can be shown that:

$$V_o = V_r(nG + 1) - G(V_1 + V_2 + \dots + V_n)$$

for an amplifier with n identical input resistors (i.e. same gain for each input). The demo circuit produced the following simulated inputs:

$$R_f = 20\text{k and } R_1 = 10\text{k} \quad \therefore R_f/R_1 = 2 = G$$
$$V_r = 1.000\text{V} \qquad V_1 = V_2 = 1\text{V}$$

The predicted output voltage is then:

$$V_{op} = \{(2 \times 2) + 1\} - 2(1 + 1) = 1\text{V}$$

The simulation output voltage actually obtained was 1.012V, suggesting that this configuration has the highest error of these test configurations. This is because it is operating with a high reference input and additional input currents.

6.2.7 Difference Amplifier

The difference amplifier shown in Figure 6.8(f) gives an output which is proportional to the difference between the input voltages. The ideal analysis predicts the output:

$$V_o = R_f/R_1 \cdot (V_2 - V_1) = G(V_2 - V_1)$$

if the resistors connected to both terminals have the same values. V_2 is the input on the $+$ terminal, V_1 on the $-$ terminal. This circuit can be used with sensors that have a positive offset on their output, to bring the output voltages into the appropriate range (0–3.5V with a +5V single supply). The simulation shows a moderate output offset.

6.2.8 Universal Amplifier

The amplifier configurations described earlier can be regarded as special cases of a universal amplifier (Figure 6.10). This has both difference and summing inputs, and can be adapted to applications where a combination of these is required.

In theory, the universal amplifier can have any number of inputs and outputs, but to simplify the analysis, we will set the following conditions:

1. *The number of inverting inputs is equal to the number of non-inverting inputs*
2. *The input resistors (R_i) are all equal*
3. *The feedback resistors (R_f) are equal*

By summing the currents at the op-amp input terminals, we can see:

$$I_f = \frac{V_o - V_x}{R_f} = \frac{(V_x - V_1)}{R_1} + \frac{(V_x - V_3)}{R_3} + \frac{(V_x - V_5)}{R_5}$$

$$I_r = \frac{V_x - 0}{R_f} = \frac{(V_2 - V_x)}{R_2} + \frac{(V_4 - V_x)}{R_4} + \frac{(V_6 - V_x)}{R_6}$$

Assume $R_1 = R_2 = R_3 = R_4 = R_5 = R_6 = R_i$

Then $V_o = R_f/R_i(V_2 + V_4 + V_6) - (V_1 + V_3 + V_5)$

Or generally $V_o = R_f/R_i(V_2 + V_4 + V_6 + \ldots) - (V_1 + V_3 + V_5 + \ldots)$

The output voltage is given by the arithmetic sum of the input voltages multiplied by the gain, where the non-inverting ($+$) inputs are even numbered and the inverting ($-$) odd. The amplifier then behaves as a combination summing and difference amplifier, allowing positive and negative signals and offset inputs to be added as required. When designing op-amp-based interfaces, the universal amplifier offers a single starting point for designs

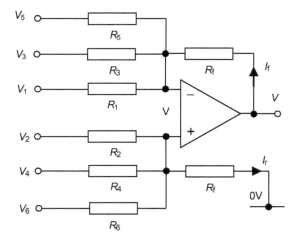

Figure 6.10
Universal amplifier.

with multiple inputs and demonstrates the principle of op-amp circuit modeling in a more general form than is usually seen.

6.3 Circuit Design

As well as using a linear stage to obtain signal gain, we may need to reduce the signal amplitude, filter or otherwise modify it in order to obtain an input suitable for the ADC. Some relevant techniques and design factors are outlined in the following sections.

6.3.1 Component Values

Most passive components can have wide range of values that tend to be supplied in preferred values, with a specified tolerance. Taking resistors, a basic set has values that are decimal multiples or sub-multiples of:

$$10, \ 12, \ 15, \ 18, \ 22, \ 27, \ 33, \ 39, \ 47, \ 56, \ 68, \ 75, \ 86, \ 91$$

Note that if each value can vary by $\pm 10\%$, approximately the whole range is covered. If a resistance requirement is calculated and it does not coincide with one of these values, the nearest preferred value (NPV) can usually be used in a general purpose (low precision) circuit. In practice, 2% ¼W metal film resistors are most often used, as they are reasonably stable with temperature, and intermediate values can be obtained if necessary. More expensive high-stability types are available at precise values.

Self-heating in resistors carrying significant current may affect the value of low-stability resistors when a circuit is active. Obviously, any resistor may be damaged by excessive power dissipation (volts × current) exceeding its rated value, but this will normally only be an issue in output load circuits.

Capacitors are usually available in a more limited range of 20% values, typically 10, 15, 22, 33, 47, 68 and 86, since their values are not so critical in most applications. The voltage rating must always be noted, as the component may break down if it is exceeded.

6.3.2 Voltage Divider

The simplest form of attenuator is the voltage divider (Figure 6.11), which reduces the signal voltage in proportion to the resistor values. Source and load resistances are shown connected, as they may affect the output, depending on their values relative to the divider resistors. If they are insignificant ($R_S = 0$, $R_L = \infty$), the output will be:

$$V_o = R_2/(R_1 + R_2) \cdot V_i$$

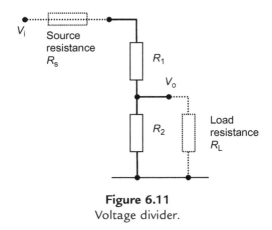

Figure 6.11
Voltage divider.

If the source and load resistance are significant, R_1 will be replaced by $R_1 + R_S$ and R_2 by $R_2//R_L$ (parallel resistances) where

$$R_2//R_L = R_2 \cdot R_L/(R_2 + R_L) \text{ (product/sum)}$$

In either case, to maintain signal accuracy, buffering with a unity gain non-inverting stage may be advisable to isolate the divider (or any other conditioning stage) from adjacent circuits.

6.3.3 Gain and Offset

A single op-amp stage can provide both amplification and level shifting, using the basic configurations outlined in the previous section. When using a single supply and all voltages are positive, the non-inverting amplifier can provide gain and offset control without affecting the signal polarity. In addition, the input is high impedance, minimising loading of the signal source, and the output low impedance, isolating the next stage.

In the demo circuit (VSM project GOFF2) in Figure 6.12, the input varies from 100 to 300mV, and the output is required to change from 1.00 to 3.00V (gain = 10). Simulation allows the component values to be easily modified to obtain the required operation. The input is set to the minimum and the offset adjusted and then set to maximum and the gain adjusted. This process is repeated until both are correct. The mid-value can then be checked (2.00V output) to confirm linearity, producing an error of 2mV, or 0.1%.

6.3.4 Amplifier Calibration

This adjustment process can be applied when the amplifier is implemented in hardware and calibrated. In the test circuit, switched inputs are used to speed up the component selection,

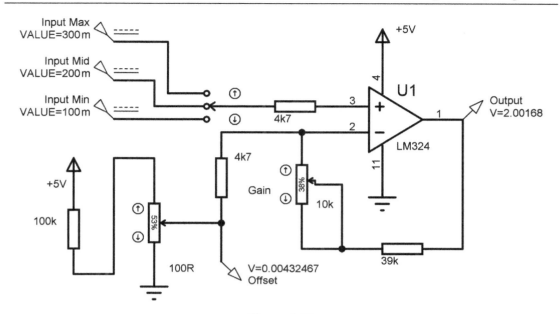

Figure 6.12
Gain and offset adjustment.

while the real input cannot necessarily be changed so easily. For example, a temperature sensor cannot be heated and cooled quickly, even if it has a calibrated output.

The LM35 IC temperature sensor, for example, provides an output of 10mV per degree Celsius, with 0mV corresponding to 0°C. 100mV then represents 10°C and 300mV represents 30°C. If we wish to measure that range using an 8-bit conversion with a reference voltage of 2.56V, a non-inverting stage could be implemented to provide the required gain and offset. This could use variable potentiometers or fixed value resistors, which are more reliable.

In simulation, the minimum output reached by the LM324 op-amp is about 80mV. Either the loss of the lowest measurements can be accepted, or the output shifted to start at a higher voltage, using the offset adjustment. For example, the voltages between 0.56 and 2.56V might be used to represent the specified temperature range within the limited output swing.

Dummy test inputs, which represent the calibrated sensor output, may be useful when testing the actual hardware. Final calibration then consists of correctly adjusting the gain and the offset at the minimum and maximum output levels, assuming that it is linear in between these values.

6.3.5 Input Resistance

The non-inverting amplifier has high input impedance, which is an advantage in that it does not load the sensor or previous stage, and minimises power consumption. However, if its input connection is long, it will be more susceptible to noise because any externally induced current will produce a significant noise voltage at the input (a small noise current will produce a large noise voltage across a large resistance). If the signal voltage is small, the noise is consequently a larger proportion of the overall signal, i.e. the signal to noise ratio is worse.

This can be alleviated by increasing the signal current and using an inverting amplifier configuration at the receiving end. It has an input impedance equal to the input resistor value, which is typically only a few kilohms. For example, if the signal is 1V and the input resistor is 1k, the current will be 1mA, swamping the noise signal. The source may need to be buffered, unless it has sufficient current driving capability. The LM35 temperature sensor, for example, which has an internal amplifier, would be able to drive a line in this manner without buffering.

An alternative is to simply attach a load resistor at a high impedance input (Figure 6.13). The sensor output must supply more power, and this may be undesirable in a battery powered system. If the sensor is remote, a current loop should be considered (see later), or some other form of serial driver. We will see sensors later with built-in serial communications that will also overcome this problem.

6.3.6 Input Capacitance

The input of the ADC has a sample and hold circuit incorporating a 120pF capacitor that is intended to hold the input voltage constant while the conversion is in progress. The input sampling switch has a resistance of about 10kΩ. The simple RC equivalent circuit is shown

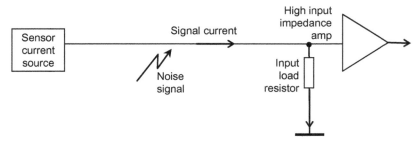

Figure 6.13
Input resistance and long input line.

in Figure 6.14(a). It acts as a low-pass filter (see Section 6.4) and affects the rise time of switched inputs.

This type of filter arrangement may be externally connected at any input to the MCU to attenuate high-frequency noise and improve reliability of input switching. We have already seen this technique applied to switch debouncing. It also represents the behaviour of any real circuit with stray capacitance and high impedance at the signal input. We will therefore analyse its behaviour in a bit more detail.

If the input to the simple RC network switches instantly between 0 and V_s (step voltage), the output will respond by rising exponentially due to the charging characteristic of the capacitor:

$$V_o = V_s(1 - e^{-t/CR})$$

The voltage rises to a final value that is the same as the input (when $t = \infty$), as seen in Figure 6.14(b). The time constant for the circuit is defined as $C \times R$ seconds, which corresponds to the time at which the voltage has reached 63.2% of the final value ($t = CR$). When $t = CR$:

$$V_o = V_s(1 - e^{-1}) = 0.632 \, V_s$$

The rise time is the time taken to reach 95% of the final value, which we can calculate by rearranging the step transfer function and substituting $V_o/V_s = 0.95$:

$$t = -CR \cdot \text{Ln}(1 - V_o/V_s) = -CR \cdot \text{Ln}(0.05) \sim 3CR$$

In practice, the final voltage is assumed to be reached after the output is over 99% of the input, when

$$t = -CR \cdot \text{Ln}(0.01) > 5CR$$

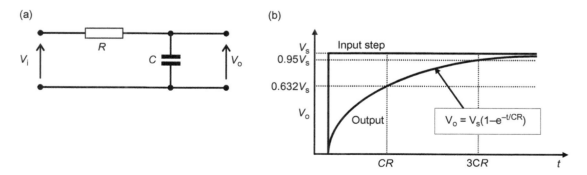

Figure 6.14
(a) RC network and (b) transient response to input step.

Therefore, a delay of $5CR$ between samples should be allowed when sampling an input with step changes through a low-pass filter. With the values of C and R in the ADC sample and hold circuit:

$$5CR = 5 \times 120 \times 10^{-12} \times 10 \times 10^{3} = 6 \times 10^{-6} = 6\mu s$$

Adding the effects of other associated components in the ADC input, a minimum conversion time of about 10μs is available. A more exact calculation is described in the MCU data sheet, which takes into account the amplifier settling time and temperature, for the selection of the optimum clock rate for the ADC. If speed is not critical, a lower clock rate may be used, as in the demo programs.

6.3.7 Transient Response

If a parallel capacitor is added to the feedback path of a linear stage (Figure 6.15(a)), a similar response is obtained to the passive RC network. When calculating the effect of feedback capacitance, remember that any internal capacitance, particularly in internally compensated op-amps (see later), must also be added to the external capacitance.

If the effective capacitance is relatively small, a similar response to the passive network is obtained. If a square wave is input with a fixed period, the output rises exponentially, and may not reach the final value before the input is reversed (Figure 6.15(b)). This shows why d.c. switching frequencies are limited in all active digital and analogue circuits − the outputs may not reach the valid logic levels if the switching is faster than the internal and external stray capacitance will allow.

If a relatively large value of capacitor is used with a large (or infinite) parallel resistance in the feedback path of the op-amp, the curve is so extended in time that it appears to be a straight line. The integrator will thus generate a sawtooth (triangular) waveform

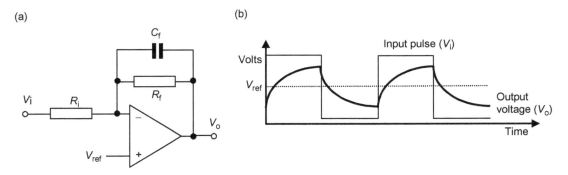

Figure 6.15
Feedback capacitance: (a) low-pass amplifier and (b) transient response of low-pass amplifier.

from a square wave input. If a (non-symmetrical) TTL signal is used, the amplifier will need a suitable offset added to the output to keep the sawtooth within the output limits.

A small capacitance in parallel with the feedback resistor is often included to provide additional stability in the operation of a linear amplifier if the speed of response is not critical. It acts as a low-pass filter, reducing or eliminating high-frequency noise and transients.

6.4 Frequency Response

Having analysed the transient response of RC networks associated with MCU input interfaces, we must also consider the frequency response, even though most sensor inputs are in the form of direct voltages. This is because even d.c. inputs may have higher frequency components in the form of noise and crosstalk. A square wave consists of a fundamental frequency (sine wave) plus a set of odd harmonics, so these are present in any digital switching signal. We may also sometimes wish to process signals such as audio for sampling by the ADC.

6.4.1 Low-Pass Filter

Where only direct voltage signals are of interest, the response time is not critical and the input resistance is high, a simple low-pass RC filter, as discussed earlier, should be included in any input path as a basic precaution against noise, and should always be considered with long input connections. The frequency response of a basic low-pass filter is shown in Figure 6.16. It consists of a 1k resistor and 1.5µF capacitor.

The reactance (a.c. resistance) of the shunt capacitor is $1/2\pi fC$ at frequency f, so treating the network as a voltage divider, it can be shown that:

$$V_o/V_i = 1/\{1+(2\pi fCR)^2\}^{\frac{1}{2}}$$

At low frequency, the term containing the frequency is small, so $V_o/V_i = 1$ and the output is the same as the input. At high frequency, the frequency term is dominant, so:

$$V_o/V_i = 1/2\pi fCR$$

and the output is inversely proportional to frequency. When plotted as attenuation (in decibels) against frequency (in decades), a curve is obtained that is level up to a break frequency and then falls away at 20dB per decade (see Appendix B for an explanation of dB measurement).

(a)

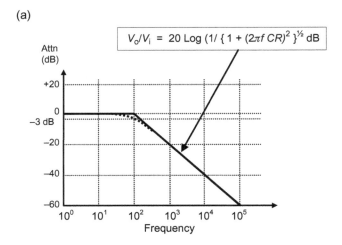

$$V_o/V_i = 20 \, Log \, (1/\{1 + (2\pi f \, CR)^2\}^{\frac{1}{2}} \, dB$$

(b)

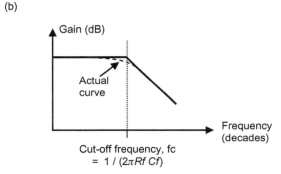

Figure 6.16
Low-pass frequency response: (a) passive RC filter and (b) d.c. amplifier.

The break frequency occurs when the impedance of the capacitor is equal to the resistance:

$$R = 1/2\pi f C \quad \text{or when} \quad f = 1/2\pi RC$$

At this frequency, the voltage ratio is 0.707 and the attenuation -3dB. In the sample plot shown, the break frequency is around 100Hz. This can be predicted from the component values:

$$fc = 1/2 \times \pi \times 10^3 \times 1.5 \times 10^{-6} = 106 \, Hz$$

The filter will reduce unwanted signal above 1kHz, with the attenuation increasing with frequency.

6.4.2 Internal Compensation

Real circuits and components have stray capacitance associated with the signal conductor components. This is a particular problem in IC amplifiers, where planar elements are formed in close proximity, and affects their response to switching and a.c. signals. The overall effect can be represented by shunt capacitance in the input stage, acting as a low-pass filter, which limits the upper frequency of operation of the amplifier.

Such capacitance will often be deliberately included within an amplifier design to improve overall stability and rejection of noise in d.c. amplifiers and to control the bandwidth in a.c. applications. If internal compensation is not incorporated, additional pins may be provided for fitting an external compensation capacitor so that the designer can set the upper frequency limit.

6.4.3 Gain/Bandwidth Product

The internally compensated op-amps, such as the traditional LM741 and LM324, are designed to have an open loop bandwidth of 10Hz. The gain then falls away at 20dB per decade of frequency, such that at 1MHz the gain is reduced to unity. This response determines the closed loop bandwidth as well, where the gain is inversely proportional to the bandwidth. This is represented by the frequency plot shown in Figure 6.17.

The bandwidth can be predicted for any value of closed loop gain by reading off the curve. The gain axis is usually scaled in decibels, so we need to be able to convert the gain as a voltage ratio into this form, using the definition of the decibel:

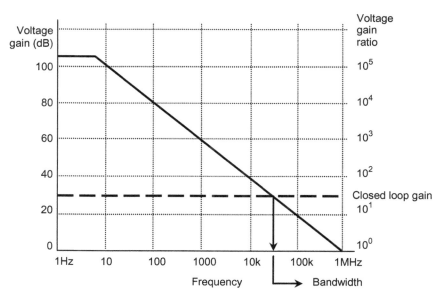

Figure 6.17
Gain versus bandwidth of d.c. amplifier.

$$\text{Gain} = 20 \cdot \text{Log}_{10}(V_\text{o}/V_\text{i})\text{dB}$$

The closed loop gain is usually between 1 and 100:

$$V_\text{o}/V_\text{i} = 1 \qquad \text{Gain} = 20 \cdot \text{Log}_{10}1 = 20 \times 0 = 0\,\text{dB}$$
$$V_\text{o}/V_\text{i} = 10 \qquad \text{Gain} = 20 \cdot \text{Log}_{10}10 = 20 \times 1 = 20\,\text{dB}$$
$$V_\text{o}/V_\text{i} = 100 \quad \text{Gain} = 20 \cdot \text{Log}_{10}100 = 20 \times 10 = 40\,\text{dB}$$

The bandwidth at any value of gain can be calculated from:

$$\text{Gain} \times \text{Bandwidth} = 1\text{MHz} = 10^6\ \text{Hz}$$

For example:

> If gain $= 30\text{dB}$, voltage gain $= 10^{30/20} = 10^{1.5} = 31.6$
> Then bandwidth $= 10^6/10^{1.5} = 10^{4.5} = 31.6\text{kHz}$

If additional capacitance is connected across the feedback resistance, the effect will be to reduce the gain/bandwidth product (GBWP), so the response rolls off at a lower frequency. However, the basic form of the first-order low-pass active filter frequency plot remains the same.

6.4.4 Alternating Current Coupling

Amplifier stages may be coupled together via a series capacitance, as seen in Figure 6.18(a) (VSM project AUDIO2). In simple terms, this allows a.c. signals to pass and block d.c. signals. In this way, the d.c. bias on a mid-band input signal can be eliminated or modified, since the d.c. level on either side of the coupling capacitor can be different.

Alternating current coupling will affect the low-frequency response, operating as a high-pass filter. Taken in conjunction with the usual low-pass characteristic of all real op-amp circuits, a band pass stage will result. That is, only signals within a given frequency range will be amplified.

A relatively large value of capacitor will provide a.c. coupling with a low cut-in frequency, while smaller values will cut in at a higher frequency. Alternating current coupling is standard in discrete component amplifiers, because the d.c. bias conditions in the output of one stage is usually different from the bias in the next.

6.4.5 Band Pass Amplifier

If the MCU is receiving analogue signals at a range of frequencies, pass band filtering is often required to eliminate the unwanted signals. For example, audio signals occupy the range of 20Hz−20kHz, and the PIC ADC is just about capable of sampling at a sufficiently high rate to capture an undistorted digitised version.

A simple band pass filter for audio input would consist of a low-pass RC filter with a cut-off frequency of 20kHz and a high-pass CR filter with a cut-off frequency of 20Hz. In the latter case, the resistor and capacitor are transposed compared with the low-pass filter, but the analysis is similar, with the break frequency calculated from the same formula ($f = 1/2\pi RC$).

The slope of the frequency response is +20dB per decade below this frequency and flat above. The simulated response of the band pass audio amplifier is shown in Figure 6.18(b). The low cut-off frequency, f_L, should be predicted by the values of the input coupling components C_1 and R_1:

$$f_L = 1/2\pi R_1 C_1 = 1/(2\pi \times 10 \times 10^3 \times 2.2 \times 10^{-6}) = 7.2\text{kHz}$$

This corresponds to the result seen in the frequency response. The high cut-off frequency on the response curve is about 20kHz. The feedback resistor is 90k, so the effective capacitance controlling this frequency is

$$C = 1/2\pi R_1 f_H = 1/(2\pi \times 90 \times 10^3 \times 20 \times 10^3) = 88\text{pF}$$

The actual feedback capacitor value is 10pF, so it can be deduced that, in this case, the internal compensation in the op-amp accounts for most of the capacitance that determines the upper frequency limit.

6.5 Op-Amp Selection

Op-amps are fabricated using bipolar transistors, field effect transistors or a combination of both. There are many types, each with their own combination of characteristics and cost; the most appropriate device for any given interfacing circuit should ideally be selected, but most designers will use a limited range with which they are familiar and keep in stock.

Many are available as single, dual or quad packages. Three have been selected for comparison in Table 6.2. The LM741 was widely used when op-amps were first developed, the LM324 was an early single supply quad device and the MCP6004 is a more recently introduced low-power CMOS quad chip.

6.5.1 Op-Amp Types

The 741 is the original, standard, general purpose single op-amp in an 8-pin package. It is based on bipolar transistor technology, with internal compensation (feedback capacitance)

(a)

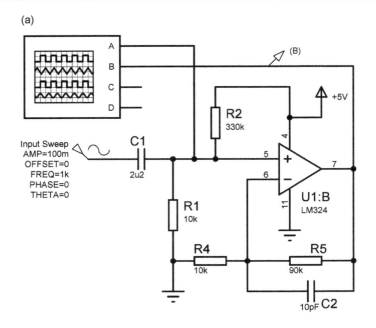

(b)

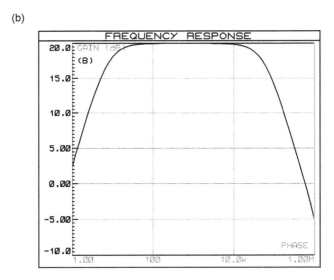

Figure 6.18
Audio amplifier: (a) band pass amplifier and (b) band pass frequency response.

to provide a stable, low bandwidth device for d.c. and audio range applications. The LM324 is a similar quad device but designed for single 5V supply operation, which is convenient in MCU systems.

Table 6.2: Op-Amp Characteristics.

Type	LM741	LM324	MCP6004
Technology	Bipolar	Bipolar	CMOS
Package	Single	Quad	Quad
Nominal supply	±15 V	+5 V	+5 V
Large signal voltage gain	200 V/mV	100 V/mV	112 dB
	=200,000	=100,000	=400,000
	=106 dB	=100 dB	=400 V/mV
Gain/bandwidth product	1 MHz	1 MHz	1 MHz
Slew rate	0.5 V/μs	0.5 V/μs	0.6 V/μs
Common mode rejection ratio	90 dB	85 dB	76 dB
Input offset voltage	±15 mV[a]	±2 mV	±5 mV
Input bias current	80 nA	45 nA	1 pA
Supply current	2 mA	1 mA	100 μA
Output voltage swing	±13 V	+5 mV to +3.5 V	+25 mV to 4.97 5V.
Maximum output current	±25 mA	+40 mA − 20 mA	±20 mA
Description	Historically most popular	Single supply quad package	Low power rail-to-rail swing

[a]External offset adjustment available.

The MOSFET op-amp uses metal oxide semiconductor field effect transistors, which have very low input current, and therefore provide low-power, high-input impedance amplifiers. However, this technology is susceptible to high voltage static electric charge in the environment and requires careful handling during assembly.

The BiFET type combines advantages of the bipolar and FET types in one chip, offering high speed with low power consumption. Protected FET inputs provide high input impedance and low input currents, while bipolar transistors provide high gain and outputs that are more robust and switch faster than FETs at high current.

6.5.2 Op-Amp Supplies

IC amplifiers can operate with dual or single supplies. Dual supplies make the circuit design easier, because the output can swing positive and negative around 0V. The default is ±15V, giving an output voltage swing of, perhaps, 28V, since the output cannot reach the supply value in many op-amps.

However, it is often convenient in microprocessor systems to use the same single supply used by the digital circuits, +5V, and avoid the need to provide separate dual op-amp supplies. Some op-amps are designed specifically to operate with a single supply, such as the LM324 type used in the examples here. The downside is that a 5V supply provides only a limited voltage swing; the LM324 output goes down to 5mV but only up to 3.5V. The MCP6004 is designed to address this limitation and swings within 25mV of both supply rails.

The MCU input range in the 8-bit demo circuits is limited to 2.56V, so output of the final amplifier stage is operating comfortably within the upper limit of the LM324, but we cannot assume that voltages near zero will be represented accurately. The conversion range can be shifted up to, say, 0.5−3.06V, or the loss of accurate conversion at the lower end tolerated.

6.5.3 Op-Amp Characteristics

Op-amps are available which offer various combinations of desirable characteristics such as high precision, low noise, low power consumption, high bandwidth, high output current and low input currents. When designing analogue signal conditioning for specific applications, an op-amp with the optimum combination of features should be selected. The comparison of different generations of op-amps in Table 6.2 defines the main electrical characteristics of representative devices.

6.5.3.1 Large Signal Voltage Gain

This is the open loop gain of the op-amp. Note that this is expressed in different ways, in volts per millivolt of input, a simple ratio or in decibels. It is mainly of academic interest, since the op-amp is mainly used in amplifiers with a gain of 100 or less, where the performance is determined by other factors.

6.5.3.2 Gain/Bandwidth Product

All the examples here are internally compensated to provide unconditional stability with direct voltage signals, and they all have the same unity gain bandwidth of 1MHz. The frequency response then follows the first-order characteristic seen in Figure 6.17. The open loop gain is maximum at low frequency but breaks at a low frequency (10Hz) to follow an inverse linear relationship between gain and frequency, with axes scaled in decades (or decibels for the gain). If the op-amp is used as a linear amplifier with a suitable feedback network, the open loop plot predicts the closed loop bandwidth. If the GBWP is 1MHz, the plot intersects the scales at decade points, so the bandwidth is easy to read off or calculate for any value of closed loop gain.

6.5.3.3 Slew Rate

The slew rate is the maximum rate of change available at the output, closely related to bandwidth since it is also controlled by the internal compensation, and is simply the corresponding transient response. Due to the high gain, the output change due to a step input is linear rather than exponential, so is quoted in volts per microsecond, or presented graphically.

6.5.3.4 Common Mode Rejection Ratio

Since the op-amps in our examples here are operating with single supply, their inputs are typically operating with a common mode voltage (relative to 0V) of at least 1V (V_{cm}). The common mode rejection ratio (CMMR) for the LM324 is quoted in the data sheet as typically 85dB. This means that the voltage at the output due to the common mode input voltage can be calculated as follows:

$$V_{co} = V_{cm}/10^{CMMR/20} = 1/10^{85/20} = 56\mu V$$

This is usually insignificant in the low precision applications outlined in this book, but may need to be taken into consideration in high precision designs, or when applying low-power op-amps such as the MCP6004 that have an inferior CMMR (76dB).

6.5.3.5 Input Offset Voltage

The inputs of the op-amp have a matched pair of transistors working as a differential pair. Since these never have identical characteristics, the mismatch is amplified and appears at the output as an offset voltage when the inputs are tied together. The input offset voltage is that which is required to zero the output for zero input. In the 741, where this characteristic is particularly poor, an external pot can be connected to balance the input stage. If an amplifier offset input is needed anyway, this can also be used to zero the output.

6.5.3.6 Input Bias Current

The ideal amplifier has infinite input resistance, resulting in zero input current in the terminals. In practice, the input transistors draw some finite current, with bipolar transistors much worse than FETs. This diversion of current from the feedback network is a significant cause of inaccuracy in bipolar amplifiers, but the current generation of low-power amplifiers can be seen to have improved this figure by several orders of magnitude; the MCP6004 has an input current of only 1pA, compared with the 741 at 80nA.

6.5.3.7 Quiescent Supply Current

The low-power Microchip MCP6004 op-amp improves on quiescent power consumption by a factor of about 200 compared with the bipolar types, with typical figure of 100μA. Its power consumption is typically 0.5mW per amplifier with a 5V supply.

The LM324 has been used extensively in the test circuits since it demonstrates the limitations frequently found in op-amp circuits. All can be improved by selecting a current device with better performance, typically with lower power consumption and better output swing and overall precision.

6.5.4 Op-Amp Selection

At the time of writing, the Microchip Analog and Interface Product Selector Guide at www.microchip.com lists an extensive range of op-amps with different combinations of characteristics, primarily classified by their GBWP. Most have a low input current around 1pA (IGFET input), while the low gain/bandwidth devices have the advantage of a typical quiescent supply current of less than $100\mu A$. They operate at a similar range of voltages to the current generation of MCUs, so that power consumption can be reduced by operating at lower voltage, and reliable battery powered operation is easier to achieve. If this is required, the minimum operating voltage and current consumption of each device must be carefully considered.

As a potential replacement for the LM324 in the designs in this book, the lowest cost quad device currently listed in the Microchip guide is the MCP6004. Its output can swing within 25mV of the supply, which can vary from 1.8 to 6V. It has the standard GBWP of 1MHz, which means that it has the similar frequency response as the internally compensated LM324. However, its CMMR is slightly inferior at 76dB, giving an output offset of 0.4mV at 1V input common mode voltage, which is still insignificant in the context of these designs.

6.6 Comparators

Linear op-amp applications use negative feedback. However, discrete op-amps can be used in switching mode, where the voltages at the inputs are compared, the output forced to its maximum or minimum voltage depending on the relative input polarity, thus operating as a comparator. In this case, there is either no feedback connected or it is positive.

When identifying circuit configurations in existing designs, positive feedback indicates a comparator, or an oscillator, while negative feedback generally indicates linear amplifier operation.

A specific type of op-amp that has an open collector output is normally used for this type of application. The output transistor switching circuit has to be completed by an external pull-up (load) resistor, in the same way as the switch input seen in earlier applications. This allows the output switching levels to be different from the comparator supply voltage.

This level shifting is useful for interfacing an MCU output to a load circuit operating at a higher voltage, 24V for example. In addition, the open collector outputs can be connected together to operate as 'wired-or' outputs, as in the window comparator described later. The comparator switching speed can also be improved by using a lower value pull-up resistor, at the cost of higher power consumption.

Three types of comparator circuit are shown in Figure 6.19 (VSM project COMPS2). The default chip type used here is the traditional TLC339, a quad comparator. The comparator is used in open loop mode (no negative feedback) to compare two input voltages connected to the + and − terminals. The output will switch high or low depending on the relative polarity; if the + terminal is positive with respect to the − terminal, the output will go high, and vice versa.

6.6.1 Simple Comparator

The basic comparator shown in Figure 6.19(a) detects whether the input is above or below the reference voltage of 2.5V applied to the negative input terminal. The transfer characteristic (Figure 6.19(b)) describes its operation by plotting the output against the input voltage. The reference voltage can be changed as required, giving a different switching level. If the inputs are reversed, so is the output polarity.

The output of the comparator is connected to an LED indicator in the load circuit, which is useful, but not essential. The open collector output provides sufficient output current to drive the LED (10mA), without any additional driver stage. If used with dual supplies and a zero reference voltage, it would be described as a zero crossing detector, which can be used to detect symmetrical digital signals on a communication line.

Many MCUs incorporate comparators at specific inputs. In the PIC 16F877A, two simple comparators are available at Port A that can be programmed to work separately or together in various combinations. A programmable reference voltage between 1.25 and 3.75V (5V supply) can be selected. Refer to the data sheet for details.

6.6.2 Trigger Comparator

The output voltage in this circuit (Figure 6.19(c)) is fed back to the positive terminal to set the reference level, which, in turn, is dependent on the output level. The switching level therefore depends on the *previous* setting of the output. As a result, the output switches at a higher voltage when increasing from low to high, and at a lower voltage when decreasing from high to low. The input is applied to the negative input terminal, so the transfer characteristic is inverted.

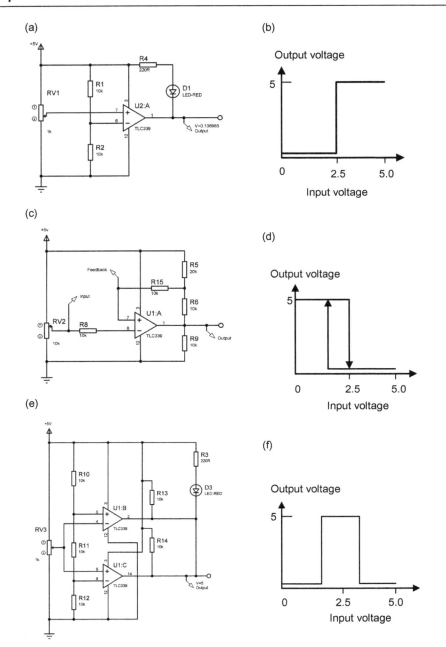

Figure 6.19
Comparators: (a) simple comparator, (b) transfer characteristic, (c) trigger comparator, (d) trigger comparator characteristic, (e) window comparator circuit and (f) window comparator characteristic.

The circuit shown has a voltage divider connected to the output that gives a high switching level (V_H) at the positive comparator input when the output of the comparator is off, and a low one (V_L) when the output is on (0V). An LED indicator is not included here as it would affect the switching levels. These can be calculated from the output resistor values:

$$V_H = 20/40 \times 5 = 2.5V$$
$$V_L = 10/30 \times 5 = 1.66V$$

A trigger logic circuit is often incorporated into digital signal paths as it helps to reduce noise (unwanted high frequencies and crosstalk). In a TTL gate, noise on a slowly changing input signal might cause multiple transitions at the output. With a Schmitt trigger input, once the gate has changed state, it does not change back unless there is a relatively large change in the input in the opposite direction. The PIC MCU has a Schmitt trigger buffer at TMR0 clock input for improved noise immunity.

6.6.3 Window Comparator

In this circuit (Figure 6.19(e)), two comparators are used together to generate an output logic high when the input voltage is within a set range, and a low when outside this range (or vice versa). The lower comparator output is near 0V when the input voltage is below 1.6V, and the upper comparator output is low when the input voltage is above 3.3V. Between these voltages, neither is low, and the output is pulled up to 5V. Open collectors allow this connection, whereas it is *not* allowed with the complementary output drivers in standard op-amps. The circuit is used to detect when a voltage is within or outside a given range, which could be used, for example, in a simple voltage tester giving a pass/fail output.

6.7 Op-Amp Applications

This section will outline some op-amp applications designed to provide accurate signal conditioning for particular sensor interfacing requirements.

6.7.1 Instrumentation Amplifier

Many sensors that need to be connected to a microcontroller analogue input have a rather small output signal. In addition, the input may only be available as a small differential voltage between two nodes with a large common mode voltage.

Strain gauges are one example; these measure small changes in the shape of a mechanical part under stress, such as a strut in a bridge. Four strain sensitive resistors are usually connected in a bridge arrangement, such that a change in their resistance due to stretching of the conductor is output as a small differential voltage, typically in the range of 0−10mV (see Chapter 9 for further details).

A sensitive amplifier is needed, with a high gain and high input resistance, which minimizes the current loading on the sensor and hence the errors. A single-stage non-inverting amplifier has a high input resistance but does not have differential inputs. The difference amplifier has these but has low input resistance. In addition, if configured for a high gain, with a high value for the feedback resistor, the feedback current is small, and thus the amplifier is more susceptible to noise and offset errors.

The solution is an instrumentation amplifier (VSM project INSTAMP2) which combines high input resistance and gain (Figure 6.20). It is made up of two stages, a pair of high-impedance inputs and an output difference amplifier. In order to highlight its limitations, our standard single supply op-amp, LM324, has been used, but the performance can be improved by selecting a higher specification op-amp, or buying the instrumentation amplifier as an integrated package.

The gain of the amplifier (100) is set by the ratio of feedback resistor chain connected between the outputs of the input stages, from the relationship:

$$G = 1 + 2R_5/R_{12} \qquad \text{where } R_5 = R_6 = 10k \text{ and } R_{12} = 202R$$
$$\therefore G = 1 + 20,000/202 = 100$$

The maximum differential input is 10.0mV, producing differential output of 1.00V. This is fed to the unity gain differential output stage that provides a single-ended output (i.e. 1.00V measured with respect to 0V). The simulation shows the output is accurate to 0.1% at full scale, but of course this does not include noise, drift and other error factors which may occur in the hardware implementation.

6.7.2 Current Loop

If a d.c. signal is to be transmitted from a remote sensor over a long connection, say more than 0.5 m, the resistance in the line may cause a voltage drop which will affect the accuracy of the received voltage. In this case, it is better to represent the measurement as a current, rather than a voltage, since d.c. loss is negligible in a closed loop.

If the operation of a simple inverting amplifier is considered ideal, the current in the feedback resistor must be the same as the current in the input resistor. If the input voltage and input resistance are constant, the feedback current will be constant, with the output voltage of the op-amp adjusting itself for any change in the feedback resistance value. This leads us to a general design for a constant current source, derived from a constant voltage at the input.

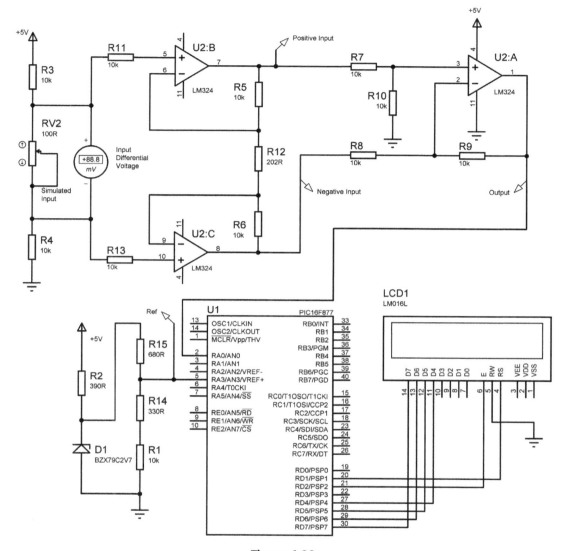

Figure 6.20
Instrumentation amplifier.

In Figure 6.21, a zener diode provides a constant voltage, and the current in the feedback path will then be (theoretically) constant and independent of the feedback resistance value. The main deviation from the ideal will probably be due to the small change in zener voltage as some of its current is diverted through the amplifier feedback loop.

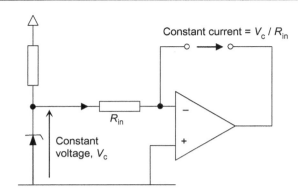

Figure 6.21
Op-amp constant current source.

This constant current principle can be adapted to a current loop which is controlled by the variable input voltage from a sensor. In Figure 6.22, a demonstration circuit (VSM project CLOOP2) is shown which will give an output change of 1.00V for an input change of 100mV, i.e. an overall gain of 10. However, the significant feature is the current loop formed by the feedback path of the line driver. A long connection between this stage and the output differential amplifier represents a line which can have a variable resistance, depending on the length and cabling type. We need the output to be independent of the variation of this resistance, which is represented by variable 10R pots. Correct operation can be confirmed by varying these pots, which should result in minimal change in the output (about 20mV maximum in simulation).

R_5 and R_6 (100R) are the input and feedback resistors in the line driver amplifier. The input stage is a simple non-inverting amplifier with a gain of 10, which feeds a voltage to the line driver which changes by 1.00V when the test switch is operated. The current switches between 0 and 10mA, to give 1.00V across the line driver load resistor, R_6. This is connected across the inputs of the unity gain output differential amplifier. The output of a standard op-amp is limited to about 25mA, so the line must operate at less than this value. On the other hand, the higher the current, the better the signal to noise ratio is likely to be.

Since the current loop is implemented using single supply op-amps running at 5V, the amplifiers are all offset by 1.50V, to avoid voltage outputs near 0V. The output switching is then between 1.50 and 2.50V. This is achieved to within about 1% in the simulation, the error being mainly due to variation in the individual amplifier offset conditions. The common offset of 1.50V is derived from a stack of two diodes supplied with a current which can be tweaked to obtain the desired voltage drop. If power diodes are used, heating effects can be reduced (the diode voltage drop changes by 2mV/°C). This offset

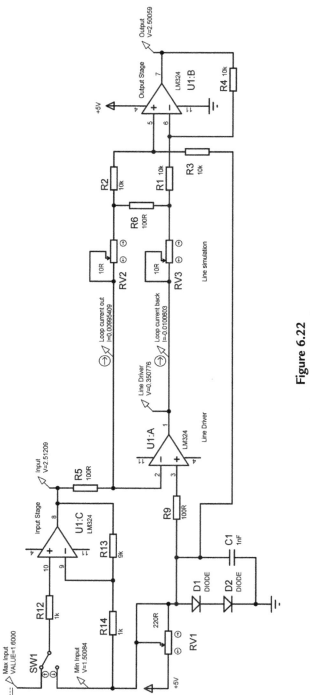

Figure 6.22
Current loop interface.

arrangement is also used in interfacing the LM35 temperature sensor which needs to go negative with respect to the reference level to measure temperatures below 0°C.

The standard current loop sensor interface operates at 4−20mA, and can provide power to the remote sensor as well as allowing it to control the current drawn from an external supply. The operating range is 16mA, a figure convenient for converting to digital form. Another advantage is that zero current can indicate a fault condition, i.e. an open circuit in the current loop.

This kind of differential driver is used extensively in data transmission. A TTL data stream is fed to a differential amplifier and transmitted as a bipolar signal on a twisted pair of conductors for reception by a difference amplifier and comparator that recover the original data bits. This will be discussed further in Chapter 8.

6.7.3 Logarithmic Amplifier

When conducting, the forward voltage drop in a semiconductor diode is approximately 0.6V, but the exact value varies with the current and temperature. Figure 6.23 shows a forward biased diode in the feedback path of an inverting stage, which allows the diode voltage drop to be simulated at input voltages and currents that increase in decades (VSM project DIODE2).

The test circuit was set up using the generic diode provided in the VSM simulator and some sample readings were taken. We can see that the diode voltage drop increases in equal steps above 30μA, within about 2%. This can be represented by the logarithmic function:

(a)

(b)

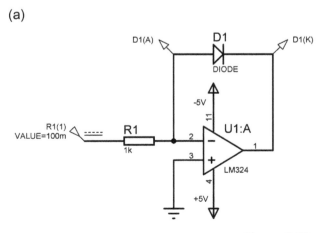

Input(V) and current (mA)	Output volts (mV)	Output step (mV)
0.003	475	-
0.03	561	86
0.30	622	61
3.00	681	59
30.0	741	60

Figure 6.23
Diode clamp: (a) test circuit and (b) simulation test results.

$$V_D = V_T \cdot \text{Ln}(I_D/I_S)$$

where I_S is diode leakage current,

$$V_T = kT/q$$

and

k = Boltzmann's constant
T = absolute temperature
q = charge on the electron

At room temperature, V_T is about 25mV. I_S depends on the type of diode. When the current was 100μA (V_{in} = 100mV) in the simulated test, the voltage drop across the diode was 591mV. I_S can then be calculated from these sample values for this device by rearranging the diode equation:

$$\begin{aligned} I_S &= I_D/\exp(V_D/V_T) \\ &= 0.0001/\exp(0.591/0.025) \\ &= 5.41 \times 10^{-15} \end{aligned}$$

A transfer characteristic for this diode is therefore:

$$V_D = 0.025 \cdot \text{Ln}(I_D/5.41 \times 10^{-15})$$

This logarithmic response can be used to measure currents and voltages over a wide range, because a decade change in the input causes a small step change in the output. The readings from the simulation show that this is about 60mV per decade of diode current at higher currents, deviating from linearity at lower currents.

The range and precision of the log amp can be improved by using a bipolar transistor in the feedback path. The base−emitter junction has the same characteristics as the diode, and the collector current is accurately proportional to the base current (transistor current gain, h_{FE}). The current gain of the transistor extends the accuracy of the base−emitter transfer function to a much greater range of currents. A demo circuit (VSM project LOGAMP2) was simulated, as seen in Figure 6.24(a). It is necessary in this case to use bipolar supplies, ±5V.

The first stage is the log amp, with a negative output developed across the transistor Q1 base−emitter junction between 0.53 and 0.83V, corresponding to an input current between 10^{-7} and 10^{-2} amps generated by the switched input voltages. An FET input quad op-amp, TL074, that has high input impedance, is used to allow lower values of current to be measured. Ideally, the op-amp input bias current should be zero, so that it does not divert any of the measured current.

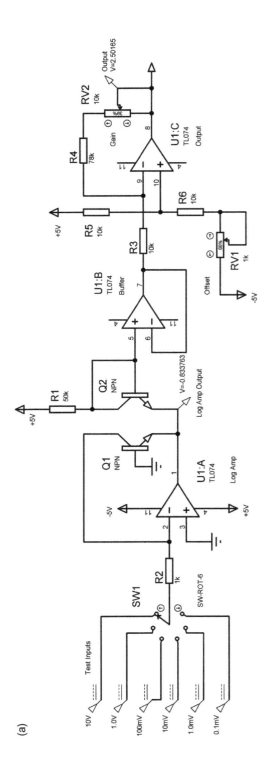

(a)

Input	Output (V)	Step (V)
0.10 mV	0.020	-
1.00 mV	0.505	0.485
10.0 mV	1.006	0.501
100 mV	1.505	0.499
1.00 V	2.003	0.498
10.0 V	2.501	0.498

(b)

Figure 6.24

Logarithmic amplifier: (a) test circuit and (b) test results.

A major factor affecting the performance of this circuit is that the output is highly sensitive to the temperature of Q1 b−e junction, due to the effect of V_T on the diode forward voltage drop. Q2, connected as a diode, compensates for this, as well as the negative output offset voltage. The transistors are a matched pair in close contact, or in the same package, so their junction temperatures and voltage drops are as similar as possible at all temperatures. The junction temperature will also be affected by the self-heating effect of the measured current, particularly at higher values. Q1 junction is biased at about 100μA, a mid-range input value. The output of Q1 is buffered with a unity gain stage, followed by an output stage that adjusts the offset and gain so that the output ranges from 0 to 2.5V in steps of 0.5V.

The simulation results are shown in Figure 6.24(b). At the higher current ranges, the output steps are uniform within about 0.5%, even though only a general purpose op-amp type is being used. If implemented using low leakage transistors, a high-performance input stage, and low-noise design techniques (particularly signal screening), the range may be extended to much lower currents. Accurate measurement of up to nine decades of current from 10^{-12} to 10^{-3} amps is then possible. One application is in the measurement of pressure in vacuum systems by ion gage.

6.8 Alternating Current Measurement

It may sometimes be necessary to measure the characteristics of a sine wave or similar symmetrical voltage, e.g. the output frequency from a sinusoidal oscillator, or the voltage output from an a.c. power supply. Amplitude can generally be measured by rectification and measurement of the resulting average or peak d.c. level. Frequency or period can be measured using a comparator, or zero crossing detector, to convert the a.c. signal to a TTL pulse waveform, whose period can then be measured using the MCU timer/counter.

6.8.1 Peak Detector

The basic rectifier circuit is shown in Figure 6.25(a). The rectifier diode only allows forward current flow, charging the capacitor up to the peak level of the a.c. signal. It then discharges slowly through the load resistance. The ripple amplitude on the output is determined by the CR time constant. A high-value resistor can be used to obtain an approximately constant d.c. output, with minimum ripple, but this is then less responsive to changes in the input level.

The basic circuit has severe limitations if accurate peak measurement is required. The input current required is high, and the output is inaccurate due to the forward voltage drop in the diode. A better performance can be obtained with the precision peak detector in Figure 6.25(b), where the input and output are buffered, and the output follows the actual capacitor voltage

(VSM project PEAK2). The circuit overcomes any reverse leakage in the rectifier diode, holding the peak voltage for longer.

The precision peak detector can be used to measure a variable a.c. voltage (50/60Hz) or signals such as audio (20Hz−20kHz). The simulation uses a TL074 FET input op-amp, but only generic components otherwise, so the output drop between samples and overall accuracy could be improved with a low leakage capacitor and diodes in the hardware implementation. The components values must be adjusted to provide the appropriate combination of response time and ripple level.

(a)

(b)

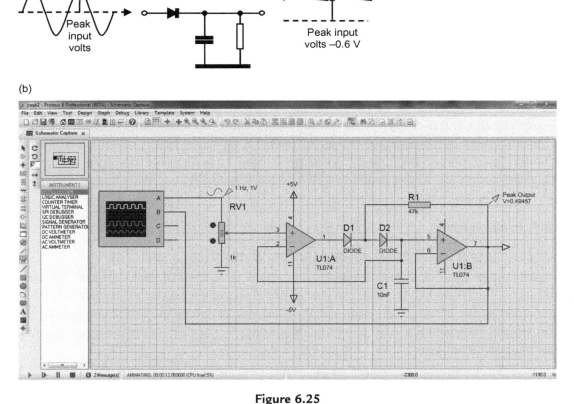

Figure 6.25
Peak detector: (a) simple rectifier circuit with input and output waveforms and (b) active peak detector simulation.

6.8.2 Frequency Measurement

The frequency of a repetitive signal can be obtained by feeding it to a comparator with a suitable switching level to produce a TTL pulse train. If the signal is a.c. coupled, with no offset, a zero crossing detector can be used. A trigger comparator is usually preferred, as it will reject noise on the signal. The resulting pulse train can then be fed to a timer/counter in a microcontroller to measure the frequency by counting pulses over a set time period (higher frequencies), or to obtain the period by timing the pulse duration (lower frequencies). Chapter 7 will provide a demo program to measure input period. This method is most suitable for signals whose frequency is changing only relatively slowly. A block diagram of the basic elements is shown in Figure 6.26.

6.8.3 Digital Sampling

Digital audio files are now the standard method for music recording, distribution and playback. For CD recording, the original audio signal is sampled at a rate of 44,100 samples per second at a resolution of 16 bits per sample, allowing all the original information in signals up to 22kHz (the limit of human hearing) to be accurately represented. Any low-frequency analogue signal can be captured, stored and processed in a similar way using the PIC MCU ADC inputs.

The basic process is illustrated in Figure 6.27. The analogue signal consists of a range of frequencies and amplitudes, represented by an irregular waveform within a variable envelope. The original signal is symmetrical, so it needs to be rectified (converted to unipolar form) and sampled as positive-going voltages. The precision rectifier described earlier could be used.

The PIC ADC has a minimum conversion time of 16μs, giving a maximum sampling rate of $1/16 \times 10^{-6} = 62,500$ per second. Therefore, in theory, full range audio could be captured,

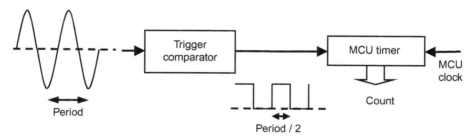

Figure 6.26
Frequency and period measurement.

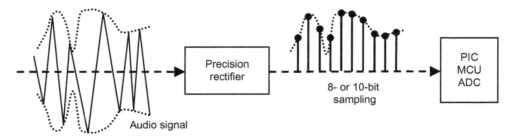

Figure 6.27
The principle of digital sampling.

but the storage process would have to be interrupt driven and fitted within the conversion time. At a maximum instruction rate of 5MHz, that is only 80 instruction cycles!

Signals at lower frequency would be easier to handle, since the processing speed would not be so critical. Additional memory will also probably be needed to store the samples, so the access time to external RAM or ROM must be considered (see Section 5.7) or even a mass storage medium (hard drive). In any case, serious signal processing would normally be implemented using a more powerful processor selected from the PIC DSP (digital signal processor) range.

6.9 Analogue Output

Analogue output from microcontrollers is less commonly required than input, because many output loads can be driven directly by a digital (pulse) signal via a suitable current amplifier. Relays and solenoids only need a switched current driver, while the output from a heater or motor can be controlled using pulse-width modulation (PWM), where a switched output current is averaged by the inductive load.

For signal processing, however, a digital to analogue converter (DAC) may be needed. It converts a binary output into the corresponding analogue voltage. A DAC is often incorporated in DSPs, where an analogue signal is converted to digital form for processing and storage and then back to analogue. Computer, MP3 and CD, and any other digital audio storage system output must use a DAC to drive the final audio amplifier stage.

6.9.1 DAC Types

The typical DAC uses a ladder network of precision resistors to produce a bit-weighted output voltage from a binary code. An output sum voltage is produced as follows: the bit connected to the most significant bit input, if set, provides half the output voltage, the second bit a quarter, the third bit an eighth and so on.

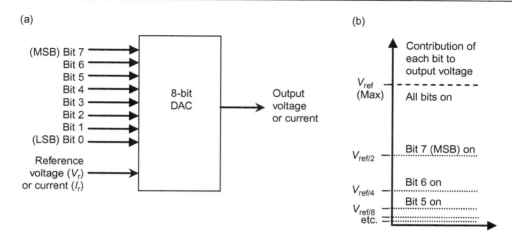

Figure 6.28

Digital to analogue converter operation: (a) general DAC hardware and (b) output voltage steps.

The passive network is fast, suitable for flash converters but needs to be buffered at its output. A summing amplifier could also be used, with the input resistor values in a power-of-two series: 1k, 2k, 4k, 8k, 16k, 32k, 64k, 128k for example, but the wide range of input currents is a potential problem.

In the general DAC shown in Figure 6.28, the output step size and maximum level are set by a reference input, as in the ADC. A reference voltage of 2.56V, for example, would give a bit step of 0.01V in an 8-bit DAC, since there are 256 (2^8) output levels. The least significant bit will produce a change of 10mV, corresponding to the resolution of the converter.

For an 8-bit DAC, the resolution is 1 part in 256, or slightly better than 0.5% at full scale, or 1% at mid range. A 10-bit DAC has a resolution of 1/1024, about 0.1%, at full scale and 1/512, about 0.2%, at the mid value. The resolution increases with the output level, since the step size is fixed.

Some converters use a current reference input and produce a current output that can be converted to a voltage by precision resistors. These resistors need to be at least as accurate as the DAC itself.

In the schematic of the DACS test circuit (Figure 6.29), a basic parallel and a serial input DAC are demonstrated (VSM project DACS2). The parallel DAC converts an 8-bit input into a corresponding analogue voltage, while the serial DAC receives its input from the SPI port.

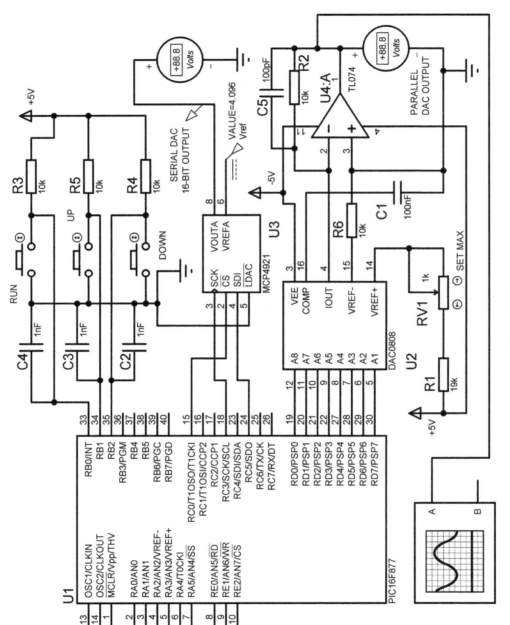

Figure 6.29
DACS schematic.

6.9.2 Parallel DAC

The parallel converter (PDAC) DAC0808 has an 8-bit digital input with a current reference and current output. The reference is derived from the supply (+5V) via a preset pot, which allows the maximum output to be adjusted to 2.55V. For greater accuracy, a stable reference voltage could be used in the current source. The PDAC therefore operates at 10mV/bit in this circuit. It also has a current output, so that a current loop output can be implemented for accurate onward signal transmission. In this circuit, a general purpose JFET (high impedance) input TL074 converts the output current into a voltage of 0–2.55V. A precision resistor should be used in the feedback path. A −5V supply allows operation down to 0V.

6.9.3 Serial DAC

The serial DAC (SDAC) is a more recently introduced device, the MCP4921 from Microchip. It uses the SPI interface, requiring a 12-bit serial input from the MCU. The output voltage range is also set by a voltage reference input. The output has 2^{12} steps (4096), so a reference voltage of 4.096V gives a conversion factor of 1mV/bit, and significant improvement on the 8-bit PDAC. The SDAC output can also reach 0V without a negative supply and needs fewer MCU I/O pins. On the other hand, the serial interface is inherently slower than the parallel.

The SPI interface uses a serial data input (2 bytes) from the MCU at the SDI output. These are simply written to the SPI output buffer in the MCU. The SDAC chip select is taken high by the MCU to trigger the data transfer, and the data strobed into the SDAC by a clock input at SCK. The most significant 4 bits of the first byte are used for control functions (0011); the low nibble contains the high 4 bits of the data, and the second byte the remaining 8. More details are given in the device data sheet of the MCP4921, and the SPI interface is described in more detail in Chapter 8.

6.9.4 DAC Program

The test program is listed in Program 6.3, and it can be seen that the software and initialisation required to drive the PDAC is relatively simple. The SDAC output in the test circuit is a simple direct voltage, controlled manually from the UP/DOWN push buttons. This increments and decrements the 12-bit data sent to the SDAC, the resulting voltage being displayed.

When the run button is pressed, the serial DAC output is disabled, and the PDAC driven with an incrementing output at maximum possible frequency, as determined by the MCU clock rate. Each output step takes three instruction cycles (INCF + GOTO). A sawtooth waveform is thus produced. If this is viewed on the oscilloscope, significant

```
;****************************************
;        DACS2.ASM MPB        18-01-13
;
;        Test program for parallel
;        and serial D/A Converters
;        DAC0808 & MCP4921
;
;        Updated for VSM v8
;
;****************************************

            PROCESSOR 16F877A
            INCLUDE 'P16F877A.INC'
            __CONFIG 0X3731

Hibyte     EQU        020          ; SPI data high byte
Lobyte     EQU        021          ; SPI data low byte

            CODE 0                  ; Load at default range
            NOP                     ; for ICD operations

; Initialise parallel and serial ports -------------

            BANKSEL    TRISD
            CLRF       TRISD                    ; Parallel port
            BCF        TRISC,5                  ; Serial data
            BCF        TRISC,3                  ; Serial clock
            BCF        TRISC,0                  ; Chip select
            CLRF       SSPSTAT                  ; default SPI mode

            BANKSEL    PORTD
            CLRF       PORTD                    ; zero PDAC
            CLRF       SSPCON                   ; default SPI mode

            MOVLW      B'00111001'              ; Initial SDAC data
            MOVWF      Hibyte                   ; and store
            MOVLW      B'11111111'
            MOVWF      Lobyte

; Check buttons ---------------------------------

up         BTFSC      PORTB,1                  ; Test UP button
           GOTO       down                     ; and jump if off
           INCF       PORTD                    ; Increment PDAC
           INCF       Hibyte                   ; Increment SDAC
waitup     BTFSS      PORTB,1                  ; Wait for..
           GOTO       waitup                   ; button release

down       BTFSC      PORTB,2                  ; Test DOWN button
           GOTO       spi                      ; and jump if off
           DECF       PORTD                    ; Decrement PDAC
           DECF       Hibyte                   ; Decrement SDAC
waitdo     BTFSS      PORTB,2                  ; Wait for..
           GOTO       waitdo                   ; button release

; Send 16-bit data to SDAC via SPI port -----------

spi        BSF        SSPCON,SSPEN             ; Enable SPI port

           BCF        PORTC,0                  ; Enable SDAC chip
           MOVF       Hibyte,W                 ; Get high data
           MOVWF      SSPBUF                   ; and send it
waithi     BTFSS      PIR1,SSPIF               ; Wait for..
           GOTO       waithi                   ; SPI interrupt
           BCF        PIR1,SSPIF               ; Reset interrupt

           MOVF       Lobyte,W                 ; Get low data
           MOVWF      SSPBUF                   ; and send it
waitlo     BTFSS      PIR1,SSPIF               ; Wait for..
           GOTO       waitlo                   ; SPI interrupt
           BCF        PIR1,SSPIF               ; Reset interrupt

           BSF        PORTC,0                  ; Disable SDAC chip

; Run output loop until reset --------------------

           BTFSC      PORTB,0                  ; Test run button
           GOTO       up                       ; and repeat loop

run        INCF       PORTD                    ; Increment PDAC
           GOTO       run

           END ;-------------------------------------
```

Program 6.3
DACS test program source code.

overshoot (ringing) can be seen on each step, and a large overshoot occurs on the falling edge. This overshoot could cause problems in subsequent stages of the system, so suitable filtering should always be considered on a digitally generated waveform. Here, the amplifier is damped with the 100pF across the feedback resistance. On the other hand, too much damping causes the waveform to lose its linearity.

Other standard waveforms can be generated in a similar way. A square wave simply requires the output to be switched between maximum and minimum output values, with a controlled delay. A triangular wave is similar to the sawtooth, except that the falling edge is decremented rather than rolling over to 0. A sine wave can be generated from a program data table, which holds pre-calculated instantaneous voltage values. Any arbitrary waveform can be generated in digital mode using a data table.

Questions 6

1. Calculate the percentage precision per bit of a 12-bit ADC at full scale. (3)
2. Explain why a 2.56V reference voltage is convenient for an 8-bit ADC input. (3)
3. State the function of the CHSx bits in ADCON0. (3)
4. Explain the difference between left and right justified ADC results. (3)
5. State the gain, input resistance and output resistance of an ideal amplifier. (3)
6. State one advantage and two disadvantages of using a single supply amplifier. (3)
7. Calculate the gain of a simple non-inverting amplifier, if the input resistor is 1k0 and the feedback resistor 19k. (3)
8. Calculate the output voltage of a two input (a) summing and (b) difference amplifier if the input voltages are 1.0V and 0.5V and each has a gain of 2, assuming a positive output is obtained. (3)
9. Describe the general effect of a capacitor across the feedback resistor in an op-amp linear stage. (3)
10. Calculate the output d.c. voltage of a simple non-inverting amplifier with a feedback voltage divider consisting of a 22k feedback resistor and a 10k offset resistor connected to a 0.5V offset voltage, if the input at the positive terminal is 1.0V d.c. (5)
11. Calculate the bandwidth of the amplifier described in Question 10 if implemented using an op-amp with a GBWP of 1MHz. (3)
12. Refer to Figure 6.19(c). If $R_5 = 10k$, $R_6 = 15k$ and R_9 remains 10k, calculate the trigger switching levels V_H and V_L, assuming that the output switches between 0 and 5V. (5)
    Total (40)

Assignments 6

6.1 Analogue Input

Modify the 8-bit conversion program so that the input measures from 0.00 to 0.64V, by right justifying the result and processing ADRESL. When the input is 0.5V, the display should show 0.500V. Calculate the resolution of the voltage measurement. Show how to detect that the input is above 0.64V, and suggest an appropriate display. Write a program outline, and write the source code if a suitable test system or simulation is available.

6.2 Amplifier Test

Run the simulation of the basic amplifier interfaces. By suitable adjustment of the input voltages, record a set of values for each amplifier input and output, and demonstrate that the expression given for the gain of each configuration is valid. Evaluate the accuracy of the outputs obtained in simulation mode, as a percentage. Construct the equivalent hardware and compare its performance with the simulated and ideal performance, and account for any discrepancies.

6.3 Summing DAC

Construct an IC summing amplifier in the circuit simulator with eight input resistors with the values 1k, 2k, 4k, 8k, 16k, 32k, 64k and 128k, and feedback resistor of 1k, using the LM324 with a single 5V supply. Connect each input to +5V via a toggle switch, and the reference (+) input to 3.5V derived from a voltage divider across the supply. Run the simulation and close the switches in reverse order (128k first). Record the output voltages obtained and demonstrate that the circuit acts as a DAC. Suggest modifications to provide a positive-going output in the range of 0−2.0V. Replace the op-amp(s) with a device model with an improved specification such as the MCP6004 and compare the performance.

CHAPTER 7

Power Outputs

Summary

- Power loads at controller outputs need a current driver interface
- Many power loads are electromagnetic, such as relays and d.c. motors
- Current switches include thyristors, triacs, bipolar transistors and FETs
- The PIC can generate PWM output and measure pulse input period
- An FET bridge can provide bidirectional current drive
- Stepper and brushless d.c. motors use a rotating magnetic field and permanent magnet rotor
- Servo controllers use feedback to control motor output position and speed

In this chapter, we will concentrate on power outputs. The microcontroller or microprocessor port only provides a limited amount of current, about 20mA in the case of the PIC and even less for standard microprocessor ports. Therefore, if we want to drive an output device that needs more current than this, some kind of current amplifier or switch is needed. We will then see how various types of motor are controlled by the MCU.

When designing applications with motors, the data supplied with a particular device must be studied in conjunction with the general principles outlined here. Motors in particular have dynamic characteristics which are not ideal or even predictable. Practical hardware testing is therefore likely to reveal issues which will not necessarily be revealed by calculation or simulation. A complete closed loop motor controller application is described in 'PIC Microcontrollers, an Introduction to Microelectronics' by the author.

7.1 Power Loads

The simplest type of power load is resistive, such as a heater or filament lamp. However, the output load in a controller system is often some kind of electromagnetic device. This is typically an actuator which uses a coil to convert electrical energy into motion, such as a solenoid, relay, loudspeaker or motor. When current is passed through a coil, the resulting magnetic field interacts with another magnet, winding or simple soft iron core to produce a mechanical output.

Interfacing PIC Microcontrollers.
DOI: http://dx.doi.org/10.1016/B978-0-08-099363-8.00007-8

A solenoid, for example, is simply a coil containing a steel pin or yoke which is attracted into the electromagnetic coil by the induction of a complementary magnetic pole. This motion can be used to operate a valve, a set of electrical contacts (relay) or any other mechanical device. In a motor, electromagnetic windings interact, or permanent magnets are used, to create torque on a drive shaft.

7.1.1 Relay

Figure 7.1 shows some common electromagnetic devices. The relay (Figure 7.1(a)) consists of an electromagnetic coil that attracts a pivoted mild steel yoke, which in turn operates a set of changeover contacts. These are used to switch an output circuit controlling a high-power load. A relay can be used for either d.c. or a.c. loads, as its switch contacts will conduct in both directions. It provides complete electrical isolation between a low-voltage control circuit (typically 24V d.c.) and a high-voltage (240V) load circuit. Three sets of contacts may be operated together to switch a three-phase load (415V a.c. contactor).

The air gap in mechanical switch contacts also provides very high off resistance, which only breaks down if there is electrical discharge (sparking). The relay is therefore generally less reliable than any of the solid-state switches described in this chapter, due to discharge and wear at the contacts. Gold-plated contacts can be used to improve durability and reliability, but its response time is also relatively slow (allow at least 10ms) because of inertia in the switching mechanism. It is therefore only suitable for infrequent, or manual, operation.

7.1.2 Direct Current Motor

The simplified direct current motor (Figure 7.1(b)) has a rectangular conductor, representing the armature windings, rotating in a magnetic field. The field is provided by permanent magnets in small motors or field windings in larger ones. A current is passed through the rotor, which produces a cylindrical magnetic field around the conductor. This interacts with the transverse magnetic field generated by the field magnets or windings, causing a tangential force on the rotor, which provides the motor torque. To allow for rotation, the current is supplied to the armature via slip rings and brushes. In order to maintain the torque in the same direction, the current has to be reversed every half revolution, so the slip ring is split to form a commutator.

A cross-section is shown in Figure 7.1(c). The current in and out of the page is represented by the cross and dot on the respective rotor conductors. The circular field caused by the rotor current is not shown (for clarity), but it is generated according to the right-hand screw rule. The stator field is distorted by interaction with the rotor field. If one imagines the stator field as elastic bands, the force is generated by the distorted field trying to straighten.

(a)

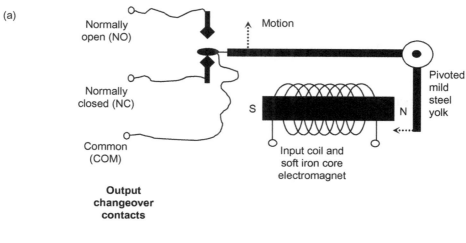

Normally
open (NO)

Motion

Normally
closed (NC)

S   N

Common
(COM)

Pivoted
mild
steel
yolk

Input coil and
soft iron core
electromagnet

**Output
changeover
contacts**

(b)

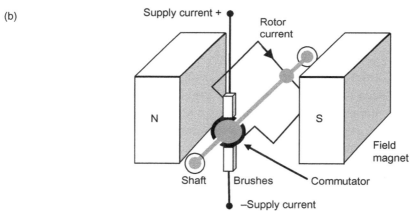

Supply current +

Rotor
current

N   S

Field
magnet

Shaft   Brushes   Commutator

–Supply current

(c)

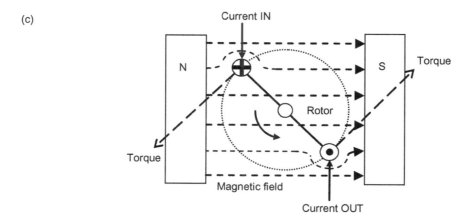

Current IN

N   S   Torque

Rotor

Torque

Magnetic field

Current OUT

Figure 7.1
Electromagnetic actuators: (a) relay operation, (b) simplified d.c. motor (3D view)
and (c) motor operating principle (cross-section).

7.1.3 Real Motors

In real motors, the armature (rotor) winding consists of many turns on a laminated soft iron former, which concentrates on the field, with multiple poles. It is informative to take apart a small, cheap, modelling d.c. motor and study its construction. It will typically have a pair of curved permanent field magnets, three armature windings and a six-segment commutator.

The asymmetric windings provide more consistent torque as the rotor moves through a complete revolution, since a pole on one side is actively driving while a gap is on the opposite side. The brushes are usually simply sprung metal contacts, but in larger motors are formed of carbon blocks which mould themselves to the cylindrical commutator to provide the maximum contact area. These sometimes need replacement in d.c. motors (e.g. motor vehicle starter motors). The brushes and commutator are therefore a weak point in the traditional d.c. motor design. Mechanical wear and sparking which occurs as the current switches between windings at the commutator means that the d.c. motor is relatively unreliable, with limited operating life.

Brushless d.c. (BLDC) motors improve on this by using a permanent magnet rotor, which eliminates the need to supply current to the armature, but these are limited in size and power. Similarly, stepper motors use a rotating magnetic field to drive a passive rotor. These can be moved one step at a time and can therefore be positioned accurately without feedback, but these are complex to drive, inefficient and limited in power output.

Larger motors tend to be three-phase a.c. motors. These use a rotating magnetic field generated by the three phases of the supply grid, resulting in high efficiency and output power in a compact unit, and an accurate, constant speed, typically 3000r.p.m. from a 50Hz supply (1 revolution per cycle). Motors and other electromechanical actuators therefore need a current drive interface and some form of controller. It may be a simple switch or a complex synthesised drive producing a rotating magnetic field with speed and torque control.

7.2 Power Interfaces

Figure 7.2 shows a selection of simple power output interfaces operating a relay, triac and oscillator (VSM project POWER2). The PIC runs a simple program which switches on each output in turn when the button is pressed.

7.2.1 Relay Interface

The relay coil is operated via a bipolar transistor switch, since the PIC output may not be able to provide enough current. When the coil is activated, the contacts change over,

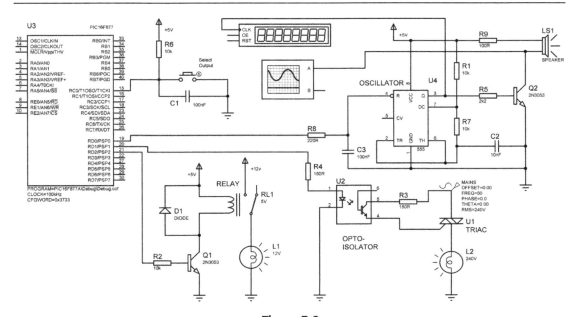

Figure 7.2
Power output interfaces.

completing the load circuit, which operates a lamp. High power loads such as heaters and motors can be easily interfaced in this way, on the condition that only infrequent on–off switching is needed. Note that relay control allows the load to be grounded and isolated from the supply when switched off.

The relay has normally open (NO) and normally closed (NC) sets of contacts. These provide flexibility in the way they are used and relays can be connected to form simple control systems without a controller. For example, latching operation can be obtained by wiring NO contacts in parallel with a push button that turns on the coil of the same relay. The relay is held on after the button is released and then switched off using an NO push button in the supply circuit. This is used to provide simple fail-safe control of a machine tool.

When used with a controller, one relay can control others via its contacts or provide feedback to confirm correct operation of the system. Relays and contactors (high current relays) are most often used in industry in conjunction with self-contained programmable controllers (PLCs), which contain a microcontroller and interfacing in a single unit.

The relay must be selected to meet load current and voltage requirements, and the interface designed to provide the necessary coil operating current. The transistor is selected for sufficient collector current, and the base resistor calculated to give sufficient base current that the transistor is in saturation when switched on. It is assumed that the current gain is 100 in this case, but transistor specification for h_{FE} (d.c. gain) should be checked.

Coil current $= 40mA =$ collector current
$\therefore$ Base current $= 40mA/100 = 400\mu A$
$\therefore$ Base resistor $< (5-0.6)/400 \times 10^{-6} = 11k\Omega \rightarrow \underline{10k\Omega}$

The relay has a rectifier diode connected across the coil to protect the drive transistor. This is a sensible precaution for all d.c. inductive loads (anything with a coil such as a motor or solenoid) because, when the coil is switched off, a large reverse voltage may be generated as the magnetic field collapses (this is the way that the spark is generated in a car ignition). The diode protects the transistor from the back EMF (voltage) by forward conduction for a brief period. In normal operation, the diode is reverse-biased and has no effect.

The relay provides an electrically controlled switch with low on resistance and high off resistance (air gap). However, it is slow, consumes a fairly large amount of power itself ($40mA \times 5V = 200mW$) and is relatively unreliable. Solid-state relays are available that are designed to switch a.c. loads directly from digital outputs with higher reliability and speed than a traditional relay or contactor. It contains TTL buffering, isolation and a triac (see below) drive in one package.

7.2.2 Thyristor and Triac

Compared to a relay, a solid-state switch is inherently more reliable, since it has no moving parts. A MOSFET may be used, but the thyristor (Figure 7.3(a)) offers the advantage of latched operation, i.e. once it is on, it stays on, until the power input is removed, or reduces to zero in the case of a.c. It is a three-terminal silicon controlled rectifier (SCR), equivalent to a pair of bipolar transistors operating in trigger mode. It allows forward current when the gate voltage is taken above 0.6V with respect to the cathode. It remains switched on until the current falls to zero, so it can be pulse triggered. However, it only passes current in one direction, providing rectified d.c. power only.

The triac is basically two thyristors connected back to back, with a common gate (trigger) input, allowing current flow in both directions (Figure 7.3(b)). The full a.c. cycle can then be utilised, usually switching at the same point in the positive and negative half cycles of the current. The gate current is positive or negative, depending on the half cycle.

A test implementation has been seen in Figure 7.2 which simply switches the triac on and off, without controlling the power level. An opto-coupler isolates the MCU from the high-voltage load circuit; the output phototransistor conducts when the light from the LED falls on its base. When the MCU output is high, the opto-switch is on, and the voltage at terminal 1 of the triac is applied to the gate, turning the triac on when the voltage passes through zero. When the switch is off, the triac does not conduct.

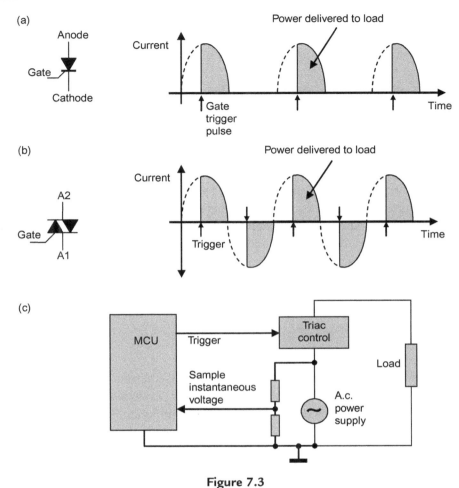

Figure 7.3
Thyristor and triac control: (a) thyristor, (b) triac and (c) MCU triac control.

A block diagram of a more complete microcontroller triac interface is shown
in Figure 7.3(c). The output power is controlled by monitoring the a.c. voltage via a
feedback voltage divider and sampling it at an analogue input. A timer controls
the delay between the zero crossing point in the cycle and the trigger point, where the
triac is switched on each half cycle. The domestic light dimmer is a commonplace
application of this type.

7.2.3 Oscillator Interface

If an output pulse signal is required at a particular frequency, it can be generated in a
variety of ways. The microcontroller can use a software delay loop to generate a pulse

output. Better, a hardware timer and interrupt driven process can be used, which will be discussed later in this chapter. Alternatively, the task can be delegated to external hardware, so that the MCU simply switches an output to enable the oscillator. This saves on MCU resources and may have other advantages, e.g. if a high frequency is needed that cannot be generated by the controller itself. A simple low-frequency oscillator can be implemented using a 555 timer chip. In Figure 7.2, it drives a loudspeaker via a bipolar transistor. Input *R* on the chip enables the oscillator, and C2 controls the frequency.

This example illustrates an important microcontroller system design principle. Any given interface needs a combination of hardware and software, but the balance between these components can vary for the same interface. The software oriented implementation will use more MCU resources in terms of available peripheral interfaces, memory, processor time and programming effort. The hardware approach saves on these resources but involves additional cost in hardware design effort and components for each system produced. Software, on the other hand, once written has a negligible reproduction cost. The optimum design mix may need careful consideration in a commercial environment, but for volume production, the software oriented solution is likely to be cheaper per unit.

7.3 Current Switches

There are two main types of transistor that can be used as a current switch. The bipolar transistor was the first to be developed, based on the p−n semiconductor junction. This consists of silicon semiconductor slices that are doped with other elements to generate extra conduction electrons (n-type) or a deficiency of electrons (p-type), forming a diode junction which conducts in one direction only. Two of these junctions back to back form a bipolar transistor, which operates as a current amplifier. The logarithmic transfer characteristic of the bipolar transistor has been described in detail in Chapter 6, but for interfacing, a simpler linear model representing it as a current amplifier is generally more appropriate.

The FET (field effect transistor) operates in a slightly different manner, where the conduction in a semiconductor channel is controlled by the voltage applied at the gate terminal, so it is basically working as voltage-controlled resistor. The input current is small or negligible, so it has higher input impedance and consumes less power in high-density circuits.

In addition, the power FET has a distinct advantage over its bipolar equivalent, in that a bipolar power transistor can suffer from thermal runaway, where the base current increases with temperature, causing even higher collector current, overheating and destruction. The FET is therefore usually preferred in high current switching applications such as motor controllers and inverters. Unlike bipolar transistors, FETs can also be simply connected with their outputs in parallel to multiply the current handling capability of a drive circuit.

7.3.1 Bipolar Transistor

One advantage of the bipolar junction transistor (BJT) is that there are only two basic types, NPN and PNP, so it is arguably easier to design with. It works as a current amplifier, i.e. a small base current controls a larger (typically $\times 100$) current in the collector. The emitter is the common terminal as far as current flow is concerned (Figure 7.4(a)). In the equivalent circuit (Figure 7.4(b)), the base behaves as a diode junction, with a forward voltage drop of about 0.6 V in normal operation. The base current controls a current source which represents the collector–emitter junction. In the NPN transistor, current flows out of the emitter, and in the PNP, into the emitter.

In the so-called common emitter configuration, a signal is input to the base of the transistor via a current-limiting resistor (Figure 7.4(c)). This then controls the larger current flow in a load connected to the collector. We are assuming that the input to the interface is coming from the MCU output port; $+5$V applied to the base resistance causes the transistor to switch on, drawing current through the load resistor and causing the voltage at the collector

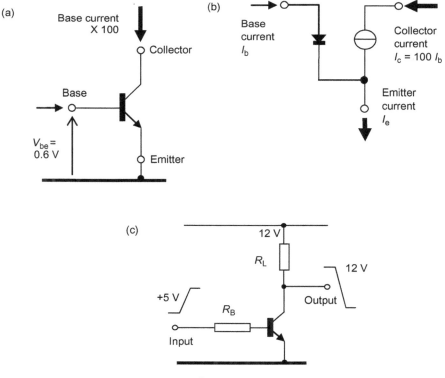

Figure 7.4
Bipolar transistor operation: (a) NPN transistor operation, (b) NPN transistor equivalent circuit and (c) simple switching transistor interface.

to go low. The supply voltage to the transistor can be some higher value (12V in this case), which allows more power to be dissipated in the load for a given collector current.

The circuit can be biased with a voltage divider on the base to operate as a linear amplifier, but this option is explained in detail in most introductory electronics texts and will not be considered further here. We will focus on the switching mode of operation where the output voltage swings over its full range, and the transistor is saturated when on. In this case, the output voltage can be close to zero. Almost the full supply voltage is applied across the load and a current flow that depends on the load resistance value. A simple resistor load will act a small heater, dissipating power, $P = V^2/R$. Alternatively, a filament lamp will convert some of this power into light or a motor into torque.

When the transistor is off, the output is pulled up to supply via the load resistance, and the load no longer dissipates power, as the voltage across it and the current through it are both low. The transistor dissipation is $P_T = V_c I_c$, where V_c and I_c are the collector voltage and current. When the transistor is off, I_c is small, and when the transistor is fully on (saturated), V_c is small, so that in both cases the transistor dissipates a relatively small amount of power.

Therefore, minimal power is wasted in the transistor, if the base current is large enough, i.e. the base resistor is small enough to ensure that the transistor is fully on. This is another advantage of pulse width modulation (PWM), where the transistor switches quickly between on and off. Since most of the power is dissipated during switchover, the transistor operating temperature rises with operating frequency. The power rating of the transistor must be selected accordingly and/or a heat sink fitted.

The PNP transistor operates in the inverse mode, with all current flows reversed. The choice of transistor depends on the circuit configuration and supply polarity. In Figure 7.5, some simulated bipolar switching circuits are shown, with operating conditions displayed using signal probes (VSM project TRANS2). All the circuits include an indicator LED so that the output state can be easily monitored.

In circuit 7.5(a), the basic common emitter switch is shown, using a +12V load supply and a generic NPN transistor. Power transistors generally have a lower current gain than signal transistors, so the circuit might need modifying for any given power transistor. A lower value base resistor will be needed, for example. The transistor current gain is specified in the data sheet as h_{FE}, the principal characteristic of the bipolar transistor. The logic input simulates an MCU output operating at TTL levels. The main disadvantage of this circuit is that the load is connected to the positive output supply, so it floats relative to ground. Even when the current switch is off, it is still 'live', in that it is permanently connected to +12V.

In circuit 7.5(b), the load is connected to ground, forming a common collector switch. When the transistor is off, both load terminals are at 0V, which is safer. The disadvantage here is that an additional stage is needed to shift the input switching level from 5 to 12V.

By changing to a PNP transistor in circuit 7.5(c), the common emitter configuration can be used with a grounded load. This allows the drive transistor to saturate, transferring more power to the load.

7.3.2 FET Switches

Figure 7.5(d) shows an equivalent FET current switch. The circuit is simplified compared with the bipolar switch because this FET is designed to accept TTL level inputs at its gate. The VN66 can provide 1A output drain current (on = 5V, off = 0V) when connected

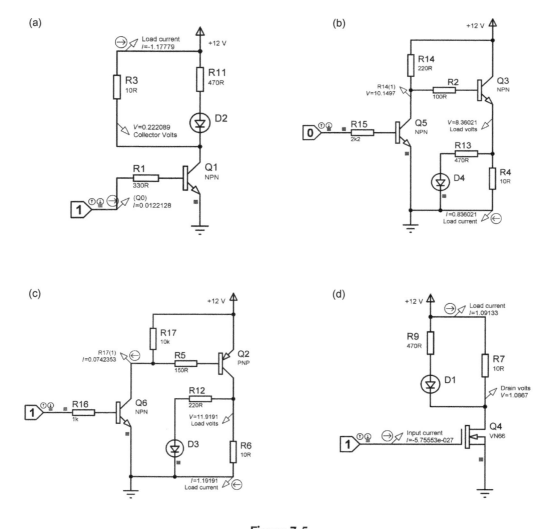

Figure 7.5
Transistor output test circuits: (a) common emitter switch, (b) grounded load switch, (c) PNP switch and (d) FET switch.

directly to a digital output. As can be seen, the input current is extremely small, around 10^{-27} A, because it is an insulated gate FET (IGFET), as indicated by the circuit symbol. The channel current is controlled electrostatically by the voltage at the gate, and negligible input current is drawn, giving almost infinite current gain. However, the power dissipation in the load is limited, because there is a significant forward resistance associated with the FET channel. Overall, the device acts as a voltage-controlled resistor and is sometimes used as such.

Equivalent devices to the VN66, with improved features such as internal overvoltage protection, are now available. For power applications, the IGFET (or MOSFET, metal oxide semiconductor) is generally used. It has two main types, P-channel and N-channel, referring to the channel charge carrier and polarity of operation, which correspond to NPN and PNP bipolar transistors. These are now generally used to implement motor drives that need bidirectional current control.

7.4 Pulse Applications

We often need a pulse output from the PIC with a variable frequency or mark/space ratio (MSR). PWM (pulse width modulation) is most often used to control the power delivered to a resistive load via a current switch. Measuring pulse feedback period or frequency allows motor speed control.

7.4.1 Pulse Output

The demo application shown in Figure 7.6 illustrates the use of a hardware timer to generate an output pulse waveform whose period and frequency can be controlled by push buttons (VSM project PULSE2).

The pulse output is generated at the output of Timer1, RC2, with a fixed 1ms positive pulse and a variable interval that can be adjusted manually. A virtual counter/timer displays the output frequency, which initially runs at 100Hz. The output is fed to a sounder, which causes the simulator to generate an audible output via the PC soundcard. The effect of pressing each button can thus be heard as well as displayed. Note that hardware debouncing has been used (capacitors across the buttons) to simplify the software.

The main purpose of this example is to illustrate the use of Timer1 compare mode. This requires preloading a register with a reference binary number, with which a count register is continuously compared in the timer hardware. When a match is detected, an interrupt is generated which calls the interrupt service routine (ISR). This allows the input to be processed immediately, preserving the accurate timing of program operation.

The Timer1 compare mode is shown as a block diagram in Figure 7.7. It uses a pair of registers, TMR1H (high byte) and TMR1L (low byte), to record a 16-bit count, driven by

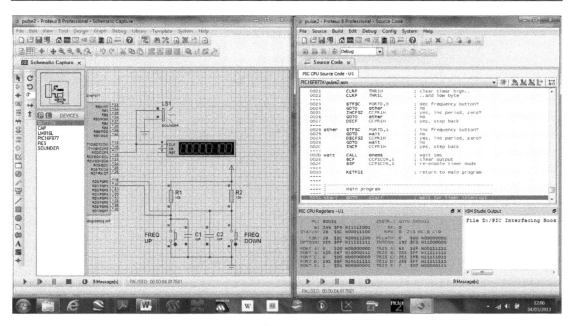

Figure 7.6
Pulse output simulation.

(a)

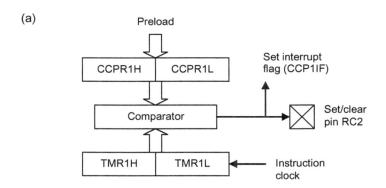

(b)

Register	Load	Effect
PIE1	0000 0100	Enable Timer 1 interrupt
INTCON	1100 0000	Enable peripheral interrupts
CCP1CON	0000 1000	Compare mode – set output pin on match
CCPR1H	027H	Initial value for high byte compare
CCPR1L	010H	Initial value for low byte compare
T1CON	0000 0001	Enable timer with internal clock

Figure 7.7
Timer1 compare mode: (a) register block diagram and (b) control register set-up.

```
PULSE2
Generates a variable interval pulse output
controlled by up/down buttons
Hardware: P16F877 (4MHz), sounder

MAIN
        Initialise
                        RC2/CCP1 = Pulse output
                        RD0,RD1 = Up/Down buttons
                        Timer1 Compare Mode & Interrupt

        Wait for interrupt

SUBROUTINE
        1ms delay

INTERRUPT SERVICE ROUTINE
        Reset interrupt
        IF Increase Frequency button pressed
                Decrement pulse interval
        IF Decrease Frequency button pressed
                Increment pulse interval
        Generate 1ms pulse
```

Figure 7.8
Pulse program outline.

the MCU instruction clock. In the system simulation, the clock is set to 4MHz, giving a 1MHz instruction clock (one instruction takes four clock cycles). The timer therefore counts in microseconds. The reference register pair, CCPR1H and CCPR1L, is preloaded with a value which is continuously compared with the 16-bit timer count (default $2710_{16} = 10,000_{10}$). With this value loaded, the compare becomes true after 10ms, the interrupt is generated and the output set high.

The ISR resets the interrupt, tests the buttons to see if the preset value should be changed, waits 1 ms and then clears the output to zero. The default output is therefore a 1ms high pulse, followed by a 9ms interval. This process repeats, giving a pulse waveform with an output period of 10 ms overall. The program is outlined in Figure 7.8.

The source code (Program 7.1) shows the initialisation required for the interrupt operation. The interrupt vector (GOTO ISR) is loaded at address 004, so the initial execution sequence has to jump over this location. The port, timer and interrupt registers are then set up. The timer is started, and the single instruction main loop then runs, waiting for the interrupt.

Timer1 counts up to 10,000, and the interrupt is triggered. The interrupt flag is first cleared, and the counter reset to zero. The next 10ms period starts immediately, because the counter runs continuously. The buttons are checked, and the compare register value incremented or decremented to change the output period if one of them is pressed. A check is also made

```
;;;;;;;;;;;;;;;;;;;;;;;;;;;;;;;;;;;;;;;;;;;;;;;;;;;;;;;;;;;;;;
;
;         PULSE2.ASM        MPB              12-01-13
;
;         Generates timed output interval using Timer 1
;         in compare mode
;
;         Updated for VSM v8
;;;;;;;;;;;;;;;;;;;;;;;;;;;;;;;;;;;;;;;;;;;;;;;;;;;;;;;;;;;;;;
           .
           PROCESSOR 16F877A        ; Select MCU
           __CONFIG 0x3731          ; Clock = XT 4MHz

;          LABEL EQUATES           ..................................

           INCLUDE "P16F877A.INC"      ; Standard register labels

Count      EQU       20              ; software timer

; Program begins ;;;;;;;;;;;;;;;;;;;;;;;;;;;;;;;;;;;;;;;;;;;;;;;;

           CODE      0               ; Place machine code
           NOP                       ; for ICD mode
           GOTO      init            ; Jump over ISR vector

           ORG       4               ; ISR vector address
           GOTO      isr             ; run ISR

init       NOP
           BANKSEL   TRISC           ; Select bank 1
           MOVLW     B'11111011'     ; RC2 = output
           MOVWF     TRISC           ; Initialise display port
           MOVLW     B'00000100'     ; Timer1 interrupt..
           MOVWF     PIE1            ; ..enable

           BANKSEL PORTC             ; Select bank 0
           CLRF      PORTC           ; Clear output
           MOVLW     B'11000000'     ; Peripheral interupt..
           MOVWF     INTCON          ; ..enable
           MOVLW     B'00001000'     ; Compare mode..
           MOVWF     CCP1CON         ; ..set output on match
           MOVLW     027             ; Initial value..
           MOVWF     CCPR1H          ; .. for high byte (10ms)
           MOVLW     010             ; Initial value..
           MOVWF     CCPR1L          ; .. for low byte (10ms)
           MOVLW     B'00000001'     ; Timer1 enable..
           MOVWF     T1CON           ; with internal clock (1MHz)

           GOTO      start           ; Jump to main program

;          SUBROUTINES.........................................

;          1ms delay with 1us cycle time (1000 cycles)

onems      MOVLW     D'249'          ; Count for 1ms delay
           MOVWF     Count           ; Load count
loop       NOP                       ; Pad for 4 cycle loop
           DECFSZ    Count           ; Count
           GOTO      loop            ; until Z
           RETURN                    ; and finish


;          INTERRUPT SERVICE ROUTINE...........................

;          Reset interrupt, check buttons, generate 1ms pulse

isr        CLRF      PIR1            ; clear interrupt flags
           CLRF      TMR1H           ; clear timer high..
           CLRF      TMR1L           ; ..and low byte

           BTFSC     PORTD,0         ; dec frequency button?
           GOTO      other           ; no
           INCFSZ    CCPR1H          ; yes, inc period, zero?
           GOTO      other           ; no
           DECF      CCPR1H          ; yes, step back
```

Program 7.1
Pulse output.

```
other    BTFSC     PORTD,1          ; inc frequency button?
         GOTO      wait             ; no
         DECFSZ    CCPR1H           ; yes, inc period, zero?
         GOTO      wait             ; no
         INCF      CCPR1H           ; yes, step back

wait     CALL      onems            ; wait 1ms
         BCF       CCP1CON,3        ; clear output
         BSF       CCP1CON,3        ; re-enable timer mode

         RETFIE                     ; return to main program


;---------------------------------------------------------------
;         Main program
;---------------------------------------------------------------
start    GOTO      start            ; wait for timer interrupt
         END                        ; of source code
```

Program 7.1
(Continued)

for zero at the upper and lower ends of the period adjustment range, to prevent the compare value rolling over or under between 00 00 and FF FF. This would cause the output frequency to jump between the minimum to maximum value, which is undesirable in this case.

The 1ms pulse period is generated as a software delay, which runs in parallel with the hardware timer count. After 1 ms, the output is cleared to zero, but the hardware count continues until the next interrupt occurs. This is an important point − the hardware timer continues independently of the program sequence, until the next interrupt is processed, allowing the timing operation and program to be executed simultaneously.

7.4.2 PWM Output

If a solid-state power switch is used to drive a current load, it can be switched on and off at reasonably high frequency. This allows the power output level to be controlled by varying the ratio of the 'on' and the 'off' time, because the load power consumption will be determined by the average level of the switched current. Thus, when using PWM, a higher mark/space ratio, or duty cycle, will result in more power delivered to the load.

Most PIC MCUs incorporate a PWM operating mode associated with one of the hardware timers. In the 16F877A, Timer2 (TMR2 register), an 8-bit counter, is used conjunction with associated registers that store the overall cycle period count and the duty cycle (high period) count. The register block diagram, register set-up and output waveform are shown in Figure 7.9. A PWM process is incorporated in Program 7.3 (see next section) that operates three different types of motor. PWM is used to drive a simple single-ended motor interface at variable speed.

In Figure 7.9, the overall period count is stored in PR2, an 8-bit register. In the demonstration program, the MCU clock is 4MHz, so the maximum TMR2 count at 1MHz

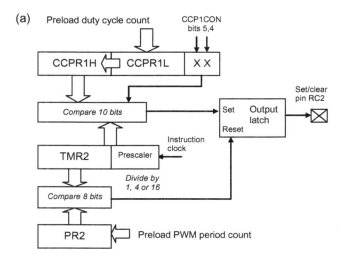

(a)

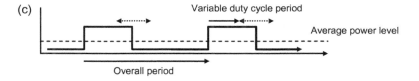

(b)

Register	Test values	Effect
PR2	249_{10}	Overall cycle time (250 μs with 4 MHz clock)
CCPR1L	124_{10}	Duty cycle time (50%) initial value
CCP1CON	00001100	Low count bits = 00, 1100 = Select PWM

(c)

Figure 7.9
Timer2 PWM mode: (a) register block diagram, (b) control register typical set-up
(see Program 7.3) and (c) PWM output.

instruction clock rate is 256μs. The count is actually initialised at 249μs. When the PWM
mode is enabled via CCP1CON (set bits 2 and 3) control register, the timer is continuously
compared with this value and timeout signalled when the contents match. The output latch
is set, indicating the start of the next cycle.

The duty cycle, during which the output is high, is timed by the contents of CCPR1L,
which must be preloaded with a suitable value. In the demo program this is 128_{10},
corresponding to half the overall period. At the start of the cycle, this value is transferred
into CCPR1H and then continuously compared with TM2 as it increments. When they
match (128 in this case), the output latch is cleared to zero until the start of the next cycle.

The buffering of the CCPR1 is designed to improve the reliability of the PWM operation.
The resolution of the duty cycle can be extended to 10 bits by using bits 4 and 5 as
CCP1CON register as the least significant bits. These are compared with the low bits of the

prescaler that function as low bits of the timer. If these bits are set to 00, as in the demo program, the PWM module effectively operates with 8-bit resolution.

To vary the duty cycle and therefore the output power delivered by a current switch at the output to a resistive load or motor, the TMR2 preload value is modified. In the demo program, it is incremented or decremented via a pair of input buttons, causing the motor to speed up or slow down. Once set up and enabled, the PWM module runs independently and continuously. An interrupt can be enabled if required, but in the demo program the control inputs are polled.

7.4.3 Pulse Input

If the speed or position of a motor is to be accurately controlled, the output shaft needs to be monitored. Usually, pulse feedback is received from a suitable sensor arrangement. An incremental encoder, for example, has an optically segmented disk or drum attached to the shaft and opto-sensor to detect its rotation. The pulse count or frequency can be processed to obtain the position, speed and acceleration of the shaft (see Section 9.1.3).

Timer1 can be set up to operate in capture mode for input measurement. A value in the 16-bit timer register (TMR1H + TMR1L) is captured in mid-count; the capture is triggered by the input RC2 changing state. A prescaler can be included between the input and capture enable so that the capture is only triggered every fourth or sixteenth pulse at the input, thereby reducing the capture rate. The timer is used here to measure the period of a pulse waveform, which is fed in at RC2 to the CCP1 module input.

In the simulation shown in Figure 7.10, a variable frequency clock signal is input at RC2 set to 500Hz. This is monitored on a virtual timer/counter set to frequency measurement mode. The 16×2 LCD is connected and driven as detailed in Chapter 4, showing the signal period in microseconds. The signal generator appears in full size when the simulation is running, as long as it is selected in the debug menu. The frequency can be adjusted as the simulation runs, and the display responds accordingly, displaying the period in microseconds.

The timer module is set up to generate an interrupt when the input changes from high to low, once per cycle. The timer counts instruction clock cycles continuously, and the count reached is stored in the CCP1 register pair on each interrupt, and the count then restarted. The timer test configuration is shown in Figure 7.11.

The program outlined in Figure 7.12 continuously converts the 16-bit contents of CCP1 to five-digit BCD and displays the result on the LCD (VSM project TIMIN2). The binary is converted to BCD using a simple subtraction loop for each digit, from ten thousands to tens (see Chapter 5). The remainder at the end is the value of the least significant digit. The source code is listed in Program 7.2.

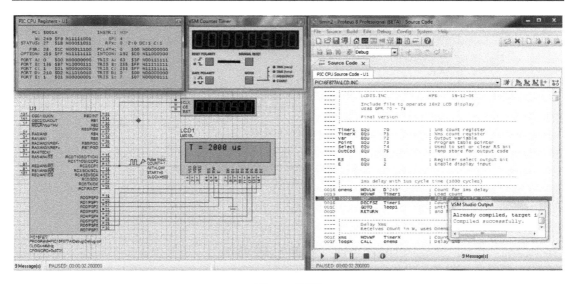

Figure 7.10
Input pulse measurement simulation.

(a)

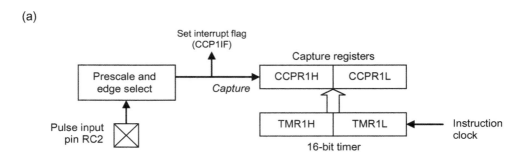

(b)

Register	Setting	Flags	Function
PIE1	0000 0100	CCP1IE	Enable CCP1 interrupt
INTCON	1100 0000	GIE, PEIE	Enable peripheral interrupts
CCP1CON	0000 0100	CCP1M0 - 3	Capture mode – every falling edge
T1CON	0000 0001	TMR1ON	Enable timer with internal clock
PIR1	0000 0X00	CCP1IF	CCP1 interrupt flag

Figure 7.11
Timer1 capture mode: (a) Timer1 capture mode block diagram and (b) Timer1 capture
mode register set-up.

TIMIN
Measure pulse waveform input period and display
P16F877A (4MHz), audio signal source, 16x2 LCD

MAIN
    Initialise
        PortD = LCD outputs
        Capture mode & interrupt
        Initalise LCD
        Enable capture interrupt

    REPEAT
        Convert 16-bit count to 5 BCD digits
        Display input square wave period
    ALWAYS

SUBROUTINES

    Convert 16-bit count to 5 BCD digits
        Load 16-bit number
        Clear registers
            Tents, Thous, Hunds, Tens, Ones
        REPEAT
            Subtract 10000 from number
        UNTIL Tents negative
        Restore Tents and remainder
        REPEAT
            Subtract 1000 from remainder
        UNTIL Thous negative
        Restore Thous and remainder
        REPEAT
            Subtract 100 from remainder
        UNTIL Hunds negative
        Restore Hunds and remainder
        REPEAT
            Subtract 10 from remainder
        UNTIL Tens negative
        Restore Tens and store remainder Ones

    Display input square wave period
        Display 'T='
        Supress leading zeros
        Display digits in ASCII
        Display 'us'

INTERRUPT SERVICE ROUTINE
        Clear Timer 1 Count Registers
        Reset interrupt flag

Figure 7.12
Input period measurement.

```
;;;;;;;;;;;;;;;;;;;;;;;;;;;;;;;;;;;;;;;;;;;;;;;;;;;;;;;;;;;;;;;
;
;          TIMIN2.ASM          MPB               12-01-13
;
;          Measure input period using Timer1 16-bit capture
;          and display in microseconds, signal input CCP1
;
;       Updated for VSM v8
;
;;;;;;;;;;;;;;;;;;;;;;;;;;;;;;;;;;;;;;;;;;;;;;;;;;;;;;;;;;;;;;;
          PROCESSOR 16F877A
          __CONFIG 0x3731

;          LABEL EQUATES
          ;;;;;;;;;;;;;;;;;;;;;;;;;;;;;;;;;;;;;;;;;

          INCLUDE "P16F877A.INC"       ; Standard register labels

;          Local label equates.................................

Hibyte    EQU       020
Lobyte    EQU       021

Tents     EQU       022
Thous     EQU       023
Hunds     EQU       024
Tens      EQU       025
Ones      EQU       026


; Program begins ;;;;;;;;;;;;;;;;;;;;;;;;;;;;;;;;;;;;;;;;;;;;;;;;;

          CODE      0               ; Place machine code
          NOP                       ; Required for ICD mode
          GOTO      init

          ORG       4               ; Interrupt vector adress
          GOTO      ISR             ; jump to service routine

init      NOP
          BANKSEL   TRISD           ; Select bank 1
          CLRF      TRISD           ; Initialise display port
          CLRF      PIE1            ; Disable peripheral interrupts

          BANKSEL PORTD             ; Select bank 0
          CLRF      PORTD           ; Clear display outputs

          MOVLW     B'11000000'     ; Enable..
          MOVWF     INTCON          ; ..peripheral interrupts
          MOVLW     B'00000100'     ; Capture mode:
          MOVWF     CCP1CON         ; ..every falling edge
          MOVLW     B'00000001'     ; Enable..
          MOVWF     T1CON           ; ..Timer 1

          GOTO      start           ; Jump to main program

; INTERRUPT SERVICE ROUTINE ;;;;;;;;;;;;;;;;;;;;;;;;;;;;;;;;;;;;;;;

ISR       CLRF      TMR1L
          CLRF      TMR1H
          BCF       PIR1,CCP1IF     ; Reset interrupt flag
          RETFIE

; SUBROUTINES
          ;;;;;;;;;;;;;;;;;;;;;;;;;;;;;;;;;;;;;;;;;;;;;;;;;;;;;

          INCLUDE "LCD.INC"   ; Include display routines

;--------------------------------------------------------------
; Convert 16 bit binary result to 5 digits
;--------------------------------------------------------------
```

Program 7.2
Input pulse period measurement.

```
conv      MOVF     CCPR1L,W          ; Get high byte
          MOVWF    Lobyte            ; and store
          MOVF     CCPR1H,W          ; Get low byte
          MOVWF    Hibyte            ; and store

          MOVLW    06                ; Correction value
          BCF      STATUS,C          ; prepare carry flag
          ADDWF    Lobyte            ; add correction
          BTFSC    STATUS,C          ; and carry
          INCF     Hibyte            ; in required

          CLRF     Tents             ; clear ten thousands register

          CLRF     Thous             ; clear thousands register
          CLRF     Hunds             ; clear hundreds register
          CLRF     Tens              ; clear tens register
          CLRF     Ones              ; clear ones register

; Subtract 10000d (2710h) and count .......................

sub10     MOVLW    010               ; get low byte to sub
          BSF      STATUS,C          ; get ready to subtract
          SUBWF    Lobyte            ; sub 10h from low byte
          BTFSC    STATUS,C          ; borrow required?
          GOTO     sub27             ; no - sub high byte

          MOVF     Hibyte,F          ; yes - check high byte
          BTFSS    STATUS,Z          ; zero?
          GOTO     take1             ; no - take borrow

          MOVLW    010               ; yes - load low byte to add
          BCF      STATUS,C          ; get ready to add
          ADDWF    Lobyte            ; restore low byte
          GOTO     subE8             ; next digit

take1     DECF     Hibyte            ; take borrow

sub27     MOVLW    027               ; get high byte to sub
          BSF      STATUS,C          ; get ready to subtract
          SUBWF    Hibyte            ; sub from high byte
          BTFSS    STATUS,C          ; borrow taken?
          GOTO     done1             ; yes - restore remainder
          INCF     Tents             ; no - count ten thousand
          GOTO     sub10             ; sub 10000 again

done1     MOVLW    010               ; restore..
          BCF      STATUS,C          ; get ready to add
          ADDWF    Lobyte            ; restore low byte
          BTFSC    STATUS,C          ; Carry into high byte?
          INCF     Hibyte            ; yes - add carry to high byte
          MOVLW    027               ; restore..
          ADDWF    Hibyte            ; ..high byte


; Subtract 1000d (03E8) and count.............................

subE8     MOVLW    0E8               ; get low byte to sub
          BSF      STATUS,C          ; get ready to subtract
          SUBWF    Lobyte            ; sub from low byte
          BTFSC    STATUS,C          ; borrow required?
          GOTO     sub03             ; no - do high byte

          MOVF     Hibyte,F          ; yes - check high byte
          BTFSS    STATUS,Z          ; zero?
          GOTO     take2             ; no - take borrow

          MOVLW    0E8               ; load low byte to add
          BCF      STATUS,C          ; get ready to add
          ADDWF    Lobyte            ; restore low byte
          GOTO     sub64             ; next digit

take2     DECF     Hibyte            ; take borrow

sub03     MOVLW    03                ; get high byte to sub
          BSF      STATUS,C          ; get ready to subtract
          SUBWF    Hibyte            ; sub from high byte
          BTFSS    STATUS,C          ; borrow taken?
```

Program 7.2
(Continued)

```
                GOTO      done2          ; yes - restore high byte
                INCF      Thous          ; no - count ten thousand
                GOTO      subE8          ; sub 1000 again

done2           MOVLW     0E8            ; restore..
                BCF       STATUS,C       ; get ready to add
                ADDWF     Lobyte         ; restore low byte
                BTFSC     STATUS,C       ; Carry into high byte?
                INCF      Hibyte         ; yes - add carry to high
byte
                MOVLW     03             ; restore..
                ADDWF     Hibyte         ; ..high byte


                ; Subtract 100d (064h) and
count................................

sub64           MOVLW     064            ; get low byte
                BSF       STATUS,C       ; get ready to subtract
                SUBWF     Lobyte         ; sub from low byte
                BTFSC     STATUS,C       ; borrow required?
                GOTO      inchun         ; no - inc count

                MOVF      Hibyte,F       ; yes - check high byte
                BTFSS     STATUS,Z       ; zero?
                GOTO      take3          ; no - take borrow

                MOVLW     064            ; load low byte to add
                BCF       STATUS,C       ; get ready to add
                ADDWF     Lobyte         ; restore low byte
                GOTO      subA           ; next digit

take3           DECF      Hibyte         ; take borrow

inchun          INCF      Hunds          ; count hundred
                GOTO      sub64          ; sub 100 again

; Subtract 10d (0Ah) and count, leaving remainder................

subA            MOVLW     0A             ; get low byte to sub
                BSF       STATUS,C       ; get ready to subtract
                SUBWF     Lobyte         ; sub from low byte
                BTFSS     STATUS,C       ; borrow required?
                GOTO      rest4          ; yes - restore byte
                INCF      Tens           ; no - count one hundred
                GOTO      subA           ; and repeat

rest4           ADDWF     Lobyte         ; restore low byte
                MOVF      Lobyte,W       ; copy remainder..
                MOVWF     Ones           ; to ones register

                RETURN                   ; done

;-----------------------------------------------------------
; Display period in microseconds
;-----------------------------------------------------------


disp            BSF       Select,RS ; Set display data mode

                MOVLW     'T'            ; Time period
                CALL      send           ; Display it
                MOVLW     ' '            ; Space
                CALL      send           ; Display it
                MOVLW     '='            ; Equals
                CALL      send           ; Display it
                MOVLW     ' '            ; Space
                CALL      send           ; Display it

; Suppress leading zeros.....................................

                MOVF      Tents,F        ; Check digit
                BTFSS     STATUS,Z       ; zero?
                GOTO      show1          ; no - show it

                MOVF      Thous,F        ; Check digit
                BTFSS     STATUS,Z       ; zero?
                GOTO      show2          ; no - show it
```

Program 7.2
(Continued)

```
                MOVF      Hunds,F          ; Check digit
                BTFSS     STATUS,Z         ; zero?
                GOTO      show3            ; no - show it

                MOVF      Tens,F           ; Check digit
                BTFSS     STATUS,Z         ; zero?
                GOTO      show4            ; no - show it

                MOVF      Ones,F           ; Check digit
                BTFSS     STATUS,Z         ; zero?
                GOTO      show5            ; no - show it

; Display digits of period.....................................

show1           MOVLW     030              ; Load ASCII offset
                ADDWF     Tents,W          ; Add digit value
                CALL      send             ; Display it

show2           MOVLW     030              ; Load ASCII offset
                ADDWF     Thous,W          ; Add digit value
                CALL      send             ; Display it

show3           MOVLW     030              ; Load ASCII offset
                ADDWF     Hunds,W          ; Add digit value
                CALL      send             ; Display it

show4           MOVLW     030              ; Load ASCII offset
                ADDWF     Tens,W           ; Add digit value
                CALL      send             ; Display it

show5           MOVLW     030              ; Load ASCII offset
                ADDWF     Ones,W           ; Add digit value
                CALL      send             ; Display it

; Show fixed characters........................................

                MOVLW     ' '              ; Space
                CALL      send             ; Display it
                MOVLW     'u'              ; micro
                CALL      send             ; Display it
                MOVLW     's'              ; secs
                CALL      send             ; Display it
                MOVLW     ' '              ; Space
                CALL      send             ; Display it
                MOVLW     ' '              ; Space
                CALL      send             ; Display it

; Home cursor .................................................

                BCF       Select,RS        ; Set display command mode
                MOVLW     0x80             ; Code to home cursor
                CALL      send             ; Do it
                RETURN                     ; done

;-------------------------------------------------------------
; MAIN LOOP
;-------------------------------------------------------------
start           CALL      inid             ; Initialise display
                BANKSEL   PIE1             ; Select Bank 1
                BSF       PIE1,CCP1IE      ; Enable capture interrupt
                BANKSEL   PORTD            ; Select Bank 0
                BCF       PIR1,CCP1IF      ; Clear CCP1 interrupt flag

loop            CALL      conv             ; Convert 16 bits to 5 digits
                CALL      disp             ; Display period in microsecs
                GOTO      loop

                END                        ;;;;;;;;;;;;;;;;;;;;;;;;;;;;;;;
```

Program 7.2
(Continued)

Note that first action in the ISR is to reset the count register and re-enable the interrupt. The data processing then proceeds at the same time as the timer counts the next cycle. The calculation cycle will run continuously, reading the captured count at the beginning of each cycle. If the calculation takes longer than one cycle of the input, it simply means that not every cycle will be measured.

The system operates successfully over the range 15Hz to 50kHz. The absolute maximum count is 65,536μs (16-bit count), and the minimum period is limited by the time taken to reset the interrupt and start counting again. This can be evaluated in the simulation, turning out to be less than 20μs.

7.5 Direct Current Motor

The basic function of a motor is to convert electrical input current into output mechanical power. A simple method of controlling any motor is to use a relay as an on/off switch, but a solid-state transistor drive is more reliable and allows control of position, speed, output power and direction. A half-bridge current driver using bipolar transistors and a full bridge using FETs are demonstrated in (VSM project BRIDGES2).

7.5.1 Bipolar Drive

The basic motor drive controller output stage based on bipolar transistors is shown in Figure 7.13(a). PNP and NPN transistors are connected as a linear output in push–pull configuration (complementary drive). As the control pot is adjusted to a positive voltage, as shown, Q1 turns on progressively and draws negative current through the motor, which rotates forward. A negative input turns on Q2 and reverses the motion. Linear (proportional) speed control is thus achieved. Generic transistors are used in the simulation, but power transistors would be used in real hardware, with suitable modification of the circuit component values. If MCU control of the linear drive is required, a DAC could replace the control pot. The op-amp stage shifts the control input to a suitable range, 0–5V.

The circuit has significant drawbacks. The main problem is that, when conducting, the transistors dissipate power equal to the product of the collector–emitter voltage and collector current. One transistor is partially on when the motor is active and will dissipate significant power when the output is at some intermediate voltage. In addition, dual supplies are needed.

7.5.2 FET Bridge

The linear drive is quite inefficient, so a PWM full bridge is usually preferred, particularly in digital systems. This operates from a single supply and lends itself more readily to control from a single MCU PWM output. A simulation of a basic circuit is shown in Figure 7.13(b) using generic transistors. It has a pair of NMOSFETs and a pair of PMOSFETs with their drains

(a)

(b)

(c)

(d)

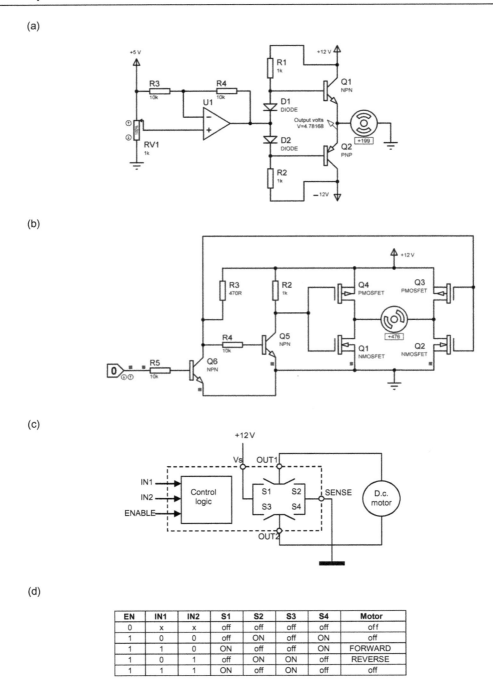

EN	IN1	IN2	S1	S2	S3	S4	Motor
0	x	x	off	off	off	off	off
1	0	0	off	ON	off	ON	off
1	1	0	ON	off	off	ON	FORWARD
1	0	1	off	ON	ON	off	REVERSE
1	1	1	ON	off	ON	off	off

Figure 7.13
Bridge drives: (a) bipolar complementary device, (b) MOSFET full-bridge drive, (c) IC bridge
driver block diagram and (d) IC bridge driver operation.

connected to the load resistor, R1. They are switched on in diagonal pairs to provide a reversible current of about 1A in R1, which represents a motor to be driven in either direction. Small motors are often designed to operate at 12V, so the MOSFETS must operate at this supply voltage.

Each NMOSFET switches on when +12V is applied at the gate relative to the source terminal, and each PMOSFET switches on when 0V is applied to its gate, or −12V relative to the source (12V supply). Thus, either the PMOSFET or the NMOSFET is switched on at any one time on either side of the bridge. A pair of bipolar transistors provides level shifting and inverts the signal between each half of the bridge. Toggling the input logic level reverses the current in the load, as indicated on the virtual ammeter. This bidirectional drive method will be extended later to drive three sets of windings in the BLDC motor.

PWM FET bridges are much more efficient than the equivalent linear drive, as the power dissipated in the drive transistor is calculated as the average drain-source voltage multiplied by the average channel current. When the FET is conducting, the voltage is low, and when off, the current is low, so the overall transistor dissipation is minimised in switch mode. This means lower power (cheaper) transistors and smaller heat sinks are needed.

Switch mode power supplies, which are more efficient than linear regulator-based supplies with the same output current, work on the same principle. The only slight problem is that the FET channel resistance is not negligible, and some power is dissipated here when conducting, but, overall, the FET is more effective in switching mode than a bipolar transistor.

7.5.3 Test Circuit

The FET bridge can be implemented as a single IC suitable for small motors, providing control logic and interfacing in one package. A block diagram based on the L6502 IC bridge is seen in Figure 7.13(c) and its control logic functions in Figure 7.13(d). It is included in the demo application described below. Larger d.c. motors usually have discrete FET drives which allow the heat from the drive transistors to be dissipated more easily via individual heat sinks.

Three low-power motor interfaces are shown in Figure 7.14, which illustrate different motor interfacing techniques (VSM project MOTORS2). The d.c. motor is controlled by a simple single-ended FET switch, the d.c. servo by an IC FET bridge and the stepper motor by a dedicated encoder and driver chipset. The motors are enabled in turn by pressing the select button. Operating parameters (speed, position, direction) can then be changed via the additional push button inputs. The control program outline is shown in Figure 7.15, and the source code in Program 7.3.

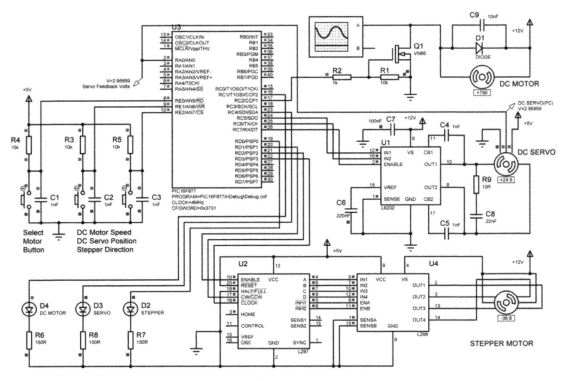

Figure 7.14
Motor interfaces schematic.

7.5.4 PWM Drive

In Figure 7.14, the d.c. motor is controlled from the PWM output of the PIC MCU, via a power FET VN66. This has an operating current of about 1A maximum, giving a maximum motor input rating of 12W at the operating voltage of 12V. The motor characteristics can be set in the simulation; a minimum motor resistance of about 10Ω is suitable.

The FET has a forward resistance of about 1Ω when switched on. The VN66 is convenient as it operates at TTL level gate voltages, switching off at 0V and on at +5V, with the threshold at about 1V. It has very high input impedance, so reliability is improved by adding shunt resistance to the gate, which improves noise immunity. The protection diode across the motor is required to cut off the back EMF from the inductive load. The shunt capacitor reduces residual noise from the motor commutator in the real motor.

When the system is started and the d.c. motor selected, a default PWM output is generated with 50% mark/space ratio (MSR). The MSR can then be increased and decreased using the up/down buttons. Note that the count preload register has to be checked each time the MSR

MOTORS
Test DC motor PWM speed, DC position step servo and
stepper motor direction with push button inputs, using P16F877 (4MHz)

Main
> Initialise
>> Port A = Analogue inputs, servo pot = RA0
>> Port C = Outputs, DC motors
>> Port D = Outputs, stepper motor
>> Port E = Digital inputs, push buttons: Select, Up, Down
>> PWM rate = 4kHz
>> Servo target value = 128

> Wait for 'Select' button
> REPEAT
> Select PWM mode, 50% MSR

>> REPEAT
>>> CALL Motor
>> UNTIL 'Select' button pressed again

> REPEAT
>> CALL Servo
>> UNTIL 'Select' button pressed again

> REPEAT
>> CALL Step
> UNTIL 'Select' button pressed again

> ALWAYS

Subroutines

> Motor
>> IF 'Up' button pressed
>>> Increment speed unless maximum
>> IF 'Down' button pressed
>>> Decrement speed unless minimum
>> RETURN

> Servo
>> IF 'Up' button pressed
>>> Add 10 to target position
>>> Move forward, until target position reached
>> IF 'Down' button pressed
>>> Subtract 10 from target position
>>> Move reverse, until target position reached
>> RETURN

> Step
>> IF 'Up' button pressed
>>> Select forward mode
>> IF 'Down' button pressed
>>> Select reverse mode
>> Output one drive pulse
>> RETURN

Figure 7.15
Motor test program outline.

```
;;;;;;;;;;;;;;;;;;;;;;;;;;;;;;;;;;;;;;;;;;;;;;;;;;;;;;;;;;;;;;
;
;         Project:                    Interfacing PICs Ed2
;         Source File Name:           MOTORS2.ASM
;
;         Devised by:                 MPB
;
;         Date:                       21-01-13
;         Status:                     Updated for VSM v8
;
;;;;;;;;;;;;;;;;;;;;;;;;;;;;;;;;;;;;;;;;;;;;;;;;;;;;;;;;;;;;;;
;
;         DC Motor PWM speed control
;         DC Servo position control
;         Stepper motor direction control
;         Select motor and direction using push buttons
;
;;;;;;;;;;;;;;;;;;;;;;;;;;;;;;;;;;;;;;;;;;;;;;;;;;;;;;;;;;;;;;

          PROCESSOR 16F877A
          __CONFIG 0x3731           ; Clock = XT 4MHz
          INCLUDE "P16F877A.INC"    ; standard register
labels

Count1    EQU       20              ; delay counter
Count2    EQU       21              ; delay counter
Target    EQU       22              ; servo target
position

;-------------------------------------------------------------
; PROGRAM BEGINS
;-------------------------------------------------------------

          CODE      0               ; Default start
address
; Port & PWM setup .........................................

init      NOP                       ; required for ICD mode
          BANKSEL   TRISB           ; Select control registers
          CLRF      TRISC           ; Output for dc motors
          CLRF      TRISD           ; Output for stepper
          MOVLW     B'00000010'     ; Analogue input setup code
                                    ; PortA = analogue inputs
                                    ; Vref = Vdd
          MOVWF     ADCON1          ; Port E = digital inputs
          MOVLW     D'249'          ; PWM = 4kHz
          MOVWF     PR2             ; TMR2 preload value

          BANKSEL   PORTB           ; Select output registers
          CLRF      PORTC           ; Outputs off
          CLRF      PORTD           ; Outputs off
          MOVLW     B'01000001'.    ; Analogue input setup code
          MOVWF     ADCON0          ; f/8, RA0, done, enable
          MOVLW     D'128'          ; intial servo position
          MOVWF     Target

;-------------------------------------------------------------
; MAIN LOOP
;-------------------------------------------------------------

but0      BTFSC     PORTE,0         ; wait for select button
          GOTO      but0

          MOVLW     B'00001100'     ; Select PWM mode
          MOVWF     CCP1CON         ;
          MOVLW     D'128'          ; PWM = 50%
          MOVWF     CCPR1L          ;

but1      BTFSS     PORTE,0         ; wait for button release

          GOTO      but1
          CALL      motor           ; check for speed change
```

Program 7.3
Motor control program.

```
                MOVLW      B'00001100'      ; Select PWM mode
                MOVWF      CCP1CON          ;
                MOVLW      D'128'           ; PWM = 50%
                MOVWF      CCPR1L           ;

but1            BTFSS      PORTE,0          ; wait for button release

                GOTO       but1
                CALL       motor            ; check for speed change

                BTFSC      PORTE,0          ; wait for select button

                GOTO       but1
                MOVLW      B'00000000'      ; deselect PWM mode
                MOVWF      CCP1CON          ;
                CLRF       PORTC            ; switch off outputs

but2            BTFSS      PORTE,0          ; wait for button release

                GOTO       but2
                CALL       servo            ; move servo cw or ccw
                BTFSC      PORTE,0          ; wait for select button
                GOTO       but2
                CLRF       PORTC            ; switch off servo

but3            BTFSS      PORTE,0          ; wait for button release
                GOTO       but3
                CALL       step             ; output one step cycle

                BTFSC      PORTE,0          ; wait for select button
                GOTO       but3
                CLRF       PORTD            ; disable stepper outputs
                GOTO       but0             ; start again
```

Program 7.3
(Continued)

is modified for the maximum (FF) or minimum (00) value, to prevent rollover and rollunder of the PWM value.

PWM drive is very commonly used for d.c. motor speed control as it can be implemented via a simple single-ended FET interface or a full bridge for bidirectional current. For this reason, the PIC timer module is specifically designed to generate PWM output.

7.6 Stepper Motor

The third sub-circuit shown in Figure 7.14 is the stepper motor interface. This uses a dedicated controller/driver chip, because current driver components would be needed in any case, and the stepper controller also incorporates sequencing logic which reduces the software burden. The stepper motor has a set of three windings distributed around the stator, and a passive rotor, with permanent (or induced) magnetic poles. It is designed for incremental, or continuous, rotation by activating the windings in a suitable sequence.

A cross-section of a stepper motor is shown in Figure 7.16(a). The stator has 16 poles and the rotor 12. When a winding is activated, it attracts the nearest poles. The varying

(a)

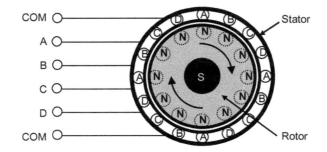

(b)

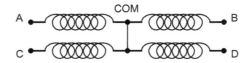

(c)

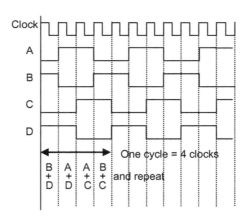

Figure 7.16
Stepper motor: (a) cross-section, (b) windings and (c) drive sequence.

offset between the rotor and stator poles allows the rotor to be moved in full or half steps. There are four sets of windings, A, B, C and D, connected in sequence around the stator, connected in pairs with centre tapped windings, giving a total of six connections. In the test circuit, the common terminals are connected to the power supply (+12V) and the individual coil terminals driven from the sequencer (active low operation).

In normal, full-step, mode, the coil sets are activated in pairs (Figure 7.16(b)) and the rotor moves half a pole per step, giving 24 steps per revolution. The step size is then $360/24 = 15°$. This mode provides full torque but lower positional resolution. In half-step mode, the rotor moves by a quarter pole per step, $7.5°$, providing twice as many steps per revolution, but less torque, since only one coil is activated at a time.

There are two chips forming the stepper drive interface. The L297 controller provides the stepping sequence on outputs A, B, C and D, and the L298 full-bridge driver provides the drive current needed by the motor windings. The drive mode (full or half step) is fixed in full step by connecting the step mode select input low. The active low reset is tied high. The MCU provides an enable signal and selects the direction of rotation (clockwise (CW) or counter-clockwise (CCW)), and the test program outputs clock pulses at a frequency of 20Hz, so that the stepping effect can be seen. When the stepper test is selected in the MCU program, the motor rotates CW by default, with the 'down' button changing the direction to CCW and the 'up' button back to CW.

If the windings are left active in any position, that position can be held against a load torque. Even when powered down, the windings have residual magnetism which holds the shaft in position against applied external torque. When actively stepping, there is a lower limit to the step time required, which translates into a maximum operating frequency and speed. If starting from stationary, the speed may need to be ramped up, until the rotor inertia gained will allow the motor to run at its maximum speed. The speed also needs to be ramped down when stopping, if correct position is to be maintained, otherwise the rotor inertia may cause overshoot. Pull-in and pull-out speeds and load torque should be specified in the stepper motor data sheet.

7.7 BLDC Motor

The brushless d.c. (BLDC) motor eliminates the main weakness of the conventional d.c. motor, the commutator and brushes, by using a permanent magnet rotor and rotating field. However, this limits its size and power and requires a complex drive controller and interface to generate the rotating field in the stator. Its operating principles are similar to the stepper motor, but it is designed for continuous rotation rather than stepwise positional control.

The BLDC motor has three sets of parallel multiple windings creating the rotating magnetic field that interacts with the permanent poles of the rotor. This arrangement can be reversed, with a circumferential rotor surrounding static windings, as found in hard disk motors, giving a useful variety of possible physical configurations, either flat or cylindrical.

Figure 7.17
Remote control helicopter.

The use of powerful rare earth alloy magnets gives a high power to weight ratio compared with conventional d.c. motors. With the development of lightweight batteries and on-board MCU-based controllers, they are so light that model aircraft can now be made with electrical propulsion using BLDC motors (Figure 7.17).

The drive control requirements for optimum performance are fairly complex, and magnetic sensors are often attached to the rotor to provide feedback to the controller. The simulated test circuit (Figure 7.18(a)) demonstrates only a basic drive system, with the motor represented by three dummy load resistors connected in a star arrangement (VSM project BLDC2). This allows the drive signals to be viewed more clearly on the virtual scope.

The motor is driven from a three-phase bridge with two IGFETs per phase to provide bidirectional current in each winding. NMOS and PMOS transistors are connected as a complementary pair in each output, the former switching on with a high input and the latter with a low input to the gate. The dummy motor windings are connected between phases, operating at 12V, so the drive transistors need level-shifting buffers (TC4469) to convert the PIC outputs to the drive voltage levels for connection to the FET gates.

The PIC generates outputs to activate the motor windings in sequence to draw the magnetised rotor around the stator poles in the same way as the stepper motor. The source code is shown in Program 7.4. The drive signals from the simulation are displayed in Figure 7.18(b). The three voltage levels seen on each signal correspond to the winding

(a)

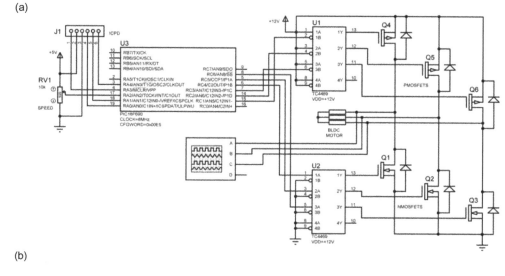

(b)

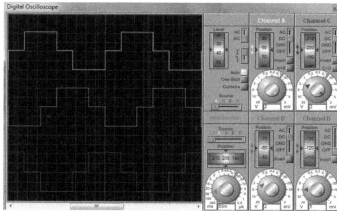

Figure 7.18
BLDC interface: (a) test schematic and (b) drive signals.

current flowing being forward, off and reverse. It can be seen that each output is driven with forward and reverse voltages in steps, and is off when the other two outputs are active. The speed can be varied between about 2 and 20 cycles per second using the pot connected to an analogue input.

This particular circuit was developed with a PIC 16F690 controller but is based on a circuit proposed in Microchip application note AN857 designed around the 16F877A. This has a full explanation of BLDC operation and demo source code and should be consulted for further work in this area.

```
;*************************************************************
;
;          BLDC2.ASM        MPB        21-01-13
;
;          BLDC motor drive
;          Speed control adjust between 2 -20 Hz
;
;          MCU = 16F690
;          Internal clock = 8MHz
;
;*************************************************************
;
;          Switching sequence
;
;          FET        Q6  Q5  Q4  Q3  Q2  Q1
;          OUT        RC6 RC5 RC4 RC3 RC2 RC1
;
;          CA         0   0   1   1   0   0
;          CB         0   1   0   1   0   0
;          AB         0   1   0   0   0   1
;          AC         1   0   0   0   0   1
;          BC         1   0   0   0   1   0
;          BA         0   0   1   0   1   0
;
;*************************************************************
          PROCESSOR 16F690     ; Specify MCU for assembler
          INCLUDE "P16F690.INC"          ; Standard labels
          __CONFIG 0x00E5                ; MCLR, PWRTE, Int
Clock

HICO      EQU       020                  ; Labels
LOCO      EQU       021

          CODE     0          ; Locate program start

; Initialize registers.................................

          BANKSEL   TRISC              ; Select Bank 1
          MOVLW     B'10000001'
          MOVWF     TRISC              ; Initialise RC1-6 for output
          MOVLW     B'00100000'        ; Analogue input setup code
          MOVWF     ADCON1             ; Fosc/32


          BANKSEL   PORTC              ; Select bank 0
          MOVLW     B'00000000'
          MOVWF     PORTC              ; Drives off
          MOVLW     B'00001001'        ; Analogue input setup code
          MOVWF     ADCON0             ; f/8, AN2, done,enable



; Start main loop.................................

start     MOVLW     B'00011000'        ; Output driver codes
          MOVWF     PORTC
          CALL      readin

          MOVLW     B'00101000'
          MOVWF     PORTC
          CALL      readin

          MOVLW     B'00100010'
          MOVWF     PORTC
          CALL      readin

          MOVLW     B'01000010'
          MOVWF     PORTC
          CALL      readin

          MOVLW     B'01000100'
          MOVWF     PORTC
          CALL      readin
```

Program 7.4
BLDC motor test program.

```
                MOVLW       B'00010100'
                MOVWF       PORTC
                CALL        readin

                GOTO        start           ; repeat output loop

; SUBROUTINE .........................................

readin   BSF        ADCON0,1        ; start ADC..
wait     BTFSC      ADCON0,1        ; ..and wait for finish
         GOTO       wait
         MOVF       ADRESH,W        ; store result

         MOVWF      HICO            ; load delay count
slow     MOVLW      d'249'          ; start 1000 cycle loop
         MOVWF      LOCO            ; = 500us
fast     NOP                        ; fill loop to 4 cycles
         DECFSZ     LOCO            ; low count
         GOTO       fast            ; loop
         DECFSZ     HICO            ; high count
         GOTO       slow            ; loop = ADC x 1000 cycles

         RETURN                     ; from variable delay
         END                        ; Terminate assembler
```

Program 7.4
(Continued)

7.8 Mechatronics Board

A very useful training and development tool for investigating motor control principles is
the microchip PICDEM mechatronics board (Figure 7.19). This has a small brushed d.c. and a
stepper motor on board, with four FET half-bridge drivers which can be connected to either (or
an external motor) to provide bidirectional drive for multiple windings. It also includes
temperature and light sensors and a 3½ digit LCD numerical display. The peripheral sub-
circuits are connected using temporary links, so that a variety of test applications can be tried
out. The brushed d.c. motor has an opto-sensor and disk producing TTL feedback pulses.

A simulation schematic is shown in Figure 7.20 (VSM project MECH2). The components
in the simulation are not necessarily identical to the actual hardware, since not all are
available as active models. The drive transistors, for example, are represented by generic
IGFETs. The physical links on the board itself are 'connected up' by simply adding
suitable labelling to the outputs of the MCU to associate them with the required peripheral
input, in the same way that the LCD connections are already in place (these are hard
wired).

The central component is the PIC 16F917, which incorporates LCD drive outputs.
These occupy a large proportion of the available I/O pins, leaving a limited number for the
other peripherals. In other designs, a serial access LCD can be used to free up I/O pins. On
the other hand, the naked LCD device is cheaper, because an alphanumeric/graphical
LCD must incorporate its own controller. The digit segments are enabled by combinations
of the LCD segment and common inputs that are defined in an include file, which must be

Figure 7.19
PICDEM board.

added to the application project. Three bias voltages are also required by the LCD at V_{cc}, $2V_{cc}/3$ and $V_{cc}/3$, which are generated in a simple resistive divider.

The push button (tactile switch) inputs on the hardware are represented by switches in the simulation, so that they can be left in the closed position when running the simulation. A bank of active high LEDs are provided for output monitoring. The temperature and light sensors are modelled as specific devices, with manual control of the set variable. They will normally be connected to an analogue input on the MCU, either a comparator or an ADC input.

The bridge control logic is represented by generic devices for the discrete CMOS gates and specific devices for the enable logic. The driver MOSFETs themselves are generic, so actual device characteristics may not represented with complete accuracy. This is not a significant issue, since the motor models are generic in any case. The PMOSFET is switched on when its gate is taken low, and the NMOSFET is switched on when its gate is logic high.

The control logic is designed to prevent the FETs being switched on together, which would damage them by overheating, and to provide a convenient set of input options. This control logic for the half-bridge that has inputs labelled P1, PMW1 and N1 and output drivers Q1 (current source) and Q2 (current sink) is given in Table 7.1.

All four half-bridge circuits operate in identical manner. Input F, operated by the current overload sensing circuit (see below), always disables the output when low. For most input combinations, the half-bridge is disabled (safe). The current source is only switched on when input P is high and PWM is low. The current sink is only switched on when input N

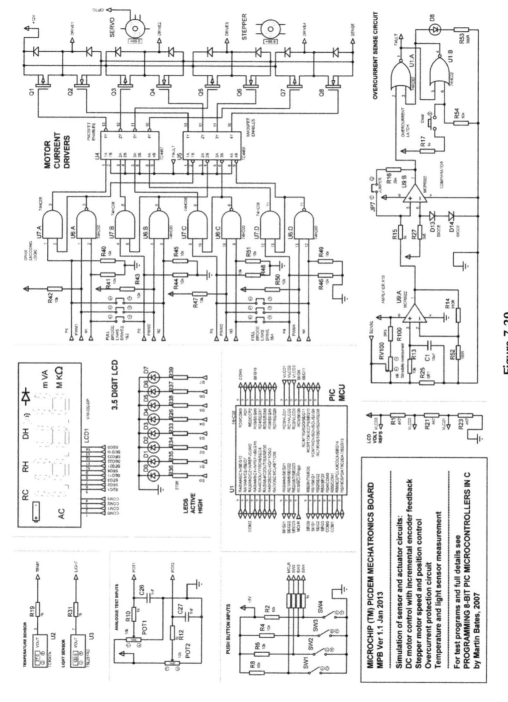

Figure 7.20
PICDEM board schematic.

Table 7.1: Half-Bridge Control Logic.

Inputs				Outputs		
P1	**PWM1**	**N1**	**F**	**Q1**	**Q2**	**Result**
X	X	X	0	Off	Off	Output disabled
0	0	X	X	Off	Off	Output disabled
0	1	0	X	Off	Off	Output disabled (default input)
1	1	0	1	Off	Off	Output disabled
1	0	X	1	On	Off	Sourcing current
X	1	1	1	Off	On	Sinking current

Input X = do not care.

is high, and PMW is high. When the inputs are all open circuit (not connected), P and N are pulled low (disabled). The output can be toggled between sink and source by holding N and P high and toggling PWM.

For full-bridge operation, P1 and N2, P2 and N1, and M1 and M2 are linked together via the six input links. Similarly driver 3 and 4 inputs can be operated together via links. Thus bidirectional drive and PMW can be applied to up to four sets of windings and could be used to operate a BLDC motor.

The board has an overcurrent protection circuit that is connected in series with the common connection of the current drivers, which trips the outputs if the total current exceeds about 1.2A. It uses a current-sensing resistor of 0.1Ω to generate 120mV at the trip current and a non-inverting amplifier with a gain of 10 to increase this to 1.2V at the comparator input. The trip level is set to this value by the pair of series diodes on the reference input of the comparator.

The comparator can be converted to trigger operation by connecting the positive feedback link. Its output operates a latch which must be reset to enable the output after an overcurrent event and on power-up. An overcurrent test circuit has been added to the simulation so that the operation of this little circuit can be studied more easily, as it is a nice illustration of some of the analogue concepts covered in Chapter 6.

7.9 Servo Systems

A servo system controls the output speed or position of a mechanical output using feedback. Originally, analogue control was used; now digital systems achieve the same result. The controller must be designed to produce the required dynamic response in conjunction with the mechanical load. Typically, this is the fastest response possible while achieving an accurate final position or speed.

7.9.1 Digital Feedback

Direct current motors cannot be positioned accurately without some kind of feedback. In applications such as printers and robot arms, the d.c. motors have pulse feedback sensors which allow the controller to monitor the motor shaft position, speed or acceleration. A slotted wheel and opto-sensor that detects each slot can be used to measure shaft rotation. A pulse count gives the position or number of revolutions of the shaft, and the pulse frequency or period, measured as described in Sections 7.2 and 7.4.3, allows the shaft speed to be calculated. An equivalent linear system is print-head control in an inkjet printer, where its position is monitored by a graduated strip attached to the traverse mechanism, and a PWM-controlled d.c. motor drives the print-head drive belt.

The block diagram shown in Figure 7.21(a) represents a generic digital rotary position or speed controller. The motor has a slotted or perforated wheel attached. If there are, say, 100 slots, then there will be 200 edges, giving a resolution of $360/200 = 1.8°$. The motor is PWM driven via a suitable current switch, and the feedback pulses measured. To improve position control, the speed can be ramped up and down to prevent the motor from overshooting the target position.

The MCU is shown acting as a slave device, receiving a position or speed command from a master controller, carrying it out, and then signalling completion of the operation. The accuracy of the system can be further improved by interpolation, where the sensor grid or disk has a variable density pattern so that each cycle can be subdivided by an analogue sensor.

7.9.2 Servo Control

A servo motor incorporates position or speed sensors for operation in a closed loop system. Direct current position servos usually have a gearbox built in to reduce the motor speed by a factor of at least 100. That is, 100 revolutions of the motor produce 1 revolution at the output shaft. This is necessary because the d.c. motor itself cannot be stopped accurately; it needs to home in on a set position.

In Figure 7.14, the d.c. servo model has a built-in pot which outputs a voltage between +5 and 0V to represent the position of the shaft. The motor is driven from an L6202 full-bridge driver. This is the IC FET bridge that provides drive to the motor in either direction under digital control. A diagram and table representing the chip operation are shown in Figure 7.13(c) and (d).

The bridge circuit contains four power FET switches connected in a bridge arrangement. The motor and supply are connected so the load current can be reversed by switching on pairs of transistors. They are controlled from a simple logic circuit (see the L6202 data sheet), as

(a)

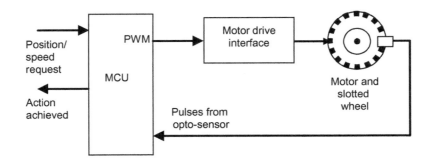

(b)

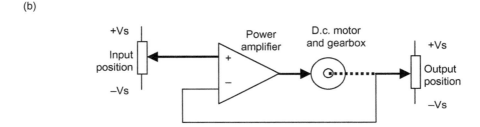

(c)

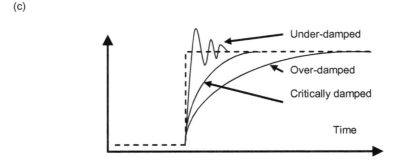

Figure 7.21
Servo systems: (a) digital servo, (b) analogue position servo and (c) step response
of position servo.

summarised in the function table. Forward and reverse are selected by setting the IN1 and IN2 inputs to opposite logic states.

The chip operates from the motor supply voltage (+12V) and the digital logic supply is derived from it, so no separate +5V supply is needed. A current-sensing resistor can be inserted in the 0V connection, so that the current flow in either direction can be monitored for control purposes. Bootstrap capacitors must be fitted as shown to ensure reliable switching of the bridge FETs. Although the FETs are protected internally with diodes, a series CR snubber network is connected across the output terminals to further protect the driver chip from current switching transients.

The test program (Program 7.3) allows the user to move the servo in steps. The required position is represented by an 8-bit number which is initially set to the mid-value of 128. If the 'up' button is pressed, the value is increased by 10, and the servo started in the forward direction. The current position is monitored from the servo pot voltage read in via AD0. When the input value matches the target value, the drive is stopped. The servo is moved in the reverse direction in the same way. Since no speed ramping is used, the output tends to overshoot. This would have to be addressed in a practical implementation using a more complex control algorithm.

7.9.3 Digital Servo

As discussed above, feedback from an incremental encoder or other pulse sensor is the most common method of position and speed measurement in processor systems, with the output from an opto-detector or magnetic sensor converted to a TTL signal. The speed can be calculated from the pulse frequency or period.

The MCU timer can be used as a counter by connecting it to an input pulse stream. The pulses must be counted over a known time period, so a second timer is used to generate an interrupt after a suitable interval, which causes the MCU to read the counter. The final count must be high enough (within the maximum count available) to obtain a reasonably precise result, since the minimum error is ± 1 bit. For example, if the count is 100, the minimum error is, by definition, 1%. With a count of 1000, it is only 0.1%. Obviously, a 16-bit count (16F877A, Timer1) can produce a more precise result than an 8-bit timer (Timer0).

In this way, the number of pulses per second can be obtained, and the speed of the shaft in revolutions per second calculated from the number of pulses per revolution produced by the shaft encoder:

$$\text{Shaft speed} = \text{pulses per second}/\text{pulses per revolution (revs/s)}$$

If the shaft speed is fairly low, it may be more convenient to measure the period of the pulse. In this case, the MCU counter register is configured as a timer, driven from the internal clock. Assuming a positive-going input pulse, the timer will be reset on the rising edge, and the count captured on the falling edge. Again, the resolution of the measurement depends largely on the magnitude of the pulse count. The accuracy depends on the

MCU clock, so a crystal clock is usually preferred in this situation. The speed is then calculated as

$$\text{Shaft speed} = 1/(\text{time per pulse} \times \text{pulses per revolution}) \ (\text{revs/s})$$

Precise speed control allows a motor to be accelerated from rest, kept at a constant output and decelerated smoothly to provide optimum performance with a given load. This is particularly important in BLDC motors, which need a fairly complex control algorithm to operate correctly. Generally, ramping the motor speed up and down avoids overshoot that can arise with a high load inertia or slow controller response.

7.9.4 Analogue Servo

Before digital controllers became available, analogue position and speed control was used in servo systems. In a basic position system, a potentiometer (pot) attached to the motor shaft provides a voltage that represents its current position (Figure 7.21(b)). The required position is set on a manual pot (or from an external controller) connected to one input of a difference amplifier, with the feedback voltage applied to the other input. A linear power amplifier (see above) then drives the motor until the voltages match and the motor stops.

An equivalent speed control system would use a tachogenerator (tacho) to measure the output shaft speed. This is a small d.c. generator that outputs a voltage or current in proportion to the speed of the shaft (a permanent magnet d.c. motor will produce this effect if the shaft is driven and the voltage measured at the terminals). As in the position servo, a difference amplifier controls the motor power until the target speed is achieved. This system illustrates the operation of a PID (proportional, integral and derivative) controller, which is still relevant to digital controllers because the mechanical load will produce the same responses due to inertia and friction.

The dynamic response is shown in Figure 7.21(c). Depending on the tuning of the amplifier and the physical characteristics of the system, the output can respond to a step change at the input in two main ways. If the slew rate of the system is slow, an under-damped response will be obtained. If too fast, the output can overshoot the target position and exhibit damped oscillation until finally settling to the target position. The ideal response is critically damped, where the response is as fast as possible without overshooting.

PID control requires the transient behaviour of the amplifier to be adjustable. The system response can be represented by a second-order differential equation, and PID control corresponds to adjusting the constants in that model to modify the transient and steady-state response of the system. This form of control can also be implemented in a digital controller using a fast DSP (digital signal processor) chip, where the PID variables are controlled in software. The motor drive amplifier would then be controlled via a high-speed DAC and the shaft speed monitored by a tachometer (speed) or pot (position) and ADC.

7.9.5 Hobby Servo

This is a self-contained position servo that has been traditionally used in remote control model craft (Figure 7.22). An actuator arm with a limited range of rotation, typically about 100°, can operate a rudder, ailerons or similar devices for directional control. It incorporates a small d.c. motor, reduction gearbox and analogue pot with a small MCU and drive interface in a lightweight package.

It has a single input which is designed to receive a variable pulse width TTL signal which is translated by the MCU into an output position. A pulse of 1.5ms provides the

(a)

(b)

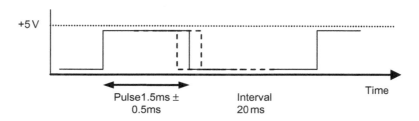

Figure 7.22
Hobby servo: (a) hobby servo connected to a PICkit2 test board and (b) hobby servo PWM input signal.

centre position, with an operating range of ± 0.5ms. The repetition rate is 20ms (50Hz), but this is not critical. It is frequently used with a radio link to a remote control unit with at least two channels over which an RF modulated version of the PWM signals is transmitted. Alternatively, it can be directly controlled from the PWM output of a PIC MCU.

Questions 7

1. State two advantages and one disadvantage of the relay as an interface device. (3)
2. Calculate the voltage drop across a 10Ω collector load resistor if the transistor has a current gain of 50, a base resistor of 1k0 and an input voltage of 4.6V. (4)
3. Describe briefly the useful characteristics of the VN66 FET when used as current switch in a digital circuit. (3)
4. Explain briefly why the d.c. motor needs a commutator and the problems this causes. (3)
5. Explain the difference between the structure and function of a thyristor and triac. (3)
6. Explain briefly how an oscillator can be implemented in hardware and software. (3)
7. Explain how PWM allows the dissipation in a power load to be controlled. (3)
8. Explain, using a suitable diagram, how a bridge driver allows a d.c. motor to drive in both directions. (3)
9. A stepper motor has a step size of $15°$. Its maximum step rate is 96Hz. Calculate the maximum speed in revolutions per second. (3)
10. Explain briefly the advantages of the BLDC motor over the d.c. motor. (3)
11. Calculate the speed of a shaft in r.p.m. if an MCU timer connected to a shaft encoder with 50 slots counts up to 200 in 100ms. (4)
12. Compare the advantages and disadvantages of the d.c. and stepper motor for position control. (5)

Total (40)

Assignments 7

7.1 Direct Current Motor Speed Control

Obtain two small d.c. permanent magnet motors, mount them in line and connect the shafts together using a suitable flexible coupling, as a motor and tachogenerator. Apply a variable voltage supply to one motor and note the output voltage from the other (tachogenerator). Interface the motor to an MCU using suitable prototyping methods (see 'PIC Microcontrollers' for guidance). Connect the drive motor to the PWM output via a suitable single-ended current switch and the tacho to the ADC input. Write a test program to operate the system under open loop control. Ensure that the motor runs correctly under 50% PWM drive. Add a potentiometer input to control the speed. Develop a closed loop control implementation that runs the motor at a consistent speed under varying load.

7.2 Stepper Motor Characteristics

In the Proteus simulation environment, select the standard stepper motor drive chip and connect up the interactive stepper motor. Operate the stepper drive clock from a push button input and note the output sequence obtained. Record the input clock and outputs accurately on a time axis and explain the significance of the sequence. Replace the input with a simulated clock and confirm correct rotation of the stepper motor. Obtain the data sheet for a stepper motor and examine the torque/speed characteristic and specifications for holding torque. Explain the significance of this characteristic in designing a robot arm with a stepper drive. Consider the maximum speed of operation and load handling for a single arm section which is rotating in a vertical plane directly driven with a stepper motor at one end and a load at the other, starting from the horizontal position.

PART 3

PIC Systems

CHAPTER 8

Serial Communications

Summary
- USART port supports RS-232 low-speed, short range communication
- RS-422 and RS-485 implement extended industrial serial links
- SPI is an on-board high-speed serial bus with hardware slave selection
- I²C is an on-board mid-speed serial bus with software slave selection
- CAN and LIN busses are network links for automotive systems
- Ethernet and Wi-Fi support remote internet monitoring and control
- Other wireless technologies are used for local control and monitoring

Serial communication links use a single signal connection for communication. This is the most practical solution for data exchange between physically remote systems and also reduces the total number of tracks on circuit boards within the local system. Serial links also simplify the wiring where there are numerous peripheral devices with which the MCU to must communicate.

Data may be transmitted in one direction only, or both directions on the same link, whereas two channels allow simultaneous communication in both directions. A simple asynchronous link can operate by dividing the data into single bytes and re-triggering the receiver at the beginning of each. A synchronous link may have a separate clock signal alongside the data for controlling the transfer, or the clock may be combined with the data on a single line.

Within the microcontroller domain, we tend to use the simpler forms of serial communications. The PIC 16F877A has the serial interfaces Universal Synchronous Asynchronous Receiver Transmitter (USART), Serial Peripheral Interface (SPI) and Inter-Integrated circuit (I²C). USB is now the standard serial port on the PC, but this is only supported by the more powerful PIC MCUs, as it is fairly complex and operates at relatively high speed.

The Master Synchronous Serial Port (MSSP) module in the PIC provides both SPI and I²C (usually pronounced I squared C) protocols. These are used for communication between processors and peripheral devices within a single system. SPI is simpler and faster, using a

Interfacing PIC Microcontrollers.
DOI: http://dx.doi.org/10.1016/B978-0-08-099363-8.00008-X

hardware peripheral select system. I²C is more complex, but with software addressing, so it does not require the additional select connections, only shared clock and data lines. The USART can be used for longer inter-system connections, such as to a PC host.

8.1 USART

The USART is a basic serial communication protocol originally developed for computer terminals to communicate with a mainframe computer. It was later adopted for the COM port of the PC to provide an interface with serial peripherals. When converted to a higher transmission voltage for distance transmission, it is traditionally known as RS-232. It can therefore, with suitable interfacing, be used by the PIC to communicate with a PC and was used in previous generations of PIC programmers for program downloading, but not via the USART port.

8.1.1 RS-232 Port

In the PIC 16F877A, the USART is accessed through pins RB6 and RB7. It has two modes of operation, synchronous (providing a separate clock signal) and asynchronous (no clock connection). In synchronous transmission, the pins are used for clock and data respectively, and the receiver uses the clock pulse to latch each data bit individually, eliminating the need for a local clock and timing circuits. Asynchronous mode allows the MCU to send and receive simultaneously on the same pins. This is the more commonly used mode, as more effective methods of local synchronous transmission are available in the PIC using the MSSP. It is usually preferred even if transmission in one direction only is needed, as we have already seen when using the serial LCD (Chapter 4). The data often consists of 7-bit ASCII character codes, as USART is most often used to send text-based messages. The eighth bit can be used for parity error detection (see below).

In the diagrams showing RS-232 operation (Figure 8.1(a)), the PIC is connected to a host system via TX and RX lines. The PIC USART output itself operates at TTL voltages, and therefore needs an external serial line driver to convert its output to a higher symmetrical line voltage required for proper RS-232 transmission. This is desirable because any data signal becomes attenuated down the line, due to the distributed resistance and capacitance of the cabling. The maximum link distance for RS-232 is about 100m with symmetrical voltages of up to ±25V, with ±12V over 10m being more typical. The line voltage signal is also inverted with respect to the TTL version. The bare TTL level signal can be used over shorter distances, less than 1m, without line drivers.

In asynchronous mode, RB6 acts as a data transmit (TX) output and RB7 as data receive input (RX) (see 16F877A data sheet, Section 10.2). Shift registers are used to transmit and

(a)

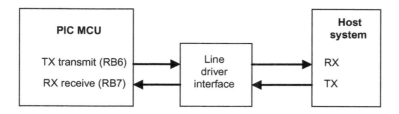

(b)

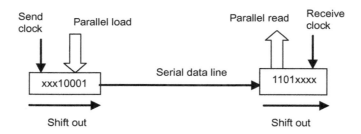

(c)

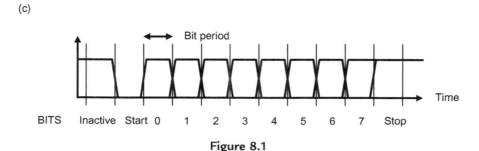

Figure 8.1

USART: (a) connections to host system, (b) shift register operation and (c) signal at PIC port.

receive (Figure 8.1(b)). At the sending end, a byte is loaded in parallel from the MCU data bus and the data shifted out onto the transmit line 1 bit at a time. At the receiving end, the line must be sampled at the correct time, i.e. in the middle of each bit, to read the correct data, as it is shifted into the register. When all bits have been received, the data can be read out onto the receiver internal data bus for storage and processing.

The format of the signal is shown in Figure 8.1(c). The receive shift in is triggered by the falling edge of the start bit. Data is usually transmitted in 8-bit words (9 is an option), with the least significant bit sent first. The receiver must sample its input at the same rate as the data is sent, so standard clock (baud) rates are used. A minimal rate of 9600 baud is used in our example here, meaning that the bits are transmitted at about 10kbits/s or 100μs per bit period.

We will assume initially that the MCU is transmitting data back to a host system. The sender and receiver have to be initialised to use the same baud rate, number of data bits (default 8) and number of stop bits (default 1). The transmit (TX) TTL output is high when idle (RS-232 line negative). When the PIC serial buffer register (TXREG) is written, the data is automatically sent, using start and stop bits to enclose the data bits.

When the falling edge of the start bit is detected, the receiver must wait 1.5-bit periods and then sample the line for the first data bit (LSB). It must then capture the next bit after a further clock cycle, and so on until the set number of bits has been read in to the receive register. The stop bit confirms the end of the byte, and another transmission can start. An interrupt flag is used to signal the receiver MCU that there is data waiting. It must be read from RCREG before the next byte arrives.

The PIC data sheet has details of the operation of the USART interface. The data is loaded into TXREG (FSR 19h) and transferred automatically to the transmit register when it is ready to send. The shift clock is derived from a baud rate generator, which uses the value in SPBRG (FSR 99h) in its counter. This counter has a post-scaler which divides the output by 16 or 64, depending on the setting of control bit BRGH, so that all the standard baud rates can be achieved (approximately) using an 8-bit counter. The value to be pre-loaded into this register to obtain a given baud rate is listed in tables in the MCU data.

The error associated with each counter value and post-scaler setting is also specified, so that the best option can be selected. The value 25d is used in the demo program, with BRGH = 1 giving an error of only 0.16% from the exact value for 9600 baud. A considerable error can be tolerated because sampling only needs to be synchronised over 10 or 11 cycles at the receiver (start + 8/9 data + stop bit). It can be seen from these tables that the error can be up to 10%, and the system should still work.

Parity checking is a simple error detection system which allows errors in the data due to poor reception quality to be detected. It can only detect the presence of errors, not correct them; this is possible using a more complex data integrity check. Parity checking is not provided within the PIC USART module but is fairly easy to implement in software. If the parity is defined in both transmitter and receiver as even, the parity bit (7 or 8) is set or cleared in the transmitted word so that the total number of 1s in the word is even (2, 4, 6 or 8).

If the data is only 7 bits (ASCII), the eighth can be used as the parity bit (bit 7 in TXREG). If the data is 8 bits, bit TX9D (bit 0 in TXSTA register) must be written with the parity value. When received, the number of 1s is counted up (including RX9D if necessary), and if it is not even, an error must be flagged up. The usual procedure is then to return a message to the sender requesting retransmission. The original specification for this protocol had additional hardware handshaking signals for the transmitter to indicate 'Ready to Send' (RTS) and the receiver to reply 'Clear to Send' (CTS), but their use is often unnecessary.

8.1.2 Test System (USART2)

A USART test system is shown in Figure 8.2 (VSM project USART2). A virtual serial terminal produces ASCII output from keyboard characters typed into the simulator host. This is connected to the USART port via a pair of MAX232 transceivers that convert the voltage levels from TTL to RS-232 levels and back again. An oscilloscope is connected to the transmission line to display the data levels, with a BCD encoded 7-segment display attached to Port D to display the data as it is received by the MCU.

When the simulation is started, the program generates a prompt on the virtual terminal, which then waits for the user to input numerical characters at the keyboard. The terminal generates the ASCII code for that key, in RS-232 format. For numbers between 0 and 9, this consists of the number plus 30_H, giving codes in the range $30_H - 39_H$. The resultant signal output is seen on the virtual scope (the virtual terminal dialogue is displayed floating on the scope screen, but is mobile).

The transmitted bits appear as a symmetrical signal of about 16V amplitude. The signal supply voltages ($\pm 8.5V$) are generated internally in the line driver chip by a charge pump. The data is inverted in the line signal, with $0 = +8V$ and $1 = -8V$, and appears in reverse order on the scope, since the bits are transmitted with bit 0 first, but are written with bit 0 last. The number 5 is represented by the code 35_H (00110101), but the data shows up as 010101100, the first 0 being the start bit. This code is received in the RXREG in the PIC serial port, the input number calculated by subtracting 30h and the digit output to the display. Only numerical characters give the correct display with this demo program.

The program is outlined in Figure 8.3 and the source code in Program 8.1. Once the USART module has been initialised, a code is transmitted and received by simply writing or reading the port buffer registers. To send, the byte is moved into TXREG, and the program then waits for the corresponding interrupt flag, TXIF, to be set. To receive, the reception enable bit, CREN, is set, and the program waits for the interrupt bit, RCIF, to be set to indicate that a byte has been

(a)

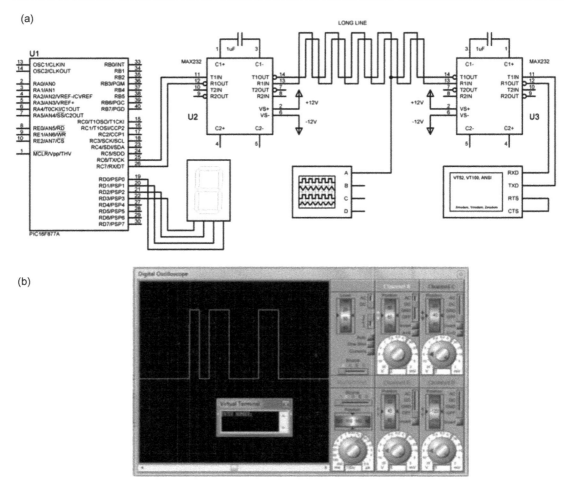

(b)

Figure 8.2
USART test: (a) RS-232 test system and (b) virtual scope and terminal display.

received in RCREG. This is then copied to a suitable storage location for processing. The USART module handles the receive and transmit operations transparently.

RS-422 uses the same data format as RS-232 but increases the range of the system by using differential signalling at ±6V on a separate twisted pair of wires for transmit and receive at up to 10Mb/s. It can operate as a multidrop system, with up to 10 receivers, with a maximum range of 1500m. It is typically used for remote programming and monitoring of programmable controllers in industrial systems. RS-485 also uses differential signaling with similar capabilities, but in a multipoint system which can handle two way communication between up to 32 nodes.

> **USART2**
> *Program to demonstrate USART operation by outputting a*
> *fixed message to a simulated terminal, reading numerical*
> *input from it, displaying it in BCD, and sending it back to the*
> *terminal.*
>
> INITIALISE
>   Port D: BCD display outputs
>   USART: 8 bits, asynchronous mode
>   9600 baud (4MHz clock)
>   Enable
>
> MAIN
>   **Write message from ASCII table to terminal**
>   REPEAT
>   **Read input, display and echo**
>   ALWAYS
>
> SUBROUTINES
>   **Write message from ASCII table to terminal**
>   REPEAT
>   Get character from table
>   Output to terminal
>   UNTIL all done
>
>   **Read input, display and echo**
>   Get input character
>   Convert to BCD and display
>   Echo character back to terminal

Figure 8.3
USART test program outline.

8.2 SPI Bus

SPI is a synchronous local link designed to allow the PIC to communicate with peripheral chips or other MCUs on the same board, or within a self-contained system, at TTL levels. It has a separate clock signal that synchronises the send and receive operations and hardware peripheral selection (Figure 8.4). Eight-bit data is clocked in and out of the SPI shift register by a set of eight clock pulses, which are either internally generated or received at the clock input from the system master. No start and stop bits are necessary, and it is faster than the USART, operating at 5MHz if the MCU clock is 20MHz.

One processor must act as a master, generating the clock. Other devices act as slaves, using the master clock for sending and receiving. The SPI signals are listed below, with the 16F877A pin allocations:

- *Serial Clock (SCK) (RC3)*
- *Serial Data In (SDI) (RC4)*

```
;;;;;;;;;;;;;;;;;;;;;;;;;;;;;;;;;;;;;;;;;;;;;;;;;;;;;;;;;;;;;;;
;
;          USART2.ASM          MPB      23-01-13
;.............................................................
;
;          Test RS232 communications using the
;          USART Asynchronous Transmit and Receive
;
;          The Proteus Virtual Terminal allows ASCII characters
;          to be displayed, and generated from the computer keys.
;          The program outputs a fixed message to the display
;          from a table, and then displays numbers input from the
;          terminal on a BCD 7-segment LED display.
;
;;;;;;;;;;;;;;;;;;;;;;;;;;;;;;;;;;;;;;;;;;;;;;;;;;;;;;;;;;;;;;;

          PROCESSOR 16F877               ; define MPU
          __CONFIG 0x3731               ; XT clock (4MHz)

          INCLUDE "P16F877A.INC"        ; Standard register labels

          Point     EQU      020
          Inchar    EQU      021

; Initialise ;;;;;;;;;;;;;;;;;;;;;;;;;;;;;;;;;;;;;;;;;;;;;;;;;;;;;

          CODE      0                   ; Place machine code
          NOP                           ; Required for ICD mode

          BANKSEL   TRISD               ; Select bank 1
          CLRF      TRISD               ; Display outputs
          BCF       TXSTA,TX9           ; Select 8-bit transmission
          BCF       TXSTA,TXEN          ; Disable transmission initially
          BCF       TXSTA,SYNC          ; Asynchronous mode
          BSF       TXSTA,BRGH          ; High baud rate

          MOVLW     D'25'               ; Baud rate counter value ..
          MOVWF     SPBRG               ; .. for 9600 baud, 4MHz clock
          BSF       TXSTA,TXEN          ; Enable transmission

          BANKSEL   RCSTA               ; Select bank 0
          BSF       RCSTA,SPEN          ; Enable serial port


; MAIN LOOP ;;;;;;;;;;;;;;;;;;;;;;;;;;;;;;;;;;;;;;;;;;;;;;;;;;;;;

          CALL      write               ; Display message on terminal
readin    CALL      read                ; Get number from terminal
          GOTO      readin              ; Keep reading until reset

; SUBROUTINES ;;;;;;;;;;;;;;;;;;;;;;;;;;;;;;;;;;;;;;;;;;;;;;;;;;;;

; Write message to terminal....................................

write     CLRF      Point               ; Table pointer = 0
next      MOVF      Point,W             ; Load table pointer
          CALL      mestab              ; Get character
          CALL      sencom              ; Output to terminal
          INCF      Point               ; Point to next
          MOVLW     D'14'               ; Number of characters + 1
          SUBWF     Point,W             ; Check pointer
          BTFSS     STATUS,Z            ; Last character done?
          GOTO      next                ; No - next
          RETURN                        ; All done
```

Program 8.1

USART serial communication.

```
; Read input numbers from terminal...........................

read     BSF      RCSTA,CREN          ; Enable reception
waitin   BTFSS    PIR1,RCIF           ; Character received?
         GOTO     waitin              ; no - wait

         MOVF     RCREG,W             ; get input character
         MOVWF    Inchar              ; store input character
         MOVLW    030                 ; ASCII number offset
         SUBWF    Inchar,W            ; Calculate number
         MOVWF    PORTD               ; display it
         RETURN                       ; done


; Transmit a character ........................................

sencom   MOVWF    TXREG               ; load transmit register
waitot   BTFSS    PIR1,TXIF ; sent?
         GOTO     waitot              ; no
         RETURN                       ; yes


; Table of message characters.................................

mestab   ADDWF    PCL                 ; Modify program counter
         RETLW    'E'                 ; Point = 0
         RETLW    'N'                 ; Point = 1
         RETLW    'T'                 ; Point = 2
         RETLW    'E'                 ; Point = 3
         RETLW    'R'                 ; Point = 4
         RETLW    ' '                 ; Point = 5
         RETLW    'N'                 ; Point = 6
         RETLW    'U'                 ; Point = 7
         RETLW    'M';                ; Point = 8
         RETLW    'B'                 ; Point = 9
         RETLW    'E'                 ; Point = 10
         RETLW    'R'                 ; Point = 11
         RETLW    ':'                 ; Point = 12
         RETLW    ' '                 ; Point = 13


         END      ;;;;;;;;;;;;;;;;;;;;;;;;;;;;;;;;;;;;;;;;;;;;;;;;;;;;;;;
```

Program 8.1
(Continued)

- *Serial Data Out (SDO) (RC5)*
- *Slave Select (!SS) (RA5)*

The test system (Figure 8.5) consists of three processors, a master, a slave transmitter and a slave receiver (VSM project SERSPI2). The slave transmitter has a BCD switch connected to Port D, which generates the test data. The binary code 0–9 is read in to the transmitter and sent to the master controller via the SPI link. The send is enabled via the !SS input of the slave by an active low signal from the master, pin RC0. The clock is supplied by the master to shift the data out of SSPSR register in the slave and into SSPSR in the master. The master then retransmits the same data to the slave receiver by the same

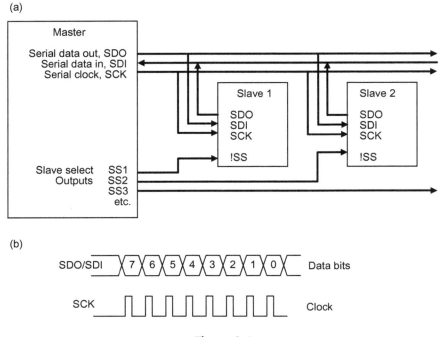

Figure 8.4
SPI operation: (a) SPI connections and (b) SPI signals.

method. The slave receiver does not need a slave select input to enable reception, as it is already initialised to expect only SPI data input.

Each chip needs its own program to operate the SPI port. The three programs are listed as shown in Program 8.2 (master, slave transmitter and slave receiver). All chips run at 4MHz, giving an SPI clock period of 1μs. The SPI outputs (SCK and SDO) need to be set as outputs in each MCU. The operation is controlled by SFRs SSPSTAT (Synchronous Serial Port Status register, address 94h) and SSPCON (Synchronous Serial Port Control register, address 14h).

In the master program, the default operating mode is selected by clearing all bits in both of these control registers. SSPSTAT bits mainly provide signal timing options. The low nibble of SSPCON sets the overall mode, master or slave (0000 = master). In the slave transmitter, the bits are set to 0100 (slave mode, slave select enabled). In the slave receiver, they are set to 0101 (slave mode, slave select disabled). Bit SSPEN enables the SPI module prior to use in all three processors.

The slave transmitter initiates the data transfer by simply writing the data read in from the switches to the SSPBUF (Synchronous Serial Port Buffer). When clock pulses are

(a)

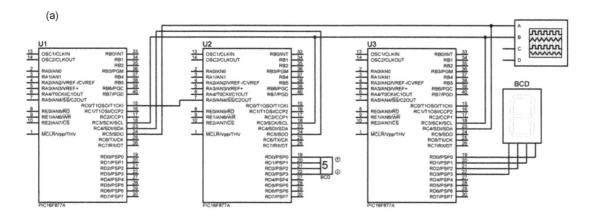

(b)

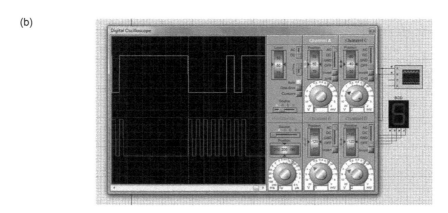

Figure 8.5
SPI test system: (a) SPI test hardware and (b) SPI test signals.

input at SCK from the master, the bits in SSPBUF are shifted out on the falling edge of each pulse. The slave transmitter program waits for the SSPIF (SSP Interrupt Flag) to be set to indicate that the data has been sent. In the master, the clock is started by a dummy write to the SSPBUF register. The master program then waits for the interrupt flag to indicate that the data has been received.

The test data is then rewritten to SSPBUF, which initiates the data output cycle. The master program again waits for SSPIF to indicate that the master transmission cycle is complete. This transmission is picked up by the slave receiver under control of the master clock. It simply waits for the interrupt flag to indicate that a data byte has been received and copies it to the BCD 7-segment display, to indicate to the user a successful data cycle.

(a)
```
;;;;;;;;;;;;;;;;;;;;;;;;;;;;;;;;;;;;;;;;;;;;;;;;;;;;;;;;;;;;;;;;;
;
;         SPIM2.ASM          MPB      24-01-13
;...............................................................
;
;         SPI Master program
;
;         Outputs clock to slave transmitter, receives BCD data
;         and sends it to slave receiver for display
;
;;;;;;;;;;;;;;;;;;;;;;;;;;;;;;;;;;;;;;;;;;;;;;;;;;;;;;;;;;;;;;;;;

              PROCESSOR 16F877           ; define MPU
              __CONFIG 0x3731            ; XT clock (4MHz)
              INCLUDE "P16F877.INC"      ; Standard register labels

Store    EQU      020

; Initialise ;;;;;;;;;;;;;;;;;;;;;;;;;;;;;;;;;;;;;;;;;;;;;;;;;;;;;

              CODE     0                 ; Place machine code
              NOP                        ; Required for ICD mode

              BANKSEL  TRISC
              BCF      TRISC,5           ; Serial data (SDO) output
              BCF      TRISC,3           ; Serial clock (SCK) output
              BCF      TRISC,0           ; Slave select (SS) output
              CLRW     SSPSTAT           ; Default clock timing

              BANKSEL  PORTD
              BSF      PORTC,0           ; Reset slave transmitter
              CLRF     SSPCON            ; SPI master mode, clock =
1MHz
              BSF      SSPCON,SSPEN      ; Enable SPI mode

; Main loop ;;;;;;;;;;;;;;;;;;;;;;;;;;;;;;;;;;;;;;;;;;;;;;;;;;;;;

again    BCF      PORTC,0           ; Enable slave transmitter
              MOVWF    SSPBUF            ; Rewrite buffer to start
clock
waitin   BTFSS    PIR1,SSPIF        ; wait for SPI interrupt
              GOTO     waitin            ; for data recieved

              BCF      PIR1,SSPIF        ; clear interrupt flag
              MOVF     SSPBUF,W          ; read SPI buffer
              MOVWF    Store             ; store BCD value
              BSF      PORTC,0           ; Disable slave transmitter
              MOVWF    SSPBUF            ; Reload SPI buffer

waits    BTFSS    PIR1,SSPIF        ; wait for SPI interrupt
              GOTO     waits             ; for data sent
              BCF      PIR1,SSPIF        ; clear interrupt flag
              GOTO     again             ; repeat main loop

              END
              ;;;;;;;;;;;;;;;;;;;;;;;;;;;;;;;;;;;;;;;;;;;;;;;;;
```

(b)
```
;;;;;;;;;;;;;;;;;;;;;;;;;;;;;;;;;;;;;;;;;;;;;;;;;;;;;;;;;;;;;;;;;
;
;         SPIT2.ASM          MPB      24-01-13
;...............................................................
;
;         SPI Slave Transmitter program
;         Waits for !SS and transmits switch BCD data
;
;;;;;;;;;;;;;;;;;;;;;;;;;;;;;;;;;;;;;;;;;;;;;;;;;;;;;;;;;;;;;;;;;

              PROCESSOR 16F877           ; define MPU
              __CONFIG 0x3731            ; XT clock (4MHz)
              INCLUDE "P16F877A.INC"     ; Standard register labels
```

Program 8.2
SPI test system source code: (a) SPI master program, (b) SPI slave transmit program
and (c) SPI slave receive program.

```
; Initialise ;;;;;;;;;;;;;;;;;;;;;;;;;;;;;;;;;;;;;;;;;;;;;;;;;;;;;;;;

            CODE     0                    ; Place machine code
            NOP                           ; Required for ICD mode

            BANKSEL  TRISC
            BCF      TRISC,5              ; Serial data output
            CLRW     SSPSTAT              ; Default clock timing

            BANKSEL  PORTD
            MOVLW    B'00000100'          ; SPI slave mode with SS
            MOVWF    SSPCON               ; SPI clock = 1MHz
            BSF      SSPCON,SSPEN         ; Enable SPI mode

; MAIN LOOP ;;;;;;;;;;;;;;;;;;;;;;;;;;;;;;;;;;;;;;;;;;;;;;;;;;;;;;;;;;

start       MOVF     PORTD,W              ; Read BCD switch
            MOVWF    SSPBUF               ; Write SPI buffer
wait        BTFSS    PIR1,SSPIF           ; wait for SPI interrupt
            GOTO     wait
            BCF      PIR1,SSPIF           ; clear interrupt flag
            GOTO     start                ; repeat main loop

            END
            ;;;;;;;;;;;;;;;;;;;;;;;;;;;;;;;;;;;;;;;;;;;;;;;;;;;;;;
```

(c)
```
;;;;;;;;;;;;;;;;;;;;;;;;;;;;;;;;;;;;;;;;;;;;;;;;;;;;;;;;;;;;;;;;;;;;;;

            PROCESSOR 16F877              ; define MPU
            __CONFIG 0x3731              ; XT clock (4MHz)
            INCLUDE "P16F877A.INC"        ; Standard register labels

; Initialise ;;;;;;;;;;;;;;;;;;;;;;;;;;;;;;;;;;;;;;;;;;;;;;;;;;;;;;;;;;

            CODE     0                    ; Place machine code
            NOP                           ; Required for ICD mode

            BANKSEL  TRISD
            CLRF     TRISD                ; Display outputs
            CLRF     SSPSTAT              ; Default clock timing

            BANKSEL  PORTD
            MOVLW    B'00000101'          ; SPI slave mode, SS
disabled
            MOVWF    SSPCON               ; SPI clock = 1MHz

; MAIN LOOP ;;;;;;;;;;;;;;;;;;;;;;;;;;;;;;;;;;;;;;;;;;;;;;;;;;;;;;;;;;

            BSF      SSPCON,SSPEN         ; Enable SPI mode
wait        BTFSS    PIR1,SSPIF           ; wait for SPI interrupt
            GOTO     wait

            MOVF     SSPBUF,W             ; get data
            MOVWF    PORTD                ; and display
            BCF      PIR1,SSPIF           ; clear interrupt flag
            GOTO     wait                 ; repeat main loop

            END
            ;;;;;;;;;;;;;;;;;;;;;;;;;;;;;;;;;;;;;;;;;;;;;;;;;;;;;;
```

Program 8.2
(Continued)

8.3 I²C Bus

I²C is a more versatile system level serial data transfer method. It only needs two bus connected signals: clock (SCL) and data (SDA) lines (Figure 8.6). These allow a master controller to be connected to up to 1023 other slave devices. These can include other MCUs, memory devices, A/D converters and so on. The example used here interfaces the PIC to external serial EEPROM memory. This expands system non-volatile data storage, as demonstrated in the general purpose PIC base module described in Chapter 10. The memory device used here is a Microchip 24AA128, which stores 16 kilobytes of data 128kbits. The system simulation is shown in Figure 8.7 (VSM project SERI2C2).

It can be seen that the signal lines are pulled up to 5V so that any one of the devices connected to it can control the line by pulling it down. This allows slaves to acknowledge operations initiated by the master. Each transmitted byte has a low start bit with an 8-bit address or data byte (MSB first) following, terminated by a low acknowledge bit from the slave. Each bit is accompanied by a clock pulse in the same way as SPI. The clock frequency is programmed by preloading a baud rate generator with a suitable value, giving speeds of up to 1MHz.

8.3.1 Control, Address and Data Format

The system uses a software addressing system, where the external device and a location within it can be selected in the same way as in a conventional address decoding system.

(a)

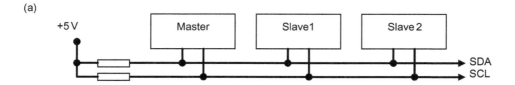

(b)

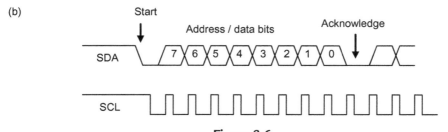

Figure 8.6
System operation: (a) connections and (b) signals.

The chip used here has three hardware address selection pins, which allow it to be allocated one of eight 3-bit addresses within the system. Thus a total of $16 \times 8 = 128$kb of memory can be installed. The location required within the chip is selected by a 14-bit address supplied by the master controller as part of the access cycle. The chip is differentiated from other I^2C devices on the bus by a 4-bit code (1010) in the control word. The data sheet for the 24AA128 memory chip details the required signalling.

The format of the data blocks for write and read are given in Table 8.1. To write a byte to memory, the control code is sent first. This alerts the memory that a write is coming, when the control code and chip select bits match its own control code and hardware address (set up on chip address inputs). The control code is 1010, the chip address is 000 and the read/not write bit is 0, to indicate a write. This chip select byte is followed by the location address to write to, which is 14 bits for this device. The high address bits are sent first, with bits A14 and A15 having no effect (X = don't care). The address is followed by the eight data bits to be written to the location specified.

The read sequence is 5 bytes in total. The first three are the same as for the write, where the chip is selected and location address is written to the address latches within the memory chip. A control byte to request a read operation is then sent, and the data returned from the selected location by the slave device.

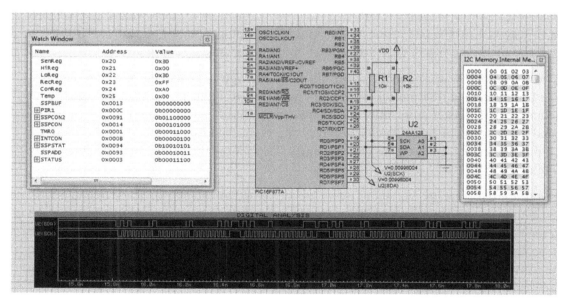

Figure 8.7
I^2C test system.

Table 8.1: I²C Data Format.

Byte	Function	Bits	Description
		Write Byte	
1	Control byte	1 0 1 0 CS2 CS1 CS0 0	Control code, chip select address, WRITE
2	Address high byte	X X A13 A12 A11 A10 A9 A8	Memory page select
3	Address low byte	A7 A6 A5 A4 A3 A2 A1 A0	Memory location select
4	Data	D7 D6 D5 D4 D3 D2 D1 D0	Data
		Read Byte	
1	Control byte	1 0 1 0 CS2 CS1 CS0 0	Control code, chip select address, WRITE
2	Address high byte	X X A13 A12 A11 A10 A9 A8	Memory page select
3	Address low byte	A7 A6 A5 A4 A3 A2 A1 A0	Memory location select
4	Control byte	1 0 1 0 CS2 CS1 CS0 1	Control code, chip select address, READ
5	Data	D7 D6 D5 D4 D3 D2 D1 D0	Data

8.3.2 Transmission Control

The clock and data lines are high when idle. The write and read sequences are initiated by a start bit sequence generated by the master controller (24AA128 data sheet, Figure 4.1). The transmission starts when the data line goes low; the clock then starts and an address or data bit output during the clock high period, which is latched into the slave receive shift register on the falling clock edge. After the eighth bit, the master (MCU) releases the data line, to allow the slave (EEPROM) to hold the line low to acknowledge the bits have been received. At the end of the next (ninth) clock high period, the slave releases the data line and the master can then transmit the first bit of the next byte (Figure 4.2).

In the memory write sequence, when the acknowledge is generated after the data byte has been received, the master stops and both lines go high. In the memory read sequence, the master stops after the address write has been sent and restarts in order to send a read control byte. It will then read the eight data bits returned by the memory chip but does not generate an acknowledge. The master then stops and the lines go idle again.

The test program reads and writes every location (16,384 addresses). The maximum write cycle time specified is 5ms (16,384 × 0.005 = 82s). It therefore takes a considerable time to complete this test (serial memory access is inherently slower than parallel). If the memory is being accessed sequentially, as is frequently the case, the overall read or write time can be reduced by using the page read and address auto-increment features of the chip, which are explained in the EEPROM data sheet.

In the test program (Program 8.3), the control operations are broken down as much as possible so that each step can be identified. The program outline (Figure 8.8) shows that

```
;;;;;;;;;;;;;;;;;;;;;;;;;;;;;;;;;;;;;;;;;;;;;;;;;;;;;;;;
;
;       I2C2.ASM          MPB        24-01-13
;
;       Test program for 24AA128 I2C 16k byte serial
;       memory with P16F877A (4MHz XT)
;       Demonstrates single byte write and read
;       with 10-bit address.
;
;       Write data from                   0x20
;       High address                      0x21
;       Low address                       0x22
;       Read data back to       0x23
;
;;;;;;;;;;;;;;;;;;;;;;;;;;;;;;;;;;;;;;;;;;;;;;;;;;;;;;;;

        PROCESSOR 16F877A
        __CONFIG 3FF1
        INCLUDE "P16F877A.INC"

; Data, address & control registers ;;;;;;;;;;;;;;;;;;;;

SenReg  EQU     0x20              ; Send data store
HiReg   EQU     0x21              ; High address store
LoReg   EQU     0x22              ; Low address store
RecReg  EQU     0x23              ; Receive data store
ConReg  EQU     0x24              ; Control byte store
Temp    EQU     0x25              ; Scratchpad location

;;;;;;;;;;;;;;;;;;;;;;;;;;;;;;;;;;;;;;;;;;;;;;;;;;;;;;;;

        CODE    0                 ; Program start address

        NOP                       ; ICPD location
        CLRF    SenReg            ; Zeroise data
        CLRF    HiReg             ; Zeroise high address
        CLRF    LoReg             ; Zeroise low address
        GOTO    begin             ; jump to main program

;----------------------------------------------------------
; SUBROUTINES ;;;;;;;;;;;;;;;;;;;;;;;;;;;;;;;;;;;;;;;;;;;

; Wait for interrupt flag SSPIF for send/recive done ...

wint    NOP                       ; BANKSEL has no address
        BANKSEL PIR1              ; Select bank
        BCF     PIR1,SSPIF        ; reset interrupt flag
win     NOP
        BTFSS   PIR1,SSPIF        ; wait for..
        GOTO    win               ; ..transmit done
        RETURN                    ; done


; Send a byte .................................

send    NOP                       ; Select..
        BANKSEL SSPBUF            ; .. bank
        MOVWF   SSPBUF            ; Send address/data
        CALL    wint              ; Wait until sent
        RETURN                    ; done

;----------------------------------------------------------
; Routines to send start, control, address, data, stop ..
;....................................................

sencon  NOP                       ; GENERATE START BIT
        BANKSEL PIR1
        BCF     PIR1,SSPIF        ; Clear interrupt flag
        BANKSEL SSPCON2          ; select register page
        BSF     SSPCON2,ACKSTAT   ; Set acknowledge flag
        BSF     SSPCON2,SEN       ; Generate start bit
        CALL    wint              ; wait till done
        MOVF    ConReg,W          ; SEND CONTROL BYTE
        CALL    send              ; Memory ID & chip address
        RETURN                    ; done
;....................................................
```

Program 8.3

I^{2}C memory access.

```
senadd    NOP
          BANKSEL   SSPCON              ; SEND ADDRESS BYTES
          MOVF      HiReg,W             ; load address high byte
          CALL      send                ; and send
          MOVF      LoReg,W             ; load address low byte
          CALL      send                ; and send
          RETURN

;.......................................................

sendat    MOVF      SenReg,W            ; Load data
          CALL      send                ; and send
          RETURN                        ; done

;.......................................................

senstop   NOP
          BANKSEL   SSPCON2             ; GENERATE STOP BIT
          BSF       SSPCON2,PEN         ; Generate stop bit
          CALL      wint                ; wait till done
          RETURN                        ; done

;.......................................................

senack    NOP
          BANKSEL   SSPCON2
          BSF       SSPCON2,ACKDT       ; Set ack. bit high
          BSF       SSPCON2,ACKEN       ; Initiate ack.sequence
          CALL      wint                ; Wait for ack. done
          RETURN                        ; done

;.......................................................

wait      NOP
          BANKSEL   TMR0                ; WAIT FOR WRITE DONE
          MOVLW     d'156'              ; Set starting value
          MOVWF     TMR0                ; and load into timer
          BANKSEL   INTCON              ; 64 x 156us = 10ms
          BCF       INTCON,T0IF         ; Reset timer out flag
wem       BTFSS     INTCON,T0IF         ; Wait 10ms
          GOTO      wem                 ; for timeout
          BANKSEL   TMR0                ; default bank
          RETURN                        ; Byte write done....

;-------------------------------------------------------
; Initialisation sequence ............................

init      NOP
          BANKSEL   SSPCON2             ;
          MOVLW     b'01100000'         ; Set ACKSTAT,ACKDT bits
          MOVWF     SSPCON2             ; Reset SEN,ACK bits
          MOVLW     b'10000000'         ;
          MOVWF     SSPSTAT             ; Speed & signal levels
          MOVLW     0x13                ; Clock = 50kHz
          MOVWF     SSPADD              ; Load baud rate count-1
          BANKSEL   SSPCON              ;
          MOVLW     b'00101000'         ;
          MOVWF     SSPCON              ; Set mode & enable
          BCF       PIR1,SSPIF          ; clear interrupt flag

; Initialise TIMER0 for write delay ..............

          BANKSEL   OPTION_REG          ;
          MOVLW     B'11000101'         ; TIMER0 setup code
          MOVWF     OPTION_REG          ; Internal clock,1/64
          BANKSEL   TMR0
          RETURN

;-------------------------------------------------------
; Write a test byte to given address ...............
```

Program 8.3
(Continued)

```
writeb     MOVLW     0xA0              ; Control byte for WRITE
           MOVWF     ConReg            ;
           CALL      sencon            ; Send control byte
           CALL      senadd            ; Send address bytes
           CALL      sendat            ; Send data byte
           CALL      senstop           ; Send stop bit
           CALL      wait              ; Wait 10ms for write
           RETURN


;---------------------------------------------------------
; Read the byte from given address ..................

readb      MOVLW     0xA0              ; Control byte to WRITE
           MOVWF     ConReg            ; address to memory
           CALL      sencon            ; Send control byte
           CALL      senadd            ; Send address bytes
           CALL      senstop           ; Stop

           MOVLW     0xA1              ; Control byte to READ
           MOVWF     ConReg            ; data from memory
           CALL      sencon            ; Send control byte
           BANKSEL   SSPCON2
           BSF       SSPCON2,RCEN      ; Enable receive mode
war        BTFSS     SSPSTAT,BF        ; Check ...
           GOTO      war               ; for read done
           CALL      senack            ; send NOT acknowledge
           CALL      senstop           ; send stop bit

           MOVF      SSPBUF,W          ; Read receive buffer
           MOVWF     RecReg            ; and store it
           RETURN

; MAIN PROGRAM          ;;;;;;;;;;;;;;;;;;;;;;;;;;;;;;;;;;;;;;;;

begin      CALL      init              ; Initialise for I2C
next       CALL      writeb            ; write the test byte
           CALL      readb             ; and read it back
           INCF      SenReg            ; next data
           INCF      LoReg             ; next location
           BTFSS     STATUS,Z   ; end of memory block?
           GOTO      next              ; no, next location
           INCF      HiReg             ; next block

           MOVF      HiReg,W           ; copy high address byte
           MOVWF     Temp              ; store it
           MOVLW     0x40              ; Last block = 3F
           SUBWF     Temp              ; Compare
           BTFSS     STATUS,Z   ; Finish if block = 40xx
           GOTO      next              ; next memory block..
           SLEEP                       ; .. unless done

           END             ;;;;;;;;;;;;;;;;;;;;;;;;;;;;;;;;;;;;;;;;;;;;
```

Program 8.3
(Continued)

the data write and read operations are carried out one after the other on the same location, but the address is re-sent for the read so that read and write sequences can be used separately in other programs. In real applications, a sequential read or write is more likely; this can be completed more quickly for a sequential data block, especially the read, because the memory has an automatic increment mode for the addressing, so only the first address needs to be sent.

I2C2   *Serial memory access using I2C serial bus*

MAIN
        Initialise
        Loop
                Write a byte to memory
                Read the same byte from memory
                IF end of memory page, increment page number
        Until end of memory

SUBROUTINES
        Initialise
                SSPCON2:   Set Acknowledge flags inactive + status bits inactive
                SSPSTAT: Slew rate control disabled + status bits inactive
                SSPADD: Load with baud rate count value
                SSPCON: Set SSP enable bit, select I2C master mode
                OPTION: TMR0: internal clock, divide by 64

        Write a byte to memory
                Generate start condition
                Send write control byte for memory chip
                Send address bytes to memory chip
                Send data byte to memory chip
                Send stop bit to memory chip
                Wait 10ms using TMR0

        Read the byte from memory
                Generate start condition
                Send write control byte for memory chip
                Send address bytes to memory chip
                Send read control byte for memory chip
                Wait until data received
                Send acknowledge and stop bits
                Store received data byte

Figure 8.8
Program outline.

The read and write operations use the same subroutines to generate the transmission control operations, which are:

- *Generate start bit*
- *Send control byte*
- *Send address bytes*
- *Send data byte*
- *Generate stop bit*
- *Generate acknowledge*
- *Wait 10ms for write completion*

Table 8.2: I²C Registers.

Register	Address	Bit/s	Bit Name	Active	Function
SSPBUF (Data)	13h	all		Data/address	SSP send/receive buffer register
SSPCON (Control)	14h	3-0	SSPMx	1000	SSP mode select bits
		5	SSPEN	1	SSP enable bit
SSPCON2 (Control)	91h	0	SEN	1	Initiate start of transmission
		2	PEN	1	Initiate stop condition
		3	RCEN	1	Receive mode enable bit
		4	ACKEN	1	Initiate acknowledge sequence
		5	ACKDT	1	Acknowledge data bit setting
		6	ACKSTAT	0	Acknowledge received from slave
SSPSTAT (Status)	94h	0	BF	1	SSP buffer is full
		2	R/W	1	Read/write bit − transmit in progress
		6	CKE	0	I²C clock mode
SSPADD (Preload)	93h	All		0×13	Baud rate count
PIR1	0Ch	3	SSPIF	1	SSP interrupt flag

Table 8.2 summarises the registers and bits used in the test program (see the master mode timing diagram in the 16F877A data sheet, Figure 9.14).

The shift register used to send and receive the data bits is not directly accessible. The buffer register SSPBUF holds the data until the shift register is ready to send it (transmit mode) or receives it when the shift in is finished in receive mode. The send operation is triggered by setting the Send Enable (SEN) bit, and the Buffer Full (BF) flag indicates that the data has been loaded. The interrupt flag (SSPIF) is automatically set to indicate start of transmission and must be cleared in software if necessary. The Acknowledge Status (ACKSTAT) bit is cleared by an acknowledge from the slave, to indicate that the byte has been received. SSPIF is then set again to indicate the end of the byte transmission, and the buffer can then be written with the next byte.

If data is to be received by the master, the read/write bit is set in the control word, and the receive mode enabled by setting the RCEN bit. The BF flag is set when the data has been received, and the buffer must be read (unloaded) before another data byte is received, or sent. Full details are provided for all I²C transmit and receive modes in the PIC data sheet. The EEPROM data sheet explains the requirements for that particular peripheral.

As can be seen, the software implementation of I²C is quite complicated, so SPI is recommended for simple systems. I²C really needs to be supported by ready-made routines, but if these are available, the hardware is simpler. The routines created for this demo were converted to a support file (sermem2.inc) format and included in the 'Base' system program in Chapter 10 to demonstrate this idea. Programming an I²C system in C is simpler, if suitable library routines are available, as described in 'Programming 8-bit PIC Microcontrollers in C' by the author.

8.4 Network Links

There are more complex communication options available in microcontroller systems, which may be built into the more powerful PICs or require external hardware for implementation. They will be outlined here so that they can be considered as options where circumstances demand, e.g. a wireless link is needed, but they generally will require further research by the reader into their implementation.

CAN and LIN busses are designed to allow MCUs to communicate in hostile environments such as motor vehicles and industrial systems. USB and Internet protocols are standard in computer systems, but only available in high-end PICs that are usually programmed in C and can accommodate the complex protocol stacks required. Wireless links need specific transmission and receiver hardware.

The more complex protocols typically use multi-byte data transmission frames containing addressing, data and error checking. A suitable hardware implementation (physical layer) must allow reliable transmission in the chosen medium, supported by a software system (protocol stack) which organises the data for transmission and reception. For each protocol, a set of C library functions or RTOS modules is usually available to support implementation.

8.4.1 CAN Bus

The Controller Area Network (CAN) bus protocol was developed primarily for communication between sub-systems in motor vehicles when the use of microcontrollers in engine control, window motors, airbags, anti-lock braking and so on became established. A two-wire bus connects all ECUs (electronic control units) using open-collector (wired-OR) outputs. A current loop through load resistors at the ends of the bus generates a reversible voltage at the terminals corresponding to logic 1 and 0 (Figure 8.9).

These is no master controller in CAN bus. Any ECU can start transmitting and take control of the bus. Each transmission starts with an 11-bit message identity code (MID) and, if there is a conflict (simultaneous transmissions), the message with the lowest value MID will take priority. The message can contain up to eight data bytes and is terminated with a cyclic redundancy check (CRC) error code that can detect multiple errors. The CRC is implemented by dividing the data block binary number by a known value and transmitting the remainder as the check code. The calculation is repeated at the receiver and an error is detected if the results do not match.

As well as the main data frame, a 'remote' frame requests data from a particular node. An 'error' frame can also be issued by any node and an 'overload' frame simply inserts a delay between frames. An extended MID format is also available. The bus is designed to work up to 1Mbit/s, but lower bit rates allow longer connections. There is no one standard

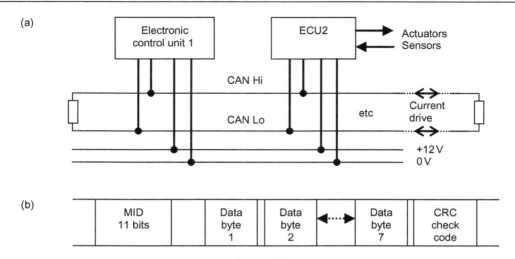

(a)

(b)

Figure 8.9
CAN bus: (a) system connections and (b) frame format.

connector, but a screened twisted pair of wires is recommended for the differential signal leads. Including two power supply connections, there are four terminals on each ECU.

CAN is used extensively for high-speed embedded applications where noise immunity and robustness are important. CAN protocol supports speeds up to 1Mbps and is highly fault-tolerant, making it ideal for safety critical applications. Microchip offers a line of products for embedded applications using the CAN protocol, including digital signal controllers with integrated CAN, standalone CAN controllers, I/O expanders and CAN transceivers.

8.4.2 LIN Bus

The Local Interconnect Network (LIN) bus (Figure 8.10) was devised as a lower cost alternative to the CAN bus. It works at a lower bit rate (up to 9.6 kbits/s) via the USART interface of the microcontroller, with a master controller and a limited number of slaves (up to 16). There is a single data line, pull-ups on each slave to the +12V supply, with active low data generated from single open-collector transceiver pins. The data is sent as RS-232 bytes with single stop and start bits.

A synchronisation byte is sent first, which allows the slaves to adjust their baud rate to match if necessary. This is followed by a frame destination identifier, consisting of up to eight data bytes and checksum error code (multi-byte parity check). There are a set of defined frame types which support master—slave communication, and the master controlled

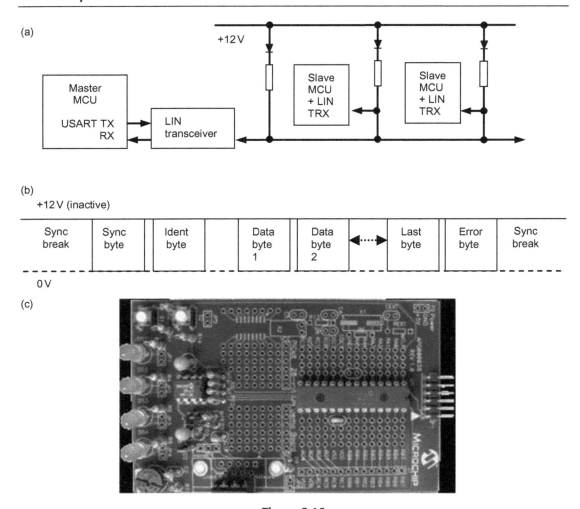

Figure 8.10
LIN controller network: (a) system connections, (b) signal format and (c) microchip LIN development board.

protocol ensures that messages are received within a specified time, which is a priority in all real-time systems.

LIN can be implemented on any PIC microcontroller (MCU) with a USART interface (nearly all). Microchip offers a physical layer interface, data link layer implementation, LIN drivers and a variety of development resources. The PICkit 28-pin LIN demo board (Figure 8.11(c)) is currently available for application development. It includes a 28-pin socket which supports various PIC16F devices, includes an LIN transceiver and prototype area with indicator LEDs and buttons.

LIN is a flexible protocol which allows the system designer a range of options to adapt it to the particular system in use and is non-proprietary. It provides an economical communication network for switched, smart sensor and actuator applications within low-end motor vehicles where the bandwidth and versatility of CAN is not necessary, or too expensive.

8.4.3 USB

A USB port is now incorporated in some of the higher performance PICs and, at the time of writing, in a small number of 16F devices. It provides a high-speed connection with PCs and other standard peripherals such as plug-in external flash ROM. It is often used as a one-to-one link but is designed to operate with a single host controlling a maximum of 127 devices in a branching network.

A logical point-to-point connection is established with each device, known as a pipe and endpoint. When initially connected, the host will establish the endpoint function and assign it a 7-bit address. The host has a set of drivers for each class of device. In the PC host, the human interface device class includes keyboard and mouse drivers, the printer class has various drivers for different types of printer, and the mass storage class provides drivers for memory sticks, audio players and digital cameras. If necessary, these are downloaded from the manufacturer. In a microcontroller, with its limited memory, only the necessary drivers will be installed as library files, and the number of endpoints will usually be strictly limited.

The USB cable has four wires, with a twisted signal pair D+ and D− operating at TTL levels, creating high and low differential voltages between the inputs at the receiver. A data 0 is represented by the voltage changing, a data 1 by the voltage remaining unchanged. The initial packet synchronisation signal consists of a sequence of full speed alternating voltages. In order that a long sequence of 1s, where the line voltage remains unchanged, is not mistaken for an error, a 0 transition is inserted after six 1s to maintain synchronisation. This is known as bit stuffing. It is removed from the data in the receiver.

In addition, there is a pair of 5V supply pins in the connector, which can power a peripheral if the current requirement is less than 500mA. There are two standard forms of USB four-pin connectors: type A is flat and B has a square profile, with miniature options.

The USB protocol is very much like a network in operation (see below), with a complex firmware stack and frame structure supporting data transmission. The data is transmitted in packets, of which there are three types: token, data and handshake. Each packet starts with a packet identifier (PID) byte which specifies the type of packet. The token packet is generated by the host and contains the target device number and a token which instructs the target device to receive or send data, or other control function. The device must respond with a handshaking packet to confirm that it is ready (or not). The host only generates an acknowledge token to confirm a successful transaction.

The data packet consists of a PID byte followed by up to 1023 bytes of data. In USB 2.0, each packet is transmitted within a 125μs time frame, with each defined by the host by a start of frame token. USB 2.0 thus operates at up to 480 Mbits/s.

The 16(L)F145X is one of the first 16 series chips to incorporate USB. A small microcontroller such as this is likely to be the slave device under the control of a PC host. It will support control, interrupt, isochronous and bulk transfers at rates up to 12 Mbits/s. The USB interface consists of the bus transceivers, serial interface engine (SIE), 512 bytes of dedicated dual access RAM and a 3.3V internal supply.

Microchip support for USB applications includes MPLAB tools for all USB PIC MCUs, peripheral applications for the 8-bit PIC16F, PIC18F family, and embedded host applications for the 16-bit PIC24F, PIC24E and dsPIC33E and 32-bit PIC32 families. Designers can use Microchip's free USB stacks, including interface drivers and file system utilities provided in source code form. Additional software support includes C and RTOS development environments.

8.4.4 Synchronous Data

In more complex networks, data tends to be transmitted in blocks, within a data frame that includes synchronisation and addressing codes at the start and error checking at the end. This produces faster, more reliable communication over longer distances, such as the Internet. The data is self-clocking, in that a signal transition occurs in each bit period. The general process for generating synchronous data is shown in Figure 8.11.

The sample data in the second row is combined with the clock signal using an XOR logic operation so that each bit period contains a transition, with a 0 represented by a high to low transition, and 1 by low to high. This process is known as Manchester encoding. The data can be recovered at the receiver by applying the same simple logical operation with a local clock which has been synchronised with the preamble of the received data frame.

The data is usually transmitted as a differential signal, as seen in the last row of the diagram. The advantages of a current loop in signal transmission have already been demonstrated in Section 6.7.2, and a differential current drive is nearly always preferred in long wired links and networks. The data is therefore typically transmitted as a sequence of long and short differential pulses on a twisted pair of conductors. The twisting ensures that any incident interference signal will affect the conductors in an equal and opposite sense and therefore will cancel out. The cable can also be shielded by a grounded foil layer. A differential amplifier and comparator circuit will receive the signal and convert it back to digital form.

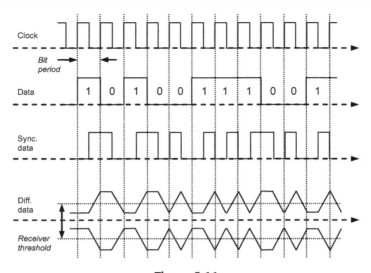

Figure 8.11
Synchronous data signal encoding and transmission.

8.4.5 Ethernet

Ethernet is the most common implementation of the local area network (LAN) for general purpose use and is now offered as a communication interface in 18F PIC microcontrollers. Ethernet operates at up 10Mbits/s in PIC-based systems, transmitting data in packets which have to be constructed for transmission via a multilevel protocol stack. Microchip also supplies a separate Ethernet controller which interfaces with the PIC SPI port.

In general, Ethernet data signal is transmitted as a differential signal in a twisted pair cable, usually terminated with an RJ45 type 8-pin connector. There is a separate pair for signals in and out. A basic cabling standard is designated 10base-T, meaning 10 Mbits/s, baseband, twisted pair. Standard networks use hubs and switches to connect all nodes to the network server, which acts as master controller, typically operating at 100 Mbits/s currently. Data rates up to 100 Gbits/s using frequency division multiplexing (FDM) are possible in optical and other media.

Each Ethernet node has a unique 48-bit MAC (media access control) address. The Ethernet data frame has the following basic elements:

- *Preamble (synchronisation sequence) 8 bytes*
- *Source MAC address 6 bytes*
- *Destination MAC address 6 bytes*
- *Frame type/length 2 bytes*

- *Data 46–1500 bytes*
- *CRC frame error check code 4 bytes*

The data frame is constructed around the original data block via a seven-layer protocol stack, where each layer has a specified function in controlling message transmission.

Ethernet works as a CSMA (carrier sense/multiple access) system, which means that a node can start transmitting at any time. Other nodes then wait until the network goes quiet again. However, since it takes a significant time for the transmission to be detected by all nodes, especially on a large network, two nodes may start at the same time, causing a collision. This is detected as corrupt data at the receiver, and both stop and wait a random time before trying again. This system only causes minor inefficiency in a lightly loaded network but causes significant delays in a more heavily loaded system.

For this reason, Ethernet does not provide a predictable response time, so alternative network types may be used in industrial control systems where a timely response is necessary. Token passing is one option, which only allows each node to transmit at a time when it possesses a virtual token. Another is scheduled communications, where each node has an allocated time slot in which to transmit.

Ethernet is usually integrated with Internet services using TCP/IP (transmission control protocol/Internet protocol). Every node originally had a 32-bit unique IP address (IPv4) which translates into a recognisable URL (universal resource locator) such as www.picmicros.org.uk. This system is now being extended to a 128-bit addressing system (IPv6).

A Microchip development board is currently available that uses the ENC28J60 Ethernet controller and the PIC18F97J60 MCU. Using a freeware TCP/IP stack, a web server can be developed to remotely monitor and control embedded applications over the Internet.

8.5 Wireless Links

Wireless technology is rapidly expanding due to low-cost implementations that are supported by the PIC and other MCUs. They are generally used for short range point-to-point control and local networking. Wireless applications include domestic and consumer electronics (thermostats, smart meters and domestic appliances, remote control, home security), industrial control systems, remote tracking, automotive systems, retail (checkout terminals, wireless price tags) and medical (wireless metering, remote patient monitoring).

8.5.1 Infrared Links

IR links are commonly used in TV remote controls and similar domestic products. The frequency of the light pulses used is slightly below the visible spectrum, so that the receiver is insensitive to ambient light. The signal is confined to line of sight, with some reflection possible from solid surfaces.

An IR interface can be used to operate a PIC application wirelessly using a standard IR remote control. Microchip application note AN946 details the operation of the interface. The signals are illustrated in Figure 8.12. The MCP2122 protocol handler forms an interface between the USART of the PIC and an IR transceiver TFDU 4100, which contains an IR LED and PIN photodiode. This is sensitive to longer wavelength light when reverse biased. These are packaged side by side, with suitable drive and detector circuits.

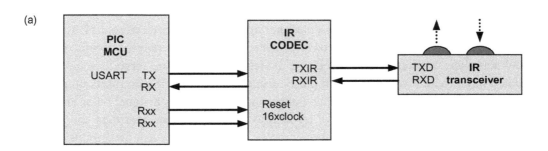

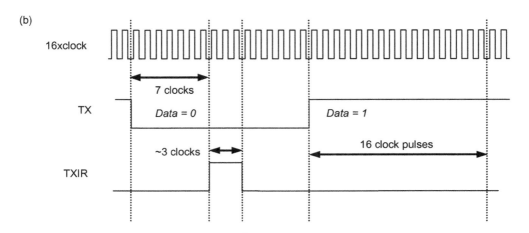

Figure 8.12
IR interface and signals: (a) IR signal connections and (b) IR signals.

The IR codec needs a clock input at 16 times the baud rate, which can be between 9600 and 115,200 bits per second. At the lowest rate, the clock required is therefore 153.6kHz, giving a period of 6.51μs. This can be generated from the PIC hardware timer in PWM mode. The bit period at the output is then 16 clock cycles. There is also a reset input which puts the codec into a low-power sleep mode.

When a low bit is input from the USART port, for example the start bit of an 8-bit word, a pulse output is generated after 7 clock cycles, which lasts about 3 cycles, producing a light pulse on the IR LED. A high bit, on the other hand, produces no pulse. Assuming a single stop bit, the receiver is resynchronised every 10-bit period, so can tolerate some timing error. Since the stop and start bits are low, a start and stop pulse will always be present, even if the data is all 1s. The receiver obviously reverses this process to produce a TTL RS-232 data word. The IrDa protocol built into the PIC codec provides data packaging, communication control and error checking.

8.5.2 Radio Links

Radio-controlled models were an early consumer application of wireless control. A basic system uses hobby grade position servos (see Section 7.9.5) to operate the control surfaces of a model aeroplane or steering and speed in a model car or boat. The servo uses a pulse width modulated input, with a variable length pulse of duration $1-2$ms, to control its output position, repeated every 20ms or so. This pulse can be received as a signal burst via a radio receiver module tuned to a frequency of (typically) 27MHz. The receiver then simply outputs a demodulated pulse to the servo. A more complex control scheme can use groups of pulses, where the number of pulses is translated into a particular command to the target system. The 2.4GHz band is now used extensively, with multichannel spread spectrum operation.

Low-cost radio frequency identification (RFID) tags are used for stock control, animal identification, secure entry systems and so on. A passive tag contains a microcontroller and transmitter, and is powered up via magnetic coils from the ID reader. It then transmits a code with unique identification data. The Microchip MCRF355 RFID chip uses a carrier frequency of 13.56MHz and data rate of 70kbits/s using simple switched modulation.

Remote radio-controlled switches are used in automatic gates, motor vehicles and similar applications. A keyfob contains the MCU and transmitter which produces a secure code at $300-400$MHz. The receiver and decoder then operate a set of relays wired up to the actuators (motors). The Microchip Keeloq™ technology supports this type of system. The PIC MCU can be interfaced to the transceiver via the I^2C bus.

8.5.3 Wi-Fi

Perhaps the most familiar type of radio data link is Wi-Fi, the wireless local area network (WLAN) standard used to provide domestic and business wireless Internet. It operates at around 2.4GHz with multiple channels spaced 5MHz apart, typically providing a 54 Mbits/s maximum bandwidth using frequency division multiplexing (FDM). Baseband network data is modulated (modem = modulator/demodulator) at the channel carrier frequency so that it can be transmitted via a radio link to its destination node. At this frequency, the signal is limited in range, as we know, to about 50m, depending on the nature of any intervening structures.

Wi-Fi implementation is governed by the IEEE (Institution of Electrical and Electronic Engineers) 802.11 standard, which is divided into sub-categories, based on technical and chronological developments, with the most commonly implemented currently in commercial and domestic modems being 802.11g.

Microchip offers a small range of Wi-Fi transceivers, such as the MRF24WB0Mx, supplied as a small surface mount package (21mm × 31mm), with options of a built-in antenna track on the PCB or a miniature antenna socket, interfacing with the SPI port.

A PIC18 MCU, or higher, must be used, because a TCP/IP stack is required to manage the data communications. The transceiver is manufactured with a unique MAC address and operates at up to 2 Mbits/s. Any such Wi-Fi enabled controller can communicate with any node on an LAN or the Internet to support remote monitoring and control.

8.5.4 Zigbee

Zigbee is a network system designed for low-cost monitoring and control, particularly in energy systems and building automation systems. It operates as a mesh network of small, self-contained, low-power, battery-operated modules communicating at 20−900 kbits/s in the industrial, scientific and medical (ISM) bands of 868MHz (Europe), 915MHz (USA) or 2.4GHz. Three device types are specified: the coordinator, router and end device. The network can operate in collision mode (see Ethernet above) or by scheduled transmissions, depending on the type of performance required (Figure 8.13).

The Microchip MRF24J40XX chip is designed to implement the PIC Zigbee interface via the SPI port (SDI, SDO, CS and SCK), operating as an IEEE 802.15.4 compliant 2.4GHz RF transceiver. It integrates the lower network layers (physical link and

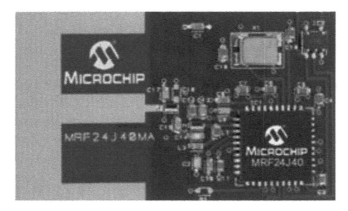

Figure 8.13
Zigbee transceiver.

MAC addressing) in a single wireless personal area network (WPAN) device at 250 or 625 kbits/s. The MRF24J40XX also supports Microchip MiWi protocol, which is its own version of the Zigbee open standard. It is designed to support development and portability of wireless applications between different Microchip RF transceivers and wireless protocols.

8.6 Comparison of Serial Protocols

A summary of types of serial links available is given in Table 8.3. It may assist in selecting the most appropriate for any planned application. Most of them are supported by PIC products that can be identified and supported by online resources. Selection will depend on the required combination of speed, complexity, cost (of hardware and software development) and overall match with the design specification. Microchip currently provides a review of communication products in 'Connectivity Solutions for Embedded Design', covering USB, Ethernet, Wi-Fi, Bluetooth, Zigbee, MiWi, CAN, LIN, IrDA and RS-485 protocols, currently at http://ww1.microchip.com/downloads/en/DeviceDoc/01181j.pdf.

Table 8.3: Comparison of Serial Links.

Link	MCU/ Port	Configuration	Protocol/Type	Signal	Data	Typical Speed (bits/s)	Interface	Error Check	Range	Typical Applications
RS-232	All PICs USART	Synchronous Asynchronous	RS-232	TTL or diff. voltage drive	1 byte	9600–115,200	Voltage driver ±12V	Parity	100m	Simple slow serial port, wireless transceiver
RS-422	USART Asynch	Up to 10 receivers	RS-232 based	Differential transceivers	1 byte	10M max	Voltage driver ±6V	Parity	1km	PLC programmer, small controller network
RS-485	USART Asynch	Up to 256 Transceivers	RS-232 based	Differential transceivers	1 byte	10M max	Voltage driver +12V/−6V	Parity	1km	Industrial control systems
SPI	MSSP	Hardware slave select	Master/slave	TTL on board	1 byte	10M max	2-wire 5V supply	None	1m	Link MCUs and peripherals Interface with RF transceivers
I²C	MSSP	Software slave address	Master/slave	TTL on board	1 byte frame	1M max	2-wire 5V Supply	None.	1m	Memory expansion Sensor interface
CAN	18FXX upwards	ECU network	Peer to peer addressing	Differential transceivers	8 byte frame	1M max	2-wire 12V supply	CRC	10m	High-performance motor vehicle control system
LIN	All PICs RS-232	Master + 16 slaves	Broadcast system	Wired OR transceivers	8 byte frame	10k max	1-wire with 12V pull-up	Checksum	10m	Low-cost motor vehicle control network
Infrared	USART	Master/slave	Point to point	IR transceiver	1 byte	9600–115,200	IR diode & photo-detector	Parity, CRC	10m	TV remote control
Radio Control	PWM	Hand-held 2/4 channel RF	Pulse width modulation	1–2ms	N/A	50 pulses per sec	2.4GHz band carrier	N/A	1km	Model craft remote control

(*Continued*)

Table 8.3: (Continued)

Link	MCU/ Port	Configuration	Protocol/Type	Signal	Data	Typical Speed (bits/s)	Interface	Error Check	Range	Typical Applications
RFID	RS-232	RF transceiver + RFID tag	Simple data frame	Simple pulse modulation	104-bit ID code	70k max	13.56MHz carrier	Checksum	0.1m	Product, customer, etc. Identification tag
Keeloq	Host I²C	Remote tag transmitter	Secure Encryption	Simple PCM	32 bits	N/A	433MHz Carrier	Block cypher	100m	Motor vehicle remote locking
USB	Selected PICs	Master +127 slaves	1 to 1 or star network	Differential transceivers	1023	V1 12M V2 480M	2-wire with 5V supply	CRC	10m	Temporary connection to peripherals, memory, etc.
Zigbee	SPI, 24F upwards	WPAN	Lower level networking	Wireless transceivers	Not specified	250k or 625k	2.4GHz transceiver	CRC	100m	Small wireless network for sensor control and data logging
Ethernet	Selected PICs	Server + LAN	Distributed network	Differential transceivers	1500	10M—1 G	2-pair	CRC	1km	Internet monitoring and control
Wi-Fi	SPI, 18F upwards	RF transceiver network + server	Full TCP/IP	Multichannel FDM	1500	2 M max	2.4GHz transceiver	CRC	100m	Internet monitoring and control

Questions 8

1. Explain why the RS-232 type protocol is described as asynchronous.   (3)
2. Explain why the signal is sent at up to 50V peak to peak on the RS-232 line.   (3)
3. Deduce the TTL RS-232 data bit sequence if the character code for number 9 is sent   (3)
    with even parity.
4. Explain why SPI signals have a more limited transmission range than RS-232.   (3)
5. Describe the function of the signal !SS in the SPI system. How does I^{2}C implement   (3)
    the same function?
6. Explain briefly why a data byte takes longer to send in the I^{2}C system, compared   (3)
    with SPI.
7. Compare briefly the features of CAN and LIN busses for automotive networking.   (3)
8. State the resulting data stream if the byte 10110010 is converted into a synchronous   (3)
    data stream. The first rising clock edge coincides with the rising edge of the first bit,
    and there is one full clock cycle per bit. The result will be 16 bits.
9. Explain briefly why many serial links use differential signals on a twisted pair of wires. (3)
10. Explain briefly why the Ethernet collision system makes it unsuitable for some   (3)
    industrial control networks.

Total (30)

Assignments 8

8.1 RS-232 Output Test

Write a test program for the PIC 16F877A, running at 4MHz. Initialise for 8 data bits, 1 stop bit at 9600 baud and output the same code AAh repeatedly to the RS-232 port. Test its simulation mode. Monitor the output signal on the virtual oscilloscope, and measure the overall time per byte, and the bit period. Compare this with the value specified in the data sheet. Change the data to 81h and show that the display is correct for this data. Change the baud rate to 19,200 and confirm that the display is correct. Compare the bit time with the specification. Try sending ASCII, and confirm that the codes are as shown in the ASCII table. Transfer the application to prototype hardware and confirm that the appropriate output is obtained on an oscilloscope.

8.2 SPI and I^{2}C Debuggers

The SPI and I^{2}C debugger are found in the list of Proteus virtual instruments. They allow these signals to be monitored when the design for a serial communication system is simulated. Use the instruments to monitor the signals in the demonstration systems. Write a full report on the interpretation of the displays obtained, and how to use these devices to aid the development of serial systems.

8.3 Greenhouse Project

A remote temperature monitoring system is required, to measure temperatures wirelessly in a large greenhouse100m long, to identify cold spots. Investigate the available technologies and outline a solution that will allow up to 16 sensors and a master controller to be completely mobile while displaying all the readings.

CHAPTER 9

Input Sensors

Summary
- Digital sensors include switches, opto-detectors and incremental encoders
- Analogue sensors produce a change in voltage, current or resistance
- Sensor characteristics include sensitivity, range, offset, accuracy and error
- Inputs include position, speed, temperature, pressure, light, strain and humidity
- Sensors are resistive, capacitive, inductive, semiconductor or voltaic
- Interface signal conditioning adjusts gain, offset and frequency response
- Integrated sensors have integrated signal conditioning and calibration

A sensor is essentially a device that responds to some environmental variable or event and converts it into an electrical output. It needs to have some characteristic that is sensitive to the environmental variable to be measured or detected. Many materials change their conductivity with temperature, including metals and semiconductors. The semiconductor (PN) junction is sensitive to temperature and light, a characteristic used in many different sensors. Alternatively, a change in capacitance or inductance may be detected. These properties form the basis for most sensors, but the most useful ones are those that can be interfaced to produce a linear, accurate and proportional output over a wide range of measurement.

Sensors measure external variables in control systems, producing a corresponding analogue (current, voltage or impedance) or digital (switched, frequency, PWM or serial data) output. If it is a switched sensor, the output may be a set of changeover contacts, open collector transistor or TTL buffer. If it is an analogue voltage or current, the signal may need to be conditioned, by shifting its voltage levels, increasing its amplitude or filtering out unwanted frequencies, to allow the MCU to receive the input in a usable form. The analogue sensor generally needs to be connected to an analogue to digital converter (ADC) or comparator input on the MCU, while digital types may produce a TTL compatible output which can be connected direct to a suitable port pin, configured accordingly. This may be a simple switched input or a serial data stream in standard or non-standard format.

Some sensors have built-in data processing so that an MCU compatible signal is produced. For example, a measured variable may be converted into a corresponding frequency. This can be fed into a digital input, and the frequency determined in software using a timer/

Interfacing PIC Microcontrollers.
DOI: http://dx.doi.org/10.1016/B978-0-08-099363-8.00009-1
© 2014 Martin Bates and Elsevier Ltd. All rights reserved.

counter to measure the number of pulses in unit time, or the period. Alternatively, the measurement may be transmitted in serial format, such as I^{2}C (see Chapter 8) (Microchip supplies a range of temperature sensors with I^{2}C output). This can provide a more accurate final result, since an analogue signal can be degraded by noise, whereas the digital code from a serial link can be error checked and is either correct or not. A variety of sensors are illustrated in Figure 9.1.

9.1 Digital Sensing

The simplest form of digital sensor is a switch. A manually operated push button or toggle switch may only need a pull-up resistor, or possibly debouncing via a parallel capacitor, hardware latch or software process, as explained in Chapter 4. Optical switching is often preferable, as there are no moving parts and is therefore more reliable.

9.1.1 Mechanical Switch

A microswitch can be attached to a mechanical system so that it detects the position of, say, the guard on a machine tool. The machine controller can then be programmed not to start until the guard is closed. Mechanical switches are inherently unreliable as they have moving parts subjected to continual wear, particularly the contacts, where electrical discharge (sparking) may occur, especially when operating inductive loads. When an inductive load is switched off, it can generate a large back EMF which appears as a high voltage at any gap in the circuit (this effect is used to operate the spark plugs in an internal combustion engine using an ignition coil).

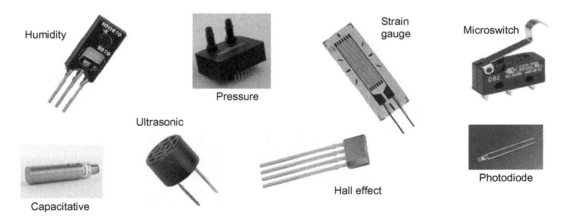

Figure 9.1
Various sensors.

There are some measures which can reduce these effects. A reed switch has a set of passive contacts enclosed in a small glass vacuum tube that are gold-plated to reduce corrosion and improve conductivity. The vacuum reduces burning at the contacts caused by switching discharge, since combustion needs oxygen. The switch is operated by the proximity of a permanent magnet, which eliminates some of the mechanical linkage.

9.1.2 Opto-Switch

A major disadvantage of any mechanical switch or relay contact is that physical wear always causes a degree of unreliability. This can be eliminated with a switched sensor which has no moving parts, such as a light-operated switch. An LED can be used in conjunction with a phototransistor that switches on when exposed to the light source and off when it is interrupted (Figure 9.2(a)).

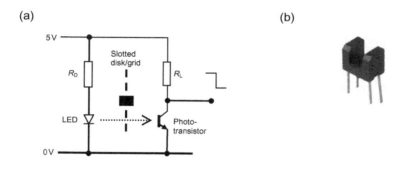

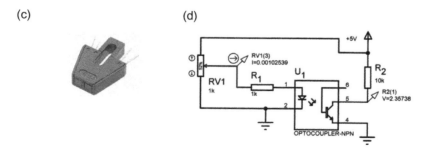

Figure 9.2
Optical detector circuit: (a) opto-detector circuit, (b) slotted photointerrupter, (c) reflective IR LED and sensor, and (d) opto-coupler test circuit.

The phototransistor has no base connection but is exposed to the incoming light via a suitable window in its housing; this generates the base charge carriers which switch on the collector current. Usually, the detector is designed to detect infrared light, with a suitable filter in the window, so that it is insensitive to visible light. The opto-detector can operate by IR transmission or reflection from a suitable surface (Figure 9.2(b) and (c)).

The LED requires a suitable current-limiting resistor, and the transistor a load resistance that will produce the required output swing while consuming as little power as possible. Assuming the LED forward voltage drop is 2V, and the minimum required LED current is 15mA, then

$$R_D = (5 - 2)/15 \times 10^{-6} = 200\Omega \quad NPV = 180R$$

NPV is the nearest preferred value in the standard 10% tolerance resistor series. These are 180R and 220R, with the lower selected because we have specified the minimum current requirement. The opto-transistor load resistor value, assuming the minimum saturation current is, say, 2mA, is

$$R_L = 5/2 \times 10^{-3} = 2.5k\Omega \quad NPV = 2k2$$

Obviously, the data sheet of the specific device must be consulted when calculating these component values. The optical switch also has a massive speed advantage over electromechanical devices; in principle, optical switching can operate at very high frequencies, as when used in communication media such as optic fibre data links.

An opto-coupler circuit is shown in Figure 9.2(d) on the point of switching with a 10k load resistor on the phototransistor (VSM project OPTO2). The LED switchover current is about 1mA, and the load current about 250μA.

9.1.3 Optical Position Sensing

The opto-detector is usually used with a slotted wheel or grating to measure position, speed or acceleration of a rotating shaft or linear axis. Light transmission through metal slots, a transparent material with a printed or engraved grating, or reflection from a surface grid may be used.

The inkjet printer provides a good example of a linear position system, which can readily be examined by dismantling a redundant unit. A plastic strip with a fine grating is used to provide position feedback for the print head. Alternatively, a pair of gratings may be used (Figure 9.3(a)), offset by 90°, so that the direction of travel can be detected by the phase relationship. The simple periodic grating can be made more precise by grading the light

(a)
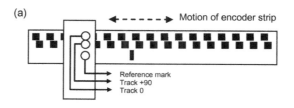

(b)

(c)

Figure 9.3
Incremental encoder: (a) linear position encoder, (b) rotary incremental encoder, and (c) Gray code rotary position encoder disk (10 bit).

transmission sinusoidally over a cycle, allowing the detection of fractions of a cycle (interpolation).

Linear axis position in machine tools can be controlled down to about 1μm by this means. If an axis is driven via a gearbox, monitoring the motor shaft before the gearbox provides maximum resolution. A rotary shaft encoder is shown that can typically generate up to 1024 pulses per revolution (Figure 9.3(b)).

To establish absolute position, a reference position is needed from which relative motion can be calculated. For example, a robot arm may need to be initialised upon start-up by physically moving it to a known 'home' position and resetting the shaft encoder count on each axis. The controller then positions the arm by counting from the reference position.

Alternatively, a Gray code disk (Figure 9.3(c)) can be attached to the output shaft. Each angular position generates a unique binary code from a set of tracks so that the absolute position of the stationary shaft can be detected. The pattern is a modified binary code which only changes one bit at a time to prevent incorrect data being sampled on the sector boundaries.

Another useful type of solid-state sensor is the Hall effect proximity sensor. It is a solid-state device that detects a changing magnetic field. Typical applications are counting the teeth on a rotating gear to drive a tachometer or detecting the piston position in a hydraulic cylinder.

9.2 Analogue Sensing

Analogue sensors produce a continuously variable output, which may be voltage, resistance or current, ideally in direct proportion to the measured quantity. In microcontroller systems, the output usually needs to be converted into a voltage within a range suitable for input to a comparator (high/low detection) or an analogue to digital converter. Suitable signal conditioning may be needed using amplifiers, filters and so on, to produce a clean signal, controlling noise, drift, interference and so on, with the required output range.

Interfacing analogue sensors requires a reasonable understanding of linear amplifier design and signal conditioning techniques, covered in Chapter 6. Connection to an MCU is simplified if the sensor contains built-in signal conditioning, such that the output is linear, has a full voltage range, and is conveniently scaled and precalibrated. The LM35 temperature sensor is a good example, giving an output of 10mV/°C, starting at 0°C = 0mV.

The behaviour of an analogue sensor can be represented by its transfer characteristic, a plot of output against input. The ideal sensor will have a plot with the transfer function $y = mx$, where x is the input, y is the output and m is the sensitivity. High sensitivity provides a large output change for a given input change. The ideal characteristic is perfectly linear with high sensitivity and zero offset, i.e. $y = 0$ when $x = 0$, so that the plot passes through the origin.

Figure 9.4 shows a reference characteristic ($y = mx$, unity sensitivity) and categorises the deviations from it in terms of the following parameters:

- *Sensitivity*
- *Offset*
- *Range*
- *Linearity*
- *Accuracy*
- *Resolution*
- *Stability*
- *Interdependence*
- *Response time*
- *Hysteresis*

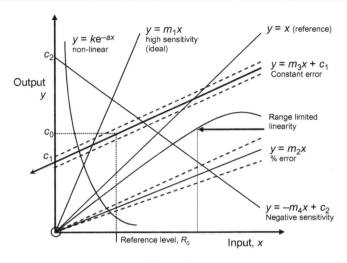

Figure 9.4
Sensor characteristics.

9.2.1 Sensitivity

An ideal sensor characteristic is shown as $y = m_1 x$. The output could be fed direct into an analogue input of the MCU, assuming that the effect of the source and input resistances are insignificant. The line goes through the origin, meaning no offset adjustment is required. A low-value linear potentiometer, using a position sensor, and connected directly across the MCU supply, would give this result.

If the sensor has low sensitivity ($y = m_2 x$), an amplifier may be needed to bring the output up to the required level at its maximum. For example, if the output of a temperature sensor is 10mV per degree, the measured range 0–50°C and the required input range 0–2.5V, we can calculated the required amplifier gain:

$$\text{Maximum sensor output} = 10 \times 50 = 500 \text{mV}$$
$$\text{Required amplifier gain} = 2.5/0.5 = 5$$

A non-inverting amplifier can be used, and the resistor ratio required calculated:

$$\text{Gain} = 1 + \frac{R_f}{R_i} = 5$$

$$\text{So,} \ \frac{R_f}{R_i} = 5 - 1 = 4$$

$$\text{Let } R_i = 1 \text{k0, then } R_f = 4 \text{k0(3k9 NPV)}$$

Precision resistors can be used for a more accurate gain performance. The full resolution of the PIC ADC is 10-bit (1 bit in 1024) or slightly better than 0.1%. To obtain this performance, a precision amplifier should be used, as well as high-precision, high-stability resistors and a circuit designed to minimise noise and drift.

Alternatively, gain and offset adjustment can be incorporated in the interface. In the above example, the feedback resistor might be replaced with a 3k6 resistor and 1k0 preset pot to adjust the gain. Bear in mind, however, that the pot introduces an element of unreliability when compared with a fixed value component.

The term responsivity may be used to describe sensitivity, while sensitivity may sometimes refer to the minimum usable signal level in a detector, so agreed use of terminology is required to avoid ambiguity.

9.2.2 Offset

Unfortunately, many sensors have considerable offset in their output. This means that, over the range for which they are useful, the output has a large positive constant added ($y = m_3x + c$). For example, many temperature characteristics start at absolute zero, $-273°C$, when we may only need to measure around normal room temperature.

In other sensors (e.g. the strain gauge), the offset is caused by a large direct current required to bring the sensor into its practical operating range. This offset has to be subtracted in the amplifier interface to bring the output back into a range where maximum resolution can be obtained. This is often achieved using a differential amplifier.

Alternatively, the sensor may have negative sensitivity as well as offset, such as the temperature characteristic of the silicon diode ($y = -m_4x + c_2$). In this case, an inverting amplifier with offset will be needed.

9.2.3 Transfer Function

If the sensitivity of a linear sensor (m) is known within the general linear function $y = mx + c$, we still need to know a value for the constant c. In thin-film metal temperature sensing resistors, this is given as the reference resistance at 25°C, for which a typical value is 1kΩ. The sensitivity may then be quoted as the 'resistance ratio', corresponding to the proportional change over 100°C. A typical value for this is 1.37. This means that at 125°C, the resistance of the 1kΩ sensor will be 1.37kΩ. The resistance (R) at any other temperature (T) may then be calculated by simple proportionality or by deriving the transfer function as follows:

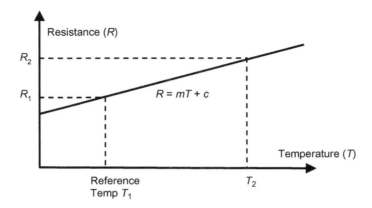

Figure 9.5
Linear temperature sensor graph.

General transfer function $\qquad R = mT + c$

Gradient(sensitivity) $\qquad m = (R_2 - R_1)/(T_2 - T_1)$
$$= (1370 - 1000/(125 - 25)$$
$$= 3.7\Omega/^\circ C$$

Offset $\qquad c = R_1 - mT_1$
$$= 1000 - (3.7 \times 25)$$
$$= 907.5\Omega$$

Transfer function of sensor $\quad R = 3.7T + 907.5\Omega$

The resistance at any temperature can then be calculated from this relationship. The graph of this transfer function is shown in Figure 9.5.

9.2.4 Linearity

The ideal sensor characteristic is a perfect straight line, where the output is exactly proportional to the input. Linearity has to be maintained through the signal conditioning and conversion processes. An example of non-linearity is seen in metal temperature sensors, which tend to deviate at higher temperatures, as their melting point is approached. The usable range then has to be defined by the part of the characteristic which is within the required limits of linearity.

The deviation from linearity is usually expressed as a maximum percentage error over the specified range, but care must be taken to establish whether this is a constant over the range, or a proportion of the output level. These two cases are illustrated by the dotted lines in Figure 9.4, indicating the possible error due to non-linearity and other factors. Constant error over the usable range means that the output is within a region that lies within a pair of lines parallel to the sensor curve ($y = m_3x + c_1$), while a percentage error of the output variable is a divergent pair of lines ($y = m_2x$).

Linear sensors are obviously easier to interface for absolute measurement purposes, but some that are non-linear may have other advantages. The thermistor, for example, has a negative exponential characteristic, but has high sensitivity, so is useful in detecting spot temperatures over a wide range. A comparator input rather than the ADC is likely to be more useful in this case.

If a continuous measurement is needed, a log amp or conversion table in the controller could be considered. In this case, the transfer function must be known precisely in order the design the interface to produce an accurate output. If necessary, correcting factors can be applied when calculating the corresponding measurement in software. This can be done by arithmetic processing (C programming preferred) or a look-up table (in assembler).

9.2.5 Other Error Factors

Many factors may contribute to sensor measurement error, including limitations in linearity, accuracy, resolution and stability. Accuracy is evaluated by comparison with a standard measurement; for example, the Celsius temperature scale is based on the known freezing and boiling points of water. The resolution of a measurement is the degree of precision achieved by the measurement system: $25.00°C$ ($\pm0.005°C$) is a more precise measurement than $25°C$ ($\pm0.5°C$).

The specified precision must be justified by the overall performance of the measurement system, and is meaningless unless it has been calibrated against an agreed standard. The SI system specifies scientific and technical standards of measurement, largely derived from the standard units of length (metre), time (second) and mass (kilogram). Electrical units are derived from these, as well as mechanical units. Accurate measurement systems have to be checked and recalibrated at specified intervals, by comparing the output with that of a standard reference system. For example, a voltmeter can be checked by connecting to a standard cell, which has a constant known voltage.

Repeatability refers to the extent to which the same sensor output is obtained on different occasions. Poor repeatability may be due to hysteresis, where the measurement differs depending on whether the input is rising or falling. Electromagnetic sensors are particularly prone to this effect, due to the residual magnetisation of ferromagnetic materials in a changing field.

Poor stability may appear as drift, a change in the sensor output over time. This may be caused by short-term heating effects when the circuit is first powered up, or the sensor performance may deteriorate over the long term, and the measurement become inaccurate; recalibration is then required.

Interdependence in the sensor may also be significant, where its output is affected by more than one environmental variable. For example, the output of a humidity sensor may change with temperature, so this incidental variable must be controlled so that the required output of the measurement compensated in processing is not affected.

Individual examples of any given sensor may vary in their response, so individual calibration data may be needed, or in system calibration. Again, some humidity sensors are supplied with individual calibration data for maximum accuracy.

9.2.6 Response Time

There is inevitably a lag between the change in the measured variable and the output of the sensor. In a temperature sensor, it may be large (several seconds) due to the thermal inertia of the detector and its substrate, or measured in nanoseconds in a photodetector. Clearly, this factor must be considered in the overall design performance of a control system.

9.3 Position Sensing

Position measurement can be implemented in many different ways. The use of switched sensors has already been introduced, where incremental opto-sensors are used for position and speed control in motorised outputs. Here we will add proximity sensing and analogue position measurement using alternative techniques.

9.3.1 Switched

The microswitch is a push button designed for chassis mounting as a proximity sensor, typically as a safety feature, to detect, for example, the closing of the safety guard in a machine tool. It may incorporate an extended actuator arm to make it less position sensitive and introduce some flex in the arm to allow for imprecise positioning of mechanical parts. The microswitch usually has changeover contacts with normally open (NO) and normally closed (NC) terminals. NO contacts wired in series with the power supply provide a very simple safety system. However, it is usually better to separate the supply and control circuits using a relay or opto-isolator. In microcontroller systems, an optical interrupter will often be preferred.

9.3.2 Resistive

A rotary potentiometer can be used as a simple three-terminal position sensor, where the voltage output represents the angular position of a shaft. A d.c. voltage is applied across the track, and the position represented by a proportional voltage (Table 9.1(a) and (b)). The linear pot is less common, but the circuit symbol is the same for both. When monitoring the output voltage, care must be taken to ensure that the current drawn from the wiper is insignificant compared with the current through the track; otherwise the position will not be accurately represented. If interfaced using an op-amp-based circuit, a non-inverting amplifier will usually ensure this condition. Logarithmic pots provide a non-linear transfer characteristic that is useful in some applications. If one end of a pot is connected to the wiper, a two-terminal variable resistance is obtained. The potentiometer can be used in a simple level measurement system, such as a fuel tank, attached to an arm and float, and connected as a variable resistance directly in series with a moving coil meter to display the level.

A basic pot has limited range (about 300°) and is subject to noise and unreliability due to the physical contact between the wiper and the track and is not particularly accurate (assume ±5%) unless a high specification or multi-turn (usually 10 turns on a spiral track) pot is used. Cheap or miniature (preset) pots have a carbon track which tolerates only a limited current at the wiper. Wire-wound pots can operate at higher currents but have a noticeable stepwise operation as the wiper moves across the track windings. There is a range of more reliable position transducers that avoid this physical contact, but they tend to be more expensive, or more difficult to interface.

9.3.3 Capacitive

Capacitive proximity sensors can detect the presence of any solid body that affects the sensor capacitance relative to the surrounding ground region. $X-Y$ position detection on a surface plane using a capacitive grid is used in touch screens.

Basic capacitor characteristics also provide opportunities to measure small distances and changes in level of insulating materials. If configured as a pair of flat plates, separated by an air gap, a small change in the gap will give a large change in the capacitance (Table 9.1(c)). These variables are inversely proportional, so if the gap is doubled, the capacitance is halved. If an insulator is partially inserted, the capacitance undergoes a corresponding change (Table 9.1(d)). This can make a simple but effective level sensor for insulating materials such as oil, powder and granules in a storage silo. A pair of vertical plates is all that is required for the sensor, so it has no moving parts, unlike the float and potentiometer.

Table 9.1: Position Sensors.

Transducer	Description	Applications	Evaluation
(a) L, d +V, 0V, V_o, $V_o = (d/L) \cdot V$	*Linear potentiometer*: Resistive track with adjustable wiper position. D.c. supply across track gives a variable voltage at the wiper representing absolute linear position	*Linear position sensing*: Faders and multi-turn presets, medium scale linear displacement	Physical wear causes unreliability but is cheap and simple
(b) +V, 0V, V_o	*Rotary potentiometer*: Rotary version senses absolute shaft position as voltage or resistance (connect one end and wiper together to form two-terminal variable resistance. Log scaling also available)	*Rotary position sensing*: Manual pots and presets, any shaft with range of movement less than 300°. May be used with float for liquid level sensing	Physical wear causes unreliability, but cheap and simple. Wire wound are more robust, but may have limited resolution
(c) d, I_{ac}, $C \propto d$	*Capacitor air gap*: Capacitance is proportional to plate separation (d is normally small). Small change in d gives a large change in C. Switching interface detects changes in rise time or capacitance	*Linear position sensing*: Sensitive transducer for small changes in position. Plate overlap can also be varied, although change may be less linear due to edge effects	No physical contact, so more reliable. Needs more complex drive and interfacing
(d) Air, Variable level, Dielectric	*Capacitor dielectric*: Capacitance depends on dielectric material, effectively producing two capacitors in parallel whose values add. Requires a high-frequency drive signal to detect changes in reactance	*Level or position sensing*: The dielectric may be any insulating material, liquid or powder. A solid dielectric can detect linear motion as its position is varied	No physical contact, so more reliable. Needs more complex drive and interfacing, involving a.c. to d.c. conversion. Simple to construct
(e) Coil and core, V_{dc}, H, Input I_{dc}, Output V_H	*Magnetic flux*: The Hall probe produces an output voltage proportional to the magnetic flux density generated by the magnetic coil. This can also be detected by another coil by mutual inductance	*Position/motion sensing*: Magnetic circuits can be used in various ways to detect position, motion or vibration with no physical contact required. LVDT, rev. counter, proximity switch	Versatile, robust sensor providing reliable pulse detection. Flux linkage types may need more complex drive and detector involving a.c. to d.c. conversion

Capacitance can be measured by various methods. If its value is reasonably large, it can be charged or discharged via a series resistance and the time taken to reach a threshold voltage measured using a comparator input and timer. If the value of the capacitor is small, a high-frequency measurement circuit may be needed, or a bridge arrangement used for greater accuracy.

9.3.4 Inductive

An electromagnetic coil can be used to implement contactless position or proximity sensing. The field produced by the current in one coil induces a corresponding current in an adjacent coil, as in a power transformer. If, however, the second coil is mobile, the induced current is reduced as the distance increases. A linear variable differential transformer (LVDT) uses electromagnetic coils to detect the position of a mild steel rod which forms a mobile core. The input coils are driven by an a.c. signal, and the rod position controls the amount of flux linked to the output coil, giving a variable peak to peak output. It needs a high-frequency a.c. supply, and is relatively complex to construct, but is reliable and accurate.

A Hall probe can also be used to measure magnetic flux, and as a position sensor in conjunction with a coil (Table 9.1(e)). It is formed from a semiconductor slice that has a current flowing across in one direction. When a magnetic field is applied perpendicular to the slice, the charge carriers are deflected and a lateral voltage generated in proportion to the field strength. This can be used to measure magnetic field strength fairly accurately, or the position of a coil, or permanent pole, producing the field.

9.3.5 Ultrasonic

Ultrasonic ranging can be used for distance measurement. The speed of a sound pulse (about 340m/s depending on conditions) travelling over a few metres and reflecting from a solid object gives a delay in milliseconds (about 3ms/m), which is easily measured by a hardware timer in a microcontroller. A short burst at about 40kHz is typically used in measurement systems, so the sound is above the range of human hearing and does not cause a nuisance or get disrupted by lower frequency sounds. It can be used in motor vehicle parking systems or bulk measurement in a silo of dry materials, for example.

9.4 Temperature Sensing

Temperature is one of the most commonly required measurements, and there is great variety of sensors for different temperature ranges and industrial applications. Some are illustrated in Table 9.2.

9.4.1 Metallic

Metals have a reasonably linear temperature coefficient of resistance over limited ranges. Metal film temperature sensing resistors operate up to about 150°C, with platinum sensors working up to 600°C (Table 9.2(a)). A typical response has already been analysed in Section 9.2.3 based on data for a 1k0 metal film resistor. The sensitivity (temperature coefficient, α) in this case was 3.7Ω/°C, with an offset at 0°C of 907.5Ω. A constant current is needed to convert the resistance change into a linear voltage change, and an accuracy of around 3% may be expected.

Self-heating needs to be considered in operating any resistive temperature sensor. A certain current is needed to allow the resistance change to be detected, but this in itself will heat the sensor. A higher current produces a better signal-to-noise ratio, i.e. a higher voltage change for a given temperature change, but will increase self-heating. Since power dissipated is proportional to the current squared, the sensor must be operated with the minimum current consistent with the required sensitivity and noise immunity.

Good thermal contact with a large target body will also reduce this problem, as any self-heating energy will be dissipated into the larger mass. In a microcontroller application, pulse width modulation could also be considered to alleviate this problem. In addition, in this type of sensing situation, a four-wire connection to the sensor may improve performance. This means connecting a separate pairs of wires to power the sensor and to measure the sensor output. This eliminates the resistance of the power leads from the measurement. Since the metal temperature sensor has low sensitivity and a large offset, connection in a bridge arrangement is indicated (see Section 9.7.1).

9.4.2 Thermocouple

Higher temperatures can be measured using a thermocouple (Table 9.2(b)). This is simply a junction of two dissimilar metals, which produces a small EMF, in a similar manner to a battery. The voltage is proportional to temperature, but has a large offset, since it is proportional to absolute temperature. This is compensated for by a cold junction, connected in series with the opposite polarity, and maintained at a known reference temperature. The difference voltage is then due to the temperature difference between the cold and hot junctions, which can be detected via a high-gain difference amplifier. Thermocouples are normally used with specially designed controllers, which incorporate the requisite cold junction temperature stabilisation.

Table 9.2: Temperature Sensors.

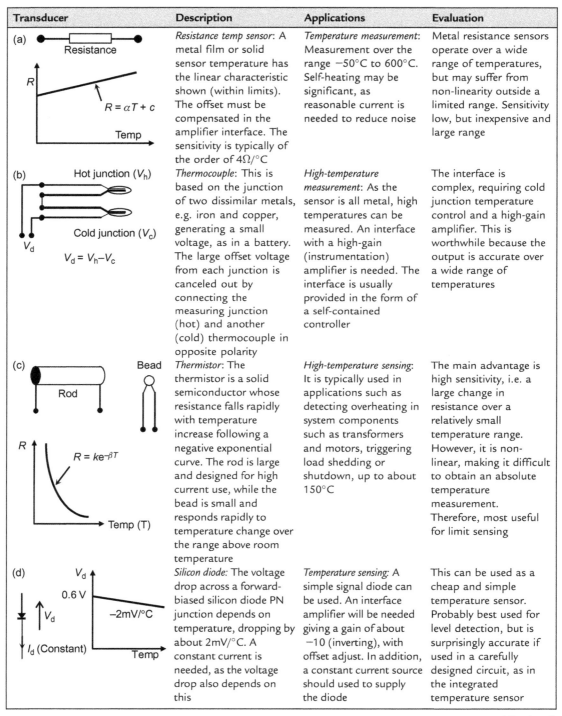

Transducer	Description	Applications	Evaluation
(a) Resistance, R, Temp, $R = \alpha T + c$	*Resistance temp sensor*: A metal film or solid sensor temperature has the linear characteristic shown (within limits). The offset must be compensated in the amplifier interface. The sensitivity is typically of the order of $4\Omega/°C$	*Temperature measurement*: Measurement over the range $-50°C$ to $600°C$. Self-heating may be significant, as reasonable current is needed to reduce noise	Metal resistance sensors operate over a wide range of temperatures, but may suffer from non-linearity outside a limited range. Sensitivity low, but inexpensive and large range
(b) Hot junction (V_h), Cold junction (V_c), V_d, $V_d = V_h - V_c$	*Thermocouple*: This is based on the junction of two dissimilar metals, e.g. iron and copper, generating a small voltage, as in a battery. The large offset voltage from each junction is canceled out by connecting the measuring junction (hot) and another (cold) thermocouple in opposite polarity	*High-temperature measurement*: As the sensor is all metal, high temperatures can be measured. An interface with a high-gain (instrumentation) amplifier is needed. The interface is usually provided in the form of a self-contained controller	The interface is complex, requiring cold junction temperature control and a high-gain amplifier. This is worthwhile because the output is accurate over a wide range of temperatures
(c) Rod, Bead, R, $R = ke^{-\beta T}$, Temp (T)	*Thermistor*: The thermistor is a solid semiconductor whose resistance falls rapidly with temperature increase following a negative exponential curve. The rod is large and designed for high current use, while the bead is small and responds rapidly to temperature change over the range above room temperature	*High-temperature sensing*: It is typically used in applications such as detecting overheating in system components such as transformers and motors, triggering load shedding or shutdown, up to about $150°C$	The main advantage is high sensitivity, i.e. a large change in resistance over a relatively small temperature range. However, it is non-linear, making it difficult to obtain an absolute temperature measurement. Therefore, most useful for limit sensing
(d) V_d, 0.6 V, $-2mV/°C$, V_d, I_d (Constant), Temp	*Silicon diode:* The voltage drop across a forward-biased silicon diode PN junction depends on temperature, dropping by about $2mV/°C$. A constant current is needed, as the voltage drop also depends on this	*Temperature sensing:* A simple signal diode can be used. An interface amplifier will be needed giving a gain of about -10 (inverting), with offset adjust. In addition, a constant current source should used to supply the diode	This can be used as a cheap and simple temperature sensor. Probably best used for level detection, but is surprisingly accurate if used in a carefully designed circuit, as in the integrated temperature sensor

(Continued)

Table 9.2: (Continued)

Transducer	Description	Applications	Evaluation
(e) +5 V 0 V 10 mV/°C	*Integrated temp sensor:* This is based on silicon junction temp sensing. A built-in amplifier gives a calibrated output of typically 10mV/°C, over the range of −50 to +150°C. The accuracy is around ±0.5°C	*Temperature measurement:* General purpose low-temperature sensing with moderate accuracy. Can be operated from +5V, so is easy to integrate into digital systems	This is a versatile sensor, and the first choice for a low-cost, low-temperature MCU-based system. It is easy to interface, does not need calibrating and is inexpensive. Response may be slow due to size

9.4.3 Thermistor

Thermistors are made from a single piece of semiconductor material in which the charge carrier mobility, and therefore the resistance, depends on temperature (Table 9.2(c)). The response is exponential, giving a relatively large change for a small change in temperature, and therefore high sensitivity. Unfortunately, the response is non-linear, so it is not simple to convert for precise measurement purposes. The thermistor therefore may be used as a simple threshold sensor to detect if a component such as a motor or transformer is overheating. The bead type could be used with a comparator to provide warning of overheating in a microcontroller output load.

For accurate temperature measurement, the transfer function of the thermistor can be represented as

$$R = R_0 \cdot \exp\{B(1/T - 1/T_0)\}$$

R_0 is the resistance at a room temperature T_0 (25°C or 298K), and B represents the negative temperature coefficient. The thermistor data sheet usually provides a table of values for the ratio R_T/R_0 so that the actual resistance can be calculated for a range of thermistors with different reference values. Alternatively, it can be calculated from the above function.

For example, the Epcos NTC bead thermistor type B57863S is available with resistance values 3k, 5k, 10k and 30k at 25°C. For the three lower values, $B = 3988$. Therefore, at 100°C, the resistance of the 10k thermistor will be

$$R = 10^4 \cdot \exp\{3988(1/373 - 1/298)\} = 678\Omega$$

This can be checked using the table in the thermistor data sheet. The function can be rearranged to obtain the temperature from the resistance:

$$T = 1/\{\mathrm{Ln}(R/R_0)/B + 1/T_0\}$$

For example, if the resistance of the thermistor same thermistor is measured as 3.6k, its temperature will be

$$T = 1/\{Ln(3.6/10)/3988 + 1/298\} = 323K \text{ or } 323 - 278 = 50°C$$

The controller will probably have to be programmed in C to implement this calculation. An assembler implementation could use a look-up table.

9.4.4 Resistive Sensor Interface

A resistive sensor can be interfaced with a microcontroller ADC input via a simple non-inverting amplifier if the range of resistance change is not too great, and the sensor is reasonably sensitive. In the prototype configuration shown in Figure 9.6, the sensor is connected as part of the feedback network, with a variable voltage at the offset terminal. By applying Ohm's law to the feedback voltage divider,

$$V_0 = V_r(R_f/R_s + 1)$$

As an example, the thermistor specified above has a resistance of 10k at 25°C and 2.5k at 60°C. If the feedback resistor is 10k, and an offset voltage $V_r = 0.5V$ is used, at 25°C the output will be 1.0V. At 60°C, the output will be 2.5V. This is within the range that can be fed direct to a PIC ADC input. As the output from the thermistor is non-linear, a look-up table is needed to obtain the temperature at selected intervals, say 1°C.

9.5 Semiconductor Sensing

The forward volt drop of a silicon diode junction is broadly taken as 0.6V, but the exact figure depends on the junction temperature (Table 9.2(d)). This temperature sensitivity is

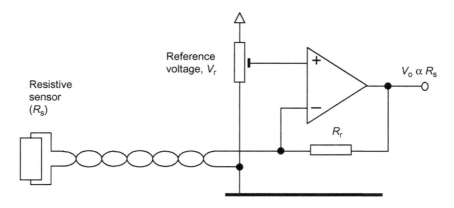

Figure 9.6
Resistive temperature sensor interface.

quite consistent, so the standard signal diode can be used as a simple, inexpensive sensor, especially if a simple high/low detector only is needed. An analysis of this characteristic is also useful in informing general design practice.

9.5.1 Semiconductor Junction

The silicon diode junction is formed of N-type and P-type semiconductors. In the N region, the silicon has been doped with a small amount of an element with an extra electron in its outer atomic shell, and the P region with an element which causes a deficiency of electrons (holes). Close to the boundary between these regions, the electrons and holes recombine to produce a depletion layer.

If the diode is reverse biased (a positive supply to the N side and negative to the P side), the depletion layer increases and no significant current flows. If forward biased, the depletion layer is reduced, allows current to flow according to an exponential relationship between the forward current (I_F) and forward voltage (V_F). An approximate version of this function, sufficient to illustrate the essentials of diode behaviour, is

$$I_F = I_S \cdot \exp(V_F/nV_T)$$

where

> I_S is the diode reverse leakage current
> n is the diode ideality factor
> V_T is the thermal voltage, where $V_T = kT/q$
> k is Boltzmann's constant $= 1.38 \times 10^{-23}$
> T is the absolute temperature of the junction ($0K = -273°C$)
> q is the charge on the electron $= 1.60 \times 10^{-19}$

At 25°C, or 298K, $V_T = 26mV$.

Rearranging the transfer function gives

$$V_F = nV_T \cdot \text{Ln}(I_F/I_S) = nV_T \cdot \text{Ln}\, I_F - nV_T \cdot \text{Ln}\, I_S$$

This shows that the diode forward voltage depends on the natural log of the forward current and the absolute temperature (variable term) with an offset determined by I_S (constant term). This characteristic is illustrated in Figure 9.7 based on data for the 1N4148 signal diode. Note that the scales are reversed when compared with the conventional representation of the transfer characteristic. The current scale is in decades, producing a linear representation of variation in the junction forward voltage with forward current.

To calculate the constants V_T, I_S and n for this particular characteristic, assuming a constant temperature (25°C), we can take two pairs of end values:

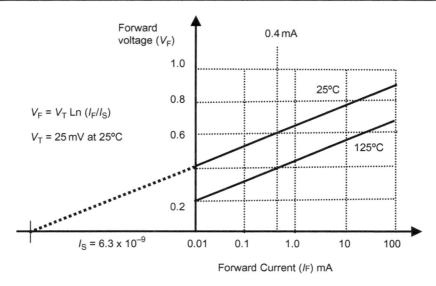

Figure 9.7
Diode characteristic (1N4148).

$$\text{When} \quad I_{F1} = 100\text{mA}, \quad V_{F1} = 0.9\text{V}$$

and when

$$I_{F2} = 0.01\text{mA}, \quad V_{F2} = 0.4\text{V}$$

So,

$$V_{F1} = nV_T \cdot \text{Ln } I_{F1} - nV_T \cdot \text{Ln } I_S$$

and

$$V_{F2} = nV_T \cdot \text{Ln } I_{F2} - nV_T \cdot \text{Ln } I_S$$

Subtract to find V_T:

$$V_{F1} - V_{F2} = nV_T \cdot \text{Ln } I_{F1} - nV_T \cdot \text{Ln } I_{F2}$$
$$0.9 - 0.4 = nV_T \cdot (\text{Ln}(0.1) - \text{Ln}(0.00001))$$
$$nV_T = 54\text{mV}$$

So,

$$n = \frac{54}{26} = 2.1$$

Divide to find I_S:

$$V_{F1}/V_{F2} = (\text{Ln } I_{F1} - \text{Ln } I_S) \cdot (\text{Ln } I_{F2} - \text{Ln } I_S)$$
$$0.9(\text{Ln } 0.00001 - \text{Ln } I_S) = 0.4(\text{Ln } 0.1 - \text{Ln } I_S)$$
$$\text{Ln } I_S = -18.9$$
$$I_S = 6.3 \times 10^{-9} \text{ A}$$

Check the value of V_F at $I_F = 0.4$ mA:

$$V_{F3} = V_T \cdot \text{Ln}(I_{F3}/I_S)$$
$$= 0.054 \, \text{Ln}(0.0004/6.32 \times 10^{-9})$$
$$= 0.6\text{V}$$

9.5.2 Temperature Sensitivity

When the temperature of the junction rises, the extra thermal energy imparted to the charge carriers eases the transition of electrons across the junction potential gradient. This causes a fall in the value of V_T, the thermal voltage, hence the value of V_F for any given value of I_F. The change per degree Celsius can be calculated, assuming a constant forward current I_{FC}:

$$V_{F1} - V_{F2} = nkT_1/q \cdot \text{Ln } I_{FC} - nkT_2/q \cdot \text{Ln } I_C = nk/q \cdot \text{Ln } I_{FC}(T_1 - T_2)$$

where

$$V_{T1} = kT_1/q$$

and

$$V_{T2} = kT_2/q$$

Now,

$$T_1 - T_2 = 1°\text{C}$$

So

$$V_{F1} - V_{F2} = nk/q \cdot \text{Ln } I_{FC}$$
$$nk/q = (2.1 \times 1.38 \times 10^{-23})/1.60 \times 10^{-19}$$
$$= 0.181\text{mV}$$

Assume $I_{FC} = 0.01\text{mA}$.
Then, $V_{F1} - V_{F2} = -2.1\text{mV}$.

If larger values of I_{FC} are used, the temperature sensitivity of the junction can be seen to be lower; for example, the change at 100mA constant current is only 0.4mV/°C. It can therefore be concluded that a higher operating current in semiconductors will improve temperature stability, as well as noise immunity, at the expense of higher power consumption. In addition, with the junction operating at higher temperature, it will be less sensitive to external temperature fluctuations, since any cooling effects in the environment are likely to be a smaller proportion of the overall energy dissipated.

9.5.3 Diode Sensor Interface

Figure 9.8 shows a basic circuit illustrating the use of a signal diode as a temperature sensor. The diode is connected in the feedback path of an inverting amplifier, with the input resistor connected to a constant voltage source. The diode therefore carries a constant forward current (I_F). The reference terminal of the amp is connected to a pot which can be used to adjust the output offset for, say, 1.0V at 25°C. The input terminals will then be at 1.4V.

Assuming a single supply at 5V in the circuit, a 2.7V zener will give a voltage of 1.3V across the input resistor, which thus needs to be 1.3V/0.01mA = 130kΩ (120k + 10k). Assuming a 2mV/°C fall in the voltage across the diode, the output will rise by the same. At 50°C, for example, it will reach approximately 1.050V. The gain could be made slightly adjustable by placing a variable resistance in series with the input resistor to tweak the diode current.

This output will probably need to be scaled to produce a greater range. Let us assume that we wish to measure temperature using a diode over the range 0−50°C using a 2.56V reference for the PIC ADC with 8-bit resolution. If we assume 2.56V corresponds to 51.2°C,

> ADC input scaling = 2.56/51.2 = 50mV/C
> Amp gain required = 50/2 = 25
> Offset = 0.950V

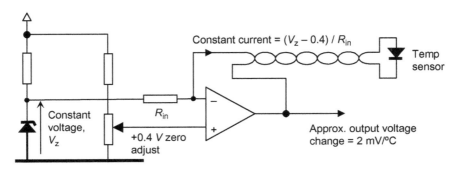

Figure 9.8
Diode tempertaure sensing circuit.

The stage outlined above would need to be followed by a difference amplifier with the required gain and offset (see Chapter 6). If working with a single supply, the minimum output available from the op-amp needs to be established. Self-heating of the sensor also needs to be considered; however, with a current of 0.01mA and forward voltage drop of around 0.4V, only 4mW will be dissipated at the junction. It should be allowed to stabilise before making a measurement.

9.5.4 Integrated Temperature Sensor

As discussed earlier, using the bare semiconductor as a temperature sensor needs careful circuit design. Using a bipolar transistor will improve matters, as it has similar characteristics as well as inherent gain which will reduce the need for further amplification. However, achieving a stable, accurate and repeatable output for a discrete d.c. design is always problematic. Therefore, for applications working around room temperature, an integrated sensor and amplifier based on the semiconductor junction temperature sensitivity described above is the most obvious choice (Table 9.2(e)).

The LM35 sensor is easy to interface and produces a calibrated output of 10mV/°C and 0mV at 0°C (zero offset). It has an input range of −50°C to +150°C giving a full output range of 2.00V, or 0.00−1.00V over the range 0−100°C, with an accuracy of ±0.5%. A negative supply is needed to take the measurement below 0°C. It can be connected directly to the PIC ADC input when operating over the full range, but for smaller ranges an amplifier might be advisable to make better use of the available resolution of the ADC. For example, to measure 0−50°C,

Temp range = 50°C
Input range used = 0−2.56V (8-bit conversion, V_{REF} = 2.56V)
Let maximum temp = 2.56 × 20 = 51.2°C
Then conversion factor = 2.56/5.12 = 50mV/°C
Output of sensor = 10mV/°C
Gain of amplifier required = 50mV/10mV = 5.0 (non-inverting)

An interface circuit for the LM35 IC sensor will be described in Section 10.5.1.

There are now various IC temperature sensors available that transmit readings in serial formats, including SPI and I²C. These are connected direct to the MCU, so the interfacing becomes a purely a programming exercise. Several types of IC temperature sensor are currently supported by models in Proteus VSM. The Maxim DS 1822 is such a device. It has a one-wire interface (plus 5 and 0 V supplies) and produces serial output using a proprietary format. It also features a unique identifier code for each sensor and alarm functions. It can be connected to any available digital input.

9.6 Light

There are several types of sensors for measuring light intensity or operating in switched mode: phototransistor, photodiode, light-dependent resistor and photovoltaic cell. Simple switched detectors (see Section 9.1.2) typically use an infrared diode and phototransistor, while the LDR can measure a wide range of visible light levels. Data links also use an infrared LED or laser diode (coherent light source) to transmit data via optical fibre transmission at high frequencies, with a PIN photodiode detector. The characteristics of light and selected other sensors are summarised in Table 9.3.

The sensitivity of the human eye is logarithmic, or measured in decades. Calibration of light sensor input is therefore not linear, as the range is large. A traditional moving coil light meter, using a selenium or CdS cell, consequently has a logarithmic scale.

9.6.1 Photodiode

At the semiconductor PN junction, the electrons and holes recombine to produce a non-conductive depletion layer. If forward biased, the depletion layer is reduced, allowing current to flow according to an exponential relationship between the current and forward voltage, as described in Section 9.5.1.

If the diode is reverse biased (a positive supply to the N side and negative to the P side), the depletion layer increases and only a small leakage current flows, due to the electrical field across the junction and residual thermal energy producing a small number of charge carriers. If light is allowed to fall on the junction, the number of charge carriers is increased in proportion to the level of illumination, producing a measurable voltage and current change. If the diode is formed as a large flat plate, a solar cell producing d.c. power is obtained.

In the PIN photodiode sensor, used in CD readers, optic fibre and remote control receivers, the speed and sensitivity are improved by adding a layer of pure (undoped) semiconductor between the P and N layers which produce the photoelectrons. A typical PIN photodiode has an operating current of $0.1-10\mu A$, linear over about four decades of light input. It has a switching time of only 2ns (maximum frequency about 200MHz) when operated with 5V reverse bias. Maximum sensitivity is at 850nm wavelength (infrared), but the visible spectrum is covered (400−700nm).

9.6.2 Phototransistor

The photodiode is insensitive at low light levels, so the phototransistor is often preferred, as it has additional inherent current gain. The bipolar transistor is a sandwich of three semiconductor layers, collector, base and emitter, either NPN or PNP, where the base layer

Table 9.3: Force, Light, Humidity Sensors.

Transducer	Description	Applications	Evaluation
(a)	*Phototransistor:* The phototransistor has no base connection, but it is exposed to light by transparent encapsulation. The base current is generated by light energy absorbed by the charge carriers. With a load resistance, the collector voltage varies with base current in the usual way	*Light sensing:* The transistor provides inherent gain (about 100), making the device quite sensitive. It is incorporated in opto-couplers and detectors, which usually use infrared light from an LED, which reduces interference from visible light sources	A high sensitivity detector, but difficult to obtain a calibrated output. It is therefore more frequently used in digital systems for isolation and position/speed measurement using a counter
(b)	*Light-dependent resistor:* The LDR uses a CdS (cadmium disulphide) cell which is sensitive to visible light over a wide range, from dark to bright sunlight. If the light input (lux) and resistance are plotted on decade scales, a straight line is obtained	*Light measurement:* The LDR is the standard cell used in light meters and cameras, since photographic exposure is also calculated on a log scale. A coarse level voltage can be obtained with a simple series resistance, e.g. dark, overcast, sun	The CdS cell provides an accurate output over a wide range, but interfacing for a calibrated output via an MCU requires conversion of the log scale, either via an accurate log amplifier or in software
(c)	*Strain gauge:* This is simply a folded conductor mounted on a flexible sheet whose resistance increases as it is stretched. Mounted in a bridge of four gauges to maximise differential voltage	*Stress, strain, position, measurement:* Typically used to measure the deformation in a mechanical component under load (e.g. crane jib) for safety monitoring purposes. A high-gain, differential (instrumentation) amplifier is needed	Relatively simple and reliable method of monitoring small mechanical deformations. The high-gain amplifier is susceptible to noise and interference and may need careful circuit design to obtain a stable output
(d)	*Pressure:* If a set of strain gauges are mounted on both sides of a diaphragm as shown, they will respond to deformation as a result of a differential pressure. The output voltages from each pair can be added to give a measurement	*Differential pressure measurement:* For measurement relative to atmosphere, one side of the gauge will be exposed to atmosphere, the other to higher pressure air or gas. If a vacuum is used on one side, absolute pressure may be gauged	Piezoresistive sensors, accurately trimmed during manufacture, and integrated amplifier provide accurate output over selected ranges

(Continued)

Table 9.3: (Continued)

Transducer	Description	Applications	Evaluation
(e) Absorbent dielectric	*Humidity*: A capacitor with an absorbent dielectric can vary in capacitance value depending on the humidity of the surrounding air	*Humidity measurement*: Environmental monitoring is the general area of application, either for weather recording, product testing or drying systems	Plain sensors requiring an HF a.c. signal to drive the detection system available. Devices with integrated signal conditioning are simpler to interface

is relatively thin. In an NPN transistor, a forward base−emitter bias voltage causes recombination of the holes and electrons to produce a base current flow. If the collector is connected to positive supply, this allows a larger number of electrons to pass directly from emitter to collector. The large collector current is proportional to the small base current, specified as the current gain (h_{FE}). The transistor thus acts as a current amplifier, where the collector−emitter current is approximately 100 times the base−emitter current.

In a phototransistor, the base is not electrically connected, but base current is generated by light falling directly on the base−emitter junction (Table 9.3(a)). The transistor action provides current gain which increases the sensitivity, allowing the output to be easily detected without further amplification by a standard MCU input, simply by connecting a suitable collector load. A typical phototransistor will produce a linear output over the same input range as the photodiode, but producing an output between 1μA and 10mA. The phototransistor is used in opto-isolators, photointerrupters, proximity detectors and incremental encoders. The opto-coupler model can be used in VSM to simulate the operation of a generic opto-switch or phototransistor sensor (see Figure 9.2(d)).

9.6.3 Light-Dependent Resistor

The light-dependent resistor (LDR) uses a cadmium disulphide (CdS) sensor that has a spectral response similar to the human eye, has high sensitivity to a wide range of values of light intensity and is relatively easy to interface. Its resistance is inversely proportional to light intensity, as shown in Figure 9.9.

Since a wide range of light intensity levels between dark and sunlight can be detected, causing a large change in resistance, the transfer characteristic is normally plotted on decade scales. The resistance changes from nearly 1MΩ in the dark to about 100Ω in bright sunlight. The transfer function is derived from the plot of the LDR response. The light level

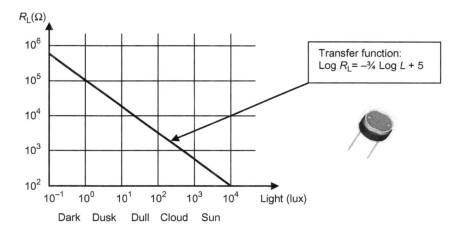

Figure 9.9
Light-dependent resistor: transfer characteristic and function.

can be obtained from the resistance by rearranging the transfer function. If R is the LDR resistance and L the light level, then

$$\text{Log } R = -\frac{3}{4} \text{ Log } L + 5$$

$$\text{so } L = 10^{\{4/3\ (5 - \text{Log } R)\}}\text{lux}$$

If a resistance of, say, 1k0 is connected in series to convert the output to volts, it could be connected directly to a PIC ADC, but the readings would have to be converted to corresponding light levels in software, or selected spot values converted using a look-up table. Alternatively, the log amplifier seen in Chapter 6 could be configured to produce a linear output. A simple interface that assumes the use of a look-up table to convert the LDR output to light level is described in Section 10.5.2.

9.6.4 Integrated Light Sensors

An IC light sensor, available in VSM, is the TSL251RD. It contains a photodiode and buffer, and outputs a voltage in the range 0–3V representing the incident visible light level. It can be connected directly to the PIC ADC but will also need calibrating in firmware for accurate measurement. An infrared option is available.

Texas Instruments TSL235 Light to Frequency Converter produces a calibrated linear output in a serial format. An input over the range $10^{-3} - 10^3$ µW/cm^2 irradiance is converted to an output of 10Hz to 500MHz (50% MSR) accurate to 0.2% with low temperature sensitivity (0.1%). This output can be received by the MCU timer/counter and converted to the corresponding temperature by counting the number of pulses per unit time.

9.7 Force Sensing

Force is often measured by its deforming effect on a sensing bar, measured as strain, the relative elastic extension, or bending effect.

9.7.1 Strain Gauge

The strain gauge is simple in principle (Table 9.3(c)). A temperature-stable alloy conductor is printed onto a flexible substrate which lengthens when the gauge is stretched (strained). The resistance increases as the conductor becomes longer and thinner. When bonded to a mechanical component, this can be used to measure small dimensional changes and hence the forces exerted upon them. They are used to measure the behaviour of, for example, bridges and cranes, under load, often to detect an overload condition for safety purposes. The strain gauge can also measure displacement force by the strain in a fixed arm, generally described as a load cell.

The strain gauge has a specified resistance at rest, which increases under strain. The ratio of the change in resistance to the strain is termed the gauge factor, having a typical value of about 2. Since this may not be specified precisely, system calibration under test is necessary. The change in the resistance when a strain gauge is active is rather small, usually less than 1%.

To detect only the change in resistance, four gauges are often connected in a bridge arrangement and the differential voltage measured between the centre terminals (Figure 9.10). The gauges are usually fixed to opposite sides of the mechanical component, say a beam subject to a bending force, so that opposite pairs are in compression or tension, giving the maximum differential voltage for a given strain. Due to their physical proximity, all the gauges should be at the same temperature, in which case its effect on the resistance of the metal conductors will be cancelled out. If necessary, the temperature coefficient of the material of the gauge may be included as an error factor.

In the example shown, a strain of 0.5% results in a change in resistance in the gauges of 1% (gauge factor = 2). The output from the potential dividers can be calculated as follows:

$$V_d = 5 \cdot \{(121/240) - (119/240)\} = 42\text{mV}$$

A high-gain instrumentation amplifier of the type described in Section 6.7.1 should be used to interface the bridge. With a gain of 100, the output will be 4.2V, more than enough for an MCU ADC input. Once the gauge has been calibrated to measure strain, displacement or

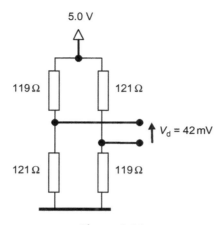

Figure 9.10
Strain gauge bridge.

stress (stress is proportional to strain), the amplifier gain should be adjusted for a convenient full-scale output, e.g. 4.096V.

Care must be taken in arranging the input connections, as the gauges will be highly susceptible to noise and interference. The amplifier should be placed as near as possible to the gauges, and connected with screened leads, and plenty of signal decoupling. A locally positioned gauge controller could be connected to the monitoring system by a standard serial link, such as RS-232.

9.7.2 Pressure Sensing

Pressure can be measured as an absolute value or relative to atmosphere. Atmospheric pressure is around 1 bar (or 1000 mbar), but this varies with the weather conditions between about 980 and 1040 mbar, with standard atmospheric pressure defined as 1013 mbar (1000 mbar = 100 kPa, kPa = kN/m^2). Small deviations from atmosphere caused by meteorological variation are inherently quite difficult to measure accurately, since this is a small change with a large offset.

In a traditional barometer, an evacuated chamber expands and contracts with atmospheric pressure, which operates a mechanical pointer. The electrical equivalent uses a diaphragm with sensitive strain gauges to measure its deformation under differential pressure (Table 9.3(d)). In a system such as a low-pressure air supply for a pneumatic system operating at, say, 5 bar, or a hydraulic system operating at 50 bar, it is usual to measure pressure relative to atmosphere. One side of the gauge diaphragm is exposed to atmosphere, while the pressurised system is connected to the other side. Any error due to atmospheric pressure variation is relatively small.

Laser trimmed piezoresistive gauge elements are used in low-cost miniature pressure sensors. Measurement with respect to atmosphere is called gauge pressure, where the diaphragm is simply fitted to a port in the target pressurised system with the other side exposed to ambient conditions. Absolute pressure measurement is made with reference to vacuum, often to measure atmospheric pressure.

The general specification for the Sensor Technics HCX Series, a representative pressure sensor range, is summarised below. Various devices are available that measure differential and absolute pressure in different ranges with span and offset adjustment of the integrated interface.

* *Range 5 mbar–5 bar max*
* *Output span 4.0V*
* *Offset 0.5V*
* *Thermal error 0.2% max*
* *Non-linearity* and hysteresis 0.5%

9.8 Humidity

There are various methods of measuring humidity, the percentage of water vapour in the atmosphere. The electrical properties of an absorbent material change with humidity, and the variation in conductivity or capacitance can be measured (Table 9.3(e)).

Humidity sensors are now typically supplied as a package with a precalibrated interface and standardised output. They typically include a temperature sensor, since both are frequently needed in humidity measurement applications, and the humidity measurement can be internally temperature compensated.

The specification for a Honeywell HIH 4000 Series humidity gauge is given below, assuming a supply of 5V and temperature 25°C:

* *Range 0–100% RH*
* *Sensitivity = 30.7mV/%RH*
* *Zero offset = 0.958V*
* *Accuracy ±3.5%*
* *Hysteresis ±3%*
* *Response time 15 s*
* *Interchangeability ±8%*

A typical domestic application is the tumble drier, which will stop when a required level of dryness in the outlet air is reached. The temperature sensor in the same detector package can regulate the heaters in the air inlet.

9.9 Integrated Sensors

Many sensors are now available in an integrated package, with built-in signal conditioning producing a calibrated output that can be connected directly to the MCU. The data sheet for such sensors must be carefully studied prior to any design work, but generally hardware interfacing of IC sensors is simple. Figure 9.11 shows a selection of such devices that are supported by Proteus VSM, connected to the 16F877A using different interfaces (VSM project ICSENS2). An LCD is included to display the sensor readings, but note that the application circuit is not in this case supported by demo firmware.

9.9.1 Distance

The Sharp GP2D12 distance measurement sensor uses an LED and photodetector to measure distances between 10 and 80cm, with an accuracy of ±3cm (10%). It produces a corresponding analogue output change of approximately 2V (negative going); some calibration may be necessary to obtain the maximum accuracy, which can be easily achieved using a metre rule or tape. This voltage can be fed directly into the PIC ADC for conversion to a distance display on the LCD. The simulated device produces 0.41V at 80cm and 2.35V at 10cm, so a 2.56V reference voltage and an 8-bit conversion would be appropriate, since the performance specification is not high precision.

9.9.2 Pressure

The Freescale MPX4115 measures absolute pressure in the range from 15 to 115kPa, using a piezoresistive transducer with internal temperature compensation and signal conditioning. Atmospheric pressure varies around 100kPa, which is usually expressed as 100 mbar in meteorology.

The minimum specified sensor output is 0.204V, and increases to a maximum of 4.794V, with a sensitivity of exactly 45.9mV/kPa and maximum error of 1.5%. In the simulated sensor, the output is 268mV at 15kPa, 4092mV at 100kPa and 4767mV at 115kPa, or 45mV/kPa, so there is a slight discrepancy which must be resolved in any software implementation.

The sensor can measure any absolute pressure up to and including the normal range of barometric pressures and can therefore be used as part of a weather monitoring system. Given that the voltage output ranges up to nearly 5V, using the supply voltage as the ADC reference is indicated, assuming it is accurate and well regulated.

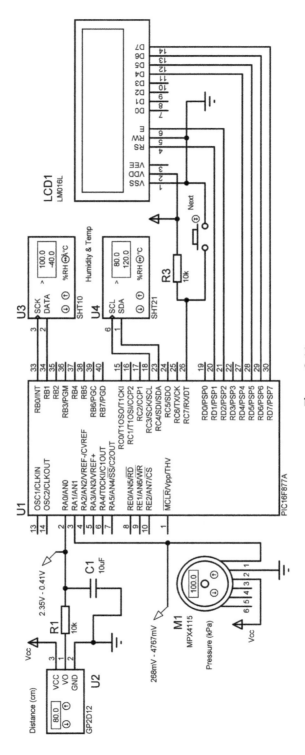

Figure 9.11
Integrated Sensors.

9.9.3 Temperature and Humidity

Two sensors from the Sensirion SHTxx sensor range are shown in the simulation. They have the same sensing elements but a different serial interface. The sensors incorporate a supply voltage regulator, capacitive polymer humidity sensor, semiconductor temperature sensor, 14-bit ADC and serial port. The humidity sensor is individually calibrated with the calibration data stored in internal memory.

The accuracy of the humidity sensor is between 2% and 5%, depending on the sensor variant, but this level of accuracy is reduced below 20% and above 80% humidity. The accuracy of the temperature measurement is at its optimum at around 25°C ($\pm 0.5\%$), but also reduces away from this central point. At 100°C, it is about $\pm 2\%$ (see sensor data sheet plots).

The SHT10 and similar sensors have a proprietary two-wire bus similar to SPI, with a separate clock (10MHz max), generated by the MCU. The open collector data line needs a pull-up resistor of 10k. The data format is similar to I^2C in that the data transmission block (from sensor to MCU) starts with a command sequence generated by the MCU, followed by a 2-byte data transfer and checksum error check code. The command requests humidity, temperature readings or a status code, and the sensor replies with data.

The SHT21 has similar sensor configuration and performance but has a standard I^2C interface. These sensors are designed for use in such applications as domestic tumble driers, automotive engine control and, medical monitoring.

Questions 9

1. Suggest three advantages of an optical switched sensor over its mechanical equivalent. (3)
2. Explain briefly why an inkjet printer contains a graduated plastic strip as part of its print head mechanism. (3)
3. Define the term sensitivity as applied to a sensor. (3)
4. Explain the difference between the terms accuracy and precision. (3)
5. Identify three sensors for measuring temperature and the main material/s from which each is constructed. (3)
6. State the gain required to obtain an output of 40mV/°C from an LM35 temperature sensor. (2)
7. From the LDR characteristic shown in Figure 9.9, state the resistance in kiloohms of the LDR at 1.0 lux illumination. (2)
8. Explain why strain gauges are normally connected as a bridge circuit. (3)
9. Explain why the instrumentation amplifier configuration is the most suitable for interfacing a strain gauge bridge. (3)
10. Calculate the sensitivity of the Sharp GP2D12 distance sensor in millivolts per centimetre and hence the distance measured if its output is 1.50V. Assume it is linear and has been calibrated in line with the VSM simulation model. (5)

Total (30)

Assignments 9

9.1 Sensor Comparison

Select the specification for three types of temperature sensor: a metal film temperature sensing resistor, a thermocouple and a thermistor. Construct a chart showing the sensitivity (if linear), range and total possible error at mid range. Investigate and establish a mathematical representation of the transfer function for each. From the function, predict the sensor output at minimum, maximum and mid-range temperature. Suggest at least one appropriate application for each sensor.

9.2 Diode Temperature Sensor Interface

The forward voltage drop across a silicon signal diode, used as temperature sensor, falls by 2mV/°C. The diode current is adjusted so that voltage is 650mV at 25°C. Design an interface that will produce an output of 0−2.50V representing diode temperatures of 0−50°C. Discuss the limitations of this measurement system.

9.3 Strain Gauge Interface

Construct and test a strain gauge bridge and interface using the instrumentation amplifier described in Chapter 6, using a high-performance op-amp. Bond the gauges onto a suitable metal bar of about 2mm thickness, fix one end and place known masses on the other to produce strain. Connect the amplifier output to an MCU with LCD display. Calibrate the system to measure these weights. Evaluate its performance in terms of accuracy, precision and other relevant error factors.

9.4 LDR Log Amp

Investigate the use of the log amp described in Chapter 6 as an interface for the LDR (the resistance of the LDR increases in decades as the light level decreases). Implement the log amp LDR interface and evaluate its performance against a standardised light meter.

9.5 IC Sensor Program

Write a control program to operate each of the sensors shown in Figure 9.11 separately, displaying the measured input in an appropriate format and resolution on the LCD (only one of the two humidity sensors needs to be implemented). Convert each into a routine that can be saved as a separate include file. Construct a program that combines these sensor functions into one program, selecting each measurement for display in turn.

CHAPTER 10

System Design

Summary
- The most suitable MCU must be selected for a particular design
- The power supply type and rating must suit the application
- An MCU project should be designed to an agreed specification
- The BASE module can be used as the basis for a range of applications
- A refrigeration controller specification illustrates the design process
- The weather station records temperature, light, humidity and pressure

Now that we have studied a range of system components and techniques, we can put them together to form a complete MCU-based hardware system. The range of PIC microcontrollers currently available will be reviewed so that the best device for any given design may be selected. A refrigeration controller will be used to illustrate the design process, a Base module described that can be used for different measurement and control applications, and a weather station designed around it.

10.1 PIC MCU Selection

For any given application, an MCU should be chosen from the PIC range which most closely meets the proposed specification. The range is constantly expanding, with new chips having additional or improved features and different combinations of existing features. The manufacturer's website www.microchip.com is an essential resource for MCU selection and design tool support.

The main MCU selection criteria are:

- *Number of I/O pins*
- *Instruction set features*
- *Program memory size*
- *Data memory size*
- *Peripheral set*
- *Clock type and speed*
- *Power consumption*

Interfacing PIC Microcontrollers.
DOI: http://dx.doi.org/10.1016/B978-0-08-099363-8.00010-8

Table 10.1 shows the main features of the different groups of 8-bit PIC MCUs (8-bit refers to the internal data bus size). An increasingly powerful set of features is provided as one progresses up the ranges, with the emphasis of new products on higher speed, additional ports and, particularly, lower power consumption. XLP (extra low power) MCU technology allows battery life of up to 20 years in mobile applications. Most newer 8-bit PICs include internal, high-speed oscillators, providing reduced instruction cycle time with low power consumption.

Table 10.2 compares the features of selected 8-bit PICs available at the time of writing. The features are compared in more detail, indicating the combination of program and data memory size, peripheral ports and maximum clock speed. The relative price is based on the reference dollar price at the time of writing, which will change over time.

Many higher power PIC MCU types are also available, including the 16-bit 24 series, running up to 70 mips (millions of instructions per second) and the dsPIC (digital signal processor) 30/33 MCUs with dedicated graphics, motor control and audio processing support features. PIC32 devices are the top of the range, offering multitasking embedded control, network support and high-performance audio—visual features. These will not be considered further here, but they share common features with the 8-bit PICs, and the whole range is designed to be as compatible as possible, with a view to design progression from the lower to the higher performance devices as the need arises.

Table 10.1: 8-Bit PIC MCUs.

MCU Family	MCU ID	Instruction Set and Program Memory	Maximum RAM and Stack	Internal and External Clock Max. Frequency	Description
Baseline	10F2XX 12F5XX	33 × 12-bit 2k	144 bytes 2 levels	4MHz 20MHz	Minimal cost and peripherals No interrupts
Mid-range	10F3XX 12F6XX 16FXXX	35 × 14-bit 8k	368 bytes 8 levels	16MHz 20MHz	General purpose Single interrupt
Enhanced mid-range	12F1XXX 16F1XXX 16LF1XXX	49 × 14-bit 32k	1.5k bytes 16 levels	32MHz 48MHz	Enhanced performance and peripherals Low power consumption Single interrupt with context saving
PIC 18 series	18FXXXX 18FXXJXX 18FXXKXX	83 × 16-bit 2M	4k bytes 32 levels	16MHz 64MHz	C optimised, hardware multiplier J = High memory, low cost K = High performance, low power Multiple interrupts with context saving

Table 10.2: Selected 8-Bit PICs.

MCU	# IC Pins	Max Instrs	RAM Bytes	EEPROM Bytes	Total I/O	# ADC Chans	# Timers 8,16 Bit	# U	# M	Int Osc (MHz)	Relative Cost
(a) Compact											
10F200	6	256	16	—	4	—	1	—	—	4	0.30
10F322	6	512	64	—	4	3	2	—	—	16	0.39
12F508	8	512	25	—	6	—	1	—	—	4	0.41
XLP											
12F1840	8	4096	256	256	6	4	2,1	—	—	32	0.78
12LF1840T48A	14	4096	256	256	6	4	2,1	1	1	32	1.12
(b) Mid-range											
16F54	18	512	25	—	12	—	1	—	—	20, N	0.39
16F628A	18	2048	224	128	16	—	2, 1	1	—	20, 4	1.47
16F690	20	4096	256	256	18	12	2, 1	1	1	20, 8	1.20
16F818	18	1024	128	128	16	5	2, 1	—	1	20, 8	1.56
16F84A	18	1024	68	64	13	—	1	—	—	20, N	3.11
[a]16F877A	40	8192	368	256	33	8	2, 1	1	1	20, N	3.71
16F887	40	8192	368	256	36	8	2, 1	1	1	20, 8	1.78
16F946	64	8196	336	256	54	8	2, 1	1	1	20, 8	2.31
XLP											
16LF1902	28	2048	128	—	25	11	1, 1	—	—	20, 16	0.78
16LF1907	40	8196	512	—	36	14	1, 1	1	—	20, 16	1.23
16LF1508	20	4096	256	—	18	12	2, 1	1	1	20, 16	0.77
16LF1527	64	16,192	1536	—	55	30	6, 3	2	2	20, 16	1.54
16F1823	14	2048	128	256	12	8	2, 1	1	1	32, 32	0.78
16F1947	64	16,192	1024	256	53	17	4, 1	2	2	32, 32	1.82
(c) High-performance 8-bit PICs											
18F1220	18	2048	256	256	16	7	1, 3	1	—	40, 8	1.96
18F4220	40	2048	512	256	36	13	1, 3	1	1	40, 8	4.46
18F8310	80	4096	768	—	70	12	2, 3	1	1	40, 8	3.01
18F96J60	100	65,536	3808	—	70	16	2, 3	1	—	42, 0	3.77
XLP											
18F13K22	20	4096	256	256	18	12	1, 3	1	1	64, 16	1.16
18F24J11	28	8096	3800	—	23	10	2, 3	2	2	48, 8	1.20
18F87K90	80	65,536	4096	1024	69	24	6, 5	2	2	64, 16	3.35

Notes: Max. # instructs = maximum number of instruction words.
Number of serial ports: #U = USART port, including A/EUSART, #M = MSSP (I^2C and SPI).
Internal oscillator = *N* if not fitted, internal oscillator only available in compact range.
Relative cost based on volume price on www.microchip.com atow (at time of writing).
[a]16F877A is a legacy product not listed in main catalogue but available by search at the time of writing.

10.1.1 Compact 8-Bit PICs

The 10Fxxx and 12Fxxx devices (Table 10.2(a)) use a simplified 12-bit instruction set with limited set of peripherals, all with internal oscillators to minimise the external component count. The smallest only have 6 pins, so in surface mount form they are among the smallest microcontrollers available. Some 8-pin devices can offer analogue inputs, EEPROM and serial interfaces, in which case the 14-bit instruction set is used. A limited number of low-power, high-speed devices are currently available in this range.

10.1.2 Mid-Range 8-Bit PICs

Mid-range (Table 10.2(b)) refers to the 16Fxxxx series of 8-bit PICs only; these all use the same 14-bit instruction set, originally with a maximum of 8k program words. The PIC 16F877A is used as the reference device for this range throughout this book, since it was one of the first to offer a full range of features. The 16F84A is still useful initially when learning to use PIC chips, since it is a minimal device which avoids the complicating features of later MCUs.

The 16F84A and the 16F877A are obsolete for new commercial applications, and their price is relatively high. For this reason, the simulation models for these chips are offered as part of a low-cost Proteus design starter package. If used for initial design investigations, the 16F877A should be replaced by the pin compatible 16F887 at the implementation stage; the minor differences between them have been outlined previously in Section 1.6. Similarly, the 16F818 (or similar) should replace the 16F84A.

The 16F54 is the most basic mid-range device available, with minimal features and a corresponding low-cost. The 16F690 chip is included in this list because it is used in the LPC demo board supplied with the PICkit3 baseline programmer. It is also a good starting point for 8-bit designs, offering a good range of features at reasonable cost. The largest on the list is the 16F946, with a total of 54 I/O channels and 8k program memory.

The original mid-range PICs are now being replaced with enhanced XLP (extra low power) devices. The 16LF1xxx chips have a 16MHz internal clock, while the 16F1xxx series has a 32MHz internal clock. The 8k program limit has been overcome in both ranges. The enhanced MCUs also have automatic context saving on interrupt, 16-level stack, an additional file select register and an extended instruction set.

The additional instructions include relative branch and indirect moves, which potentially improve the code efficiency of C programs. Hardware enhancements include a built-in temperature sensor that can trigger a warning or cooling system if the chip overheats, and an internal accurate reference voltage for the ADC. Self-programming means that program flash memory can be rewritten without an external programmer, and linear data memory addressing is also implemented, eliminating page boundary problems.

The number of timers and serial ports has also been boosted in some of the larger chips in the enhanced range. The 16LF1902 provides an LCD display driver interface in addition to the usual peripheral support. The 16LF1454/5/9 chips are just being introduced with USB ports at the time of writing. Some chips such as the 16F1508 have a configurable logic cell (CLC) which has a small set of logic gates that can take external and internal inputs and produce a logic result more quickly than a programmed operation.

The low power consumption of the XLP range typically results in a standby current of 20–30nA at a supply voltage of 1.8V in sleep mode. The operating current is quoted as 30µA per MHz for the 16F1823 (power consumption is generally proportion to clock rate). If running at a maximum of 32MHz, the current will be $30 \times 32 = 960$µA, or less than 1 mA, and the power consumption will be less than 2mW.

10.1.3 Power 8-Bit PICs

The high-power 8-bit PIC chips are designated 18Fxxxx (Table 10.2(c)). These have a more extensive 16-bit instruction set, originally ran at 40MHz, and are generally optimised for programming in 'C'. All had an 8×8 hardware multiplier to speed up arithmetic operations, and an 8MHz internal oscillator. There are now two main types of newer 18Fxxxx PICs. The J series includes some XLP devices with relatively large RAM capacity. Others offer specialist interfaces: LCD, Touch Sensing, USB and Ethernet. The K series is optimised for low power consumption with high speed, running at up to 64MHz with a program memory capacity of 64k instructions.

10.2 Power Supplies

The power supply is an important consideration when designing a PIC application. An ideal d.c. supply has a constant voltage and infinite current with no ripple or noise. Most regulated supplies come reasonably close to this. The main options are a mains powered supply with d.c. regulator, or a battery.

10.2.1 Mains Supplies

Electronic systems have traditionally been powered from a mains supply with a transformer and bridge rectifier producing an unregulated low-voltage output at about 12V. An on-board 5V IC regulator will then eliminate the mains ripple (50 or 60Hz) and deliver a smooth output voltage with sufficient current. Alternatively, a plug-top regulated supply with coaxial connector is often more convenient for small, stand-alone MCU boards, supplying up to 1A while a workshop bench supply will provide higher current for prototype systems.

A conventional linear regulator uses a series power transistor and output sensing amplifier to control the voltage, while producing current up to a specified limit. A large reservoir electrolytic capacitor provides additional smoothing of the voltage, helping to overcome any transient (short-term) current demands, such as when a peripheral current driver switches over. In a rack system with many boards or controller units, a central power supply can deliver a regulated voltage to all units, or an unregulated supply to individual on-board regulators. The linear regulator is simple and effective but inefficient. If the input voltage is 12V and the output 5V, more than half the input power is wasted.

The switch mode power supply is now preferred in commercial designs, as it is more efficient than the linear regulator, though more complex. It contains an oscillator that drives a PWM output that is averaged out to a regulated d.c. level in the output stage. Feedback voltage from the output controls its power output − as the load current increases and the voltage tends to drop, the output is increased to compensate. Small switch mode d.c. to d.c. converters can be used with a central unregulated supply.

10.2.2 Battery Supplies

Portable applications need a battery supply. Much effort has been devoted to developing improved battery technologies, as mobile applications have become widespread. The battery needs to supply the necessary current for as long as possible, and the MCU circuit needs to tolerate a variable supply voltage. The lithium-ion battery is the current standard for laptops, smartphones and most mobile technology. It can be recharged many times, from any existing charge state, is reasonably compact and has a good power to weight ratio.

Prototype systems and small hobby applications will probably still use less expensive dry cells or lithium button cells. These have to be used in combinations that produce a suitable total voltage and current. For example, $3 \times 1.5V = 4.5V$ dry cells can be used to power a PIC board, since the PIC can work down to 2V as the batteries discharge. In small applications, with minimal current requirements, a single 3V lithium cell can be used.

10.2.3 PIC Supplies

Much effort has also been directed to developing low-power microprocessors and peripherals. Not only does this extend battery life, it also reduces problems caused by power dissipation in large systems. The first generation of digital chips were based on bipolar transistors, which, although fast, had relatively high power consumption. CMOS (FET-based) chips were developed to overcome this and allowed the first microprocessors to be developed. XLP (extra low power) technology in the PIC MCU is a further step on this path.

MCU power dissipation increases with frequency, so the power supply current needs to be specified accordingly. The 16F877A, for example, will operate at between 0Hz and 20MHz with a supply between 4.0 and 5.5V (data sheet Figure 17). The minimum supply voltage required at 10MHz is 3.0V, and 2.0V at 4MHz. The replacement 16F887 improves on this somewhat and will run at up to 8MHz with a 2.0V supply.

The 16F877A draws about 2mA at 4MHz with a 5V supply (HS mode). This increases to just under 6mA at 20MHz, while 16F887 draws only 4mA at this speed. Using its internal oscillator, the 16F887 uses just over 2mA at 8MHz, and in LP mode at 32kHz, only about 30μA. A low-power version of the 16F877A is available, the 16LF877A, whose maximum supply current with a 32kHz clock is quoted as 35μA. The electrical characteristics of both variants are listed in Appendix A.

10.2.4 System Supplies

The power supply must obviously be selected to provide sufficient current with a margin of, say, 50% to cope with transients, contingencies and, maybe, modest circuit expansions or powered peripherals. The overall current demand must be estimated for the whole circuit, and the mains supply current rating chosen accordingly or the battery rated capacity (amp hours) used to predict the battery life.

All interfacing circuits must be included in the power supply current budget and, if necessary, designed to minimise power consumption, and low-power peripherals selected. For example, the LCD module used in the demo circuits may draw up to about 10mA, as it has its own processor. On the other hand, a plain LCD display driven direct from the main MCU, as seen in the mechatronics board in Chapter 7, has very low power consumption, while an LED display has comparatively high power consumption.

In most digital systems, it is advisable to decouple the supply lines adjacent to any IC component, because fast signal edges by definition contain high-frequency components which radiate energy to adjacent tracks. Clock or other high-frequency signals superimposed on an IC supply may cause it to malfunction. The power supply smoothing electrolytic capacitor will help to suppress low-frequency transients, and 10pF ceramic capacitors across the supply pins close to each active device are standard practice, providing a shunt path for high-frequency interference and crosstalk.

Since the supply pins are not shown explicitly in the circuit schematic, supply decoupling components must be introduced at the layout stage. The success of such precautions is one aspect of circuit design that can only be confirmed by final hardware testing.

10.3 System Design

When designing a single microcontroller application, we can start with a specification of the functions that the system is intended to perform, and then select the most suitable chip. In theory, all types should be considered, but the developer is likely to be committed to a particular system because of previous investment in knowledge and resources. Alternative types will probably only be considered if the default range cannot provide the features required, or there is some other reason to change, such as designing for a customer who uses a different range.

10.3.1 Application Specification

Here, our default choice is obviously the PIC. We have to identify the features required for the MCU, its interfacing and select any sensors, transducers and communication links needed. Here is a typical specification:

> *A control system is required for a refrigeration unit which will maintain the temperature within an insulated enclosed space, such as a temperature controlled shipping container, at a selected temperature between 1°C and 9°C. The controller will connect to the refrigeration unit via a suitable contactor, which switches the refrigeration unit on and off. The temperature will be monitored by suitable sensors, and controlled to within ± 0.5°C, pre-settable using up/down push buttons, and displayed on a self-illuminating display which is readable from a maximum of 2m. When the unit is powered up, the previous temperature setting must be used by default. If the temperature deviates from the set temperature by more than 2°C, or any other significant fault occurs, an alarm must sound within the unit, and remotely (in the lorry cab) with a flashing indicator. The design must be highly reliable, robust, moisture proof and low maintenance. It will be powered from a vehicle 12 or 24V d.c. supply (selectable).*

10.3.2 Design Outline

The first step in the development process is to draw a block diagram, so that the system requirements can be clearly visualised (Figure 10.1). At this point, it might be useful to suggest some rules for the construction of block diagrams in embedded system design:

- *The main elements are shown in block form and labelled accordingly*
- *These are connected by arrow segments indicating the nature of the signal and the principal direction of information flow*
- *Serial and parallel data are represented by single and block arrows, respectively*
- *If the signal is not digital, it should be labelled accordingly, specifying voltage levels and signal type, with simple signal diagram if necessary*
- *The block diagram allows the I/O requirements to be identified, and a suitable microcontroller selected*
- *The block diagram can subsequently be expanded into a circuit schematic*

For reliable operation, it is suggested that a set of four temperature sensors are installed near the four corners of the storage space. In normal operation, an average of these will be displayed. If a single sensor goes faulty, we will assume that its reading will go out of

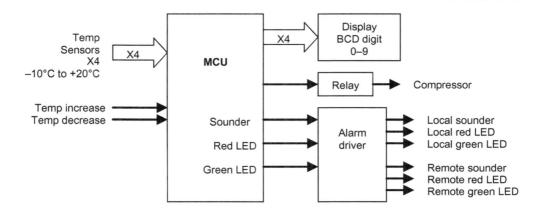

Figure 10.1
Block diagram of refrigeration controller.

range. A 'sensor faulty' alarm can be generated, say a short beep and flashing indicator. As long as the other three sensors agree within 2°C, the controller will continue to operate, taking an average of these three only and ignoring the faulty sensor. If more than one sensor goes out of range, a fault condition will be indicated and the system shut down.

Good quality push buttons with moisture-proof housings will be selected. The contactor will be the default control interface for the compressor. This may be an independent diesel unit that will have its own control unit, whose interfacing requirements must be established. The display can be a single 7-segment LED type, which is cheap, simple to interface and self-illuminating. A larger than standard size can be used for good visibility (say 2″), and these are not expensive. Since different frequencies will be used for different alarm conditions, small loudspeakers will be used, with the drive signals generated in software. Red LEDs for the alarm will be used, with a high brightness LED in the remote monitor unit. A green power LED indicating normal operation will also be incorporated into both the main and the remote alarm units.

The temperature sensors will be housed in aluminium boxes, bonded to the face of one side, for good thermal contact. The LM35 covers the range with sufficient accuracy (just), and can be used in the initial design, but a more robust industrial grade temperature sensor might be desirable that has a smaller range and greater resolution. A current-driven link is more reliable in harsh environments and mitigates the effect of any voltage drop over the length of the sensor connectors, which could be several meters. Screened screw connectors will be used at both ends of the connecting cables for electrical and mechanical robustness.

The regulated 5V supply from the main unit will be provided to the remote sensors, with all connections within an armoured screened cable. The signal 0V will be separate and screened, to minimise the possibility of interference and false alarms due to the compressor

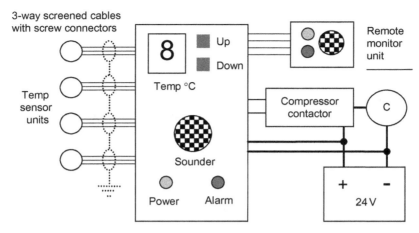

Figure 10.2
Refrigeration control system design.

switching currents or the vehicle ignition system. The same connectors and cabling can be used for the remote alarm unit, since it also needs three signal wires, and aluminium boxes can be used for all units. Connectors and cabling must be waterproof if exposed to the elements. The system overall design is visualised in Figure 10.2.

10.3.3 Implementation

The implementation of the refrigeration controller will not be completed here, but a similar system implementation will be detailed in Section 10.4. However, some general implementation issues will be considered.

We are assuming a single supply of $+5V$, derived from the 12/24V vehicle battery, to which it must be permanently connected. A regulator or d.c./d.c. converter providing sufficient current must be selected. The remote sensing units can be connected to the control unit using the current loop amplifier shown in Figure 6.12, which provides a gain of 10 overall. The 10mV/°C at the input will give 100mV/°C at the output, or a range of $0.1 \times 16 = 1.6V$. This is suitable for conversion by the ADC at 10mV/bit, or resolution of 0.1°C.

The MCU needs four 8-bit analogue inputs for the temperature sensors. At mid-range, 8-bit conversion will give a resolution of about 1%, which is more than adequate. A total of 10 digital I/O pins are needed. Program memory of 1k will be assumed initially, but this will be reviewed when the code is complete. An accurate clock is not needed, so an internal oscillator will be used, reducing the component count and improving reliability. The PIC 16F818 seems to fit the bill, with 16 I/O pins in total, including 5 analogue inputs.

Remember, we will need an extra analogue input for the reference voltage. The 16F818 has 1k program memory, and EEPROM for storing the previous set temperature. The hardware timers will be useful for generating the timed outputs. If we run out of program memory, the 16F819 (2k) can be substituted at slightly higher cost. Considering the cost of failure of the unit, this will not be significant. Both chips have an 8-MHz internal oscillator, and ICD programming and debugging. A low-power device is not necessary since the controller is running from the vehicle battery.

The arithmetic processing should be straightforward, as only single digit numbers are in use. The temperatures will be read in as 8-bit numbers in the range of 0−160. The readings should be checked for out of range values, indicating a faulty sensor. If the remaining three sensors agree (within limits), this should be taken as a correct reading and the faulty input excluded.

The temperature readings should be averaged and the result divided by 10 (see Chapter 5) to obtain a number in the range of 0−16, and 4 subtracted to calculate the temperature in the range of −4 to +12. The appropriate digit can be selected from the table for display (Chapter 4) for the range of 0−9°C. Any temperature below 0 can be classified as faulty, and a 'Temperature too low' alarm generated. Similarly, a result above 9°C should generate a 'Temperature too high' alarm.

The display (BCD input) can use a program look-up table for the digit display codes 0−9. If the temperature goes to low, 'L' could be displayed as well as the alarm operating. Similarly, 'H' could be displayed if too high. The alarm sounds should use the hardware timers to generate suitable frequencies on the outputs, and the delay times for the flashing LED warnings.

The set temperature should be displayed when the up or down button is pressed and incremented by 1°C for each press. When released, the display should revert to the measured temperature. The set temperature should be stored in EEPROM when the button is released. This figure can then be recalled during program initialisation on power-up. During normal operation, the average reading will be compared with the set temperature, and the compressor switched on and off accordingly. A program outline is shown in Figure 10.3.

Once the application specification has been converted into a block diagram, a suitable MCU provisionally selected, a program outlined and an overall hardware design established, a controller PCB can be designed. Proteus VSM can be used to create a schematic, as detailed in Chapter 3. A PCB can then be produced and constructed from the resulting circuit netlist. Hardware prototyping and testing is covered in 'PIC Microcontrollers' by the author.

```
COLD1
Refrigeration controller:
Averages input from four temperature sensors
Checks for faulty sensor and averages
Switches output to the compressor

Initialise
                Analogue inputs (5)
                        4 channels + Vref
                Digital Inputs (2)
                        Up, Down
                Digital Outputs (8)
                        Compressor
                        Display (4)
                        Power, Alarm LEDs
                        Alarm Sounder
                Analogue control
                Timers

Main
                Recall stored SetTemp
                REPEAT
                        Read inputs
                        Check for faulty sensor
                                IF fault, set alarm
                        Average inputs
                        Store Temp

                        Check Temp
                                IF too high or too low, set alarm
                        Check buttons
                                IF 'up' pressed
                                        Increment SetTemp & store
                                IF 'down' pressed
                                        Decrement SetTemp & store

                        Display Temp
                        Compare Temp with SetTemp
                        Switch compressor on/off
                ALWAYS
```

Figure 10.3
Refrigeration controller program outline.

10.4 BASE System (Project BASE2)

The BASE system is a simple, general purpose PIC controller board, using subsystems and interfaces described in previous chapters. It could be used as the basis for implementation of the refrigeration controller specified earlier. It was originally designed around a 16F877A MCU, and this is retained in the simulation version, but replacement with the 16F887 chip is recommended for implementation.

The BASE system has a 4MHz clock external circuit (which can be eliminated if using the 16F887), ICD interface, ADC reference voltage, and test input circuits and sounder output.

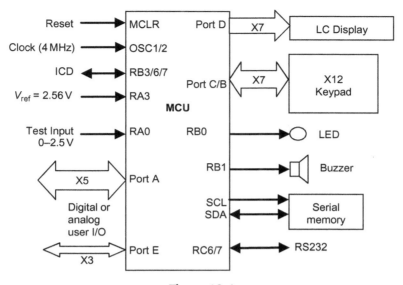

Figure 10.4
BASE module block diagram.

The on-board peripherals are a 12-button keypad, alphanumeric LCD display and serial flash memory, with an RS232 serial link. The latter allows connection to a wireless transceiver or other compatible peripheral unit. The block diagram in Figure 10.4 shows the main system features and the interfacing to the MCU.

10.4.1 BASE Hardware

The base board schematic is shown in Figure 10.5 (VSM project BASE2). The circuit is built around the 16F877A, with Ports A and E brought out to an in-line connector for the external interface circuits. Port D is allocated to the serial LCD, with Port C interfacing with the serial memory and RS232 driver. Port B programming pins are brought out to the ICD connector, and remaining Ports B and C pins used for the keypad, an LED indicator and buzzer. A test program (see later) will exercise all the main features and allows the design to be initially checked by simulation.

10.4.1.1 Reset

A manual reset is included, so that programs can be restarted when the board is running independently. If the program appears to be malfunctioning, a hardware reset is usually the first remedy. The reset timers in the PIC MCU should ensure a reliable start on power up without the need for a manual reset, but if brown-out precautions are not included in the firmware, it may be useful.

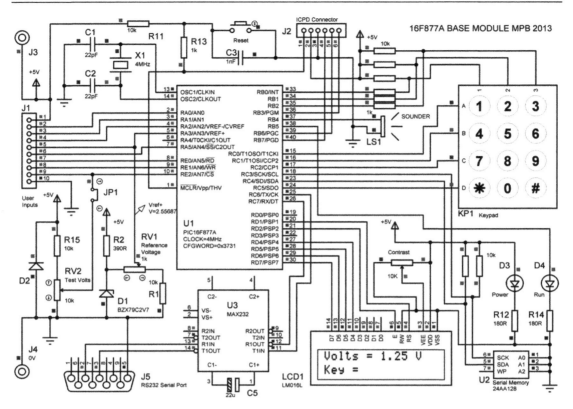

Figure 10.5
BASE module schematic.

10.4.1.2 Clock

A standard crystal circuit is used, running at 4MHz. This gives a 1µs instruction execution time, which is convenient for analysing program timing. The crystal needs to be physically near the MCU, so that additional track capacitance does not prevent the crystal from oscillating, or affect the resonant frequency. The maximum frequency possible is 20MHz, giving a 200ns instruction cycle, or 5 million instruction cycles per second. At this speed, the XT crystal must be replaced with a HS (High Speed) type.

10.4.1.3 ICD

The MCU ICD pins are brought out to a connector for the programmer module that provides program downloading and final debugging in hardware. Program creation is reviewed later, but its main objective is to create a machine code file for downloading to the chip in VSM. However, we must switch back to MPLAB for firmware downloading.

The ICD interface allows in-circuit debugging in the target hardware using the same debugging techniques as those used in initial simulated testing.

10.4.1.4 User I/O

Ports A and E are attached to a connector for external analogue or digital I/O. RA3 is used for the ADC reference voltage, 2.56V for 8-bit conversion. If 10-bit conversion is required, a 4.096V reference can be substituted, as shown in Chapter 6. An adjustable test voltage is connected to AN0; when the test program is running, its value will be displayed on the LCD (0.00−2.50V). In the test program, the 'Run' LED flashes as the simulation runs, and the buzzer provides an audible tick when a button is pressed on the keypad.

10.4.1.5 Keypad

The keypad interface is detailed in Chapter 4. The 12-button keypad is connected to pins in Ports B and C. In the test program, the outputs to the rows (ABCD) and the inputs from the columns (123) are initially all high. The rows are taken low in turn and the columns tested for 0. When a button is detected, the corresponding ASCII code is returned and processed. Note that in simulation there may be a delay between 'pressing' the button and the contact closing, to simulate switch bounce, which may cause confusion in single-step mode.

10.4.1.6 LCD

The operation of the 16×2 character LCD is also detailed in Chapter 4. It is configured in 4-bit mode, i.e. ASCII codes are fed to it in two stages (high nibble then low nibble) via four data lines connected to the high bits of Port D. Low bits of the same port provide the control lines RS (Register Select) and E (Enable). The RW (Read/Write) line is hard-wired low for writing only − it is not necessary to use the LCD handshaking which would require a change in data direction and make the software more complex.

10.4.1.7 Serial Memory

The memory chip locations are accessed via the I^2C serial interface (RC3, RC4), as detailed in Chapter 8. Data is transferred in one byte packets on SDA, preceded by addressing bytes to select the chip and the location. SCK provides a clock pulse with each bit to latch it into the destination device. The hardware address pins are connected low to assign the default address 0. WP (Write Protect) allows the chip write to be disabled to prevent accidental overwriting of important data, but it is not connected here. The serial memory is exercised in the test program by storing the analogue input voltage after it has been displayed. This can be seen in simulation mode by displaying the contents of serial memory.

10.4.1.7.1 RS232 Port

The RS232 port is connected to a 9-pin D-type connector via a standard MAX232 chip. This converts the signal level between a bipolar voltage of about 18V ($\pm$ 9V) for communication with a host system or other peripheral via a long link, and TTL levels for the MCU. The line voltage is generated by an internal charge pump from the single 5V supply, using the externally fitted capacitors. The hardware handshaking lines (RTS, CTS) are not implemented.

The operation of the RS232 interface is described in more detail in Chapter 8. For communication with a host PC, an external USB converter may be required. For communication with a wireless transceiver, the MAX232 chip can be eliminated. If the MSSP port is required for communication purposes, the serial memory may be omitted, or a separate address created for the communication port on the I^2C bus. If SPI is to be used, RD0 is available as a hardware select output.

10.4.1.8 Power Supply

A standard 5V, 1A linear regulator chip is the default choice for a small PIC board and would be suitable for the BASE board. Any high-power loads attached to the board, such as motors, can then use the unregulated supply (12V). Standard linear IC regulators will provide $+5V \pm 0.25V$, with low noise and ripple. Alternatively, an on-board d.c. to d.c. converted may be used for greater efficiency.

10.4.2 BASE Test Program

A test program that exercises all parts of the hardware, while being as simple as possible, is always useful for prototyping. If the hardware can be proved to function correctly, subsequent software development can then be undertaken in the knowledge that any problems must be caused by program bugs. The base board test program (Program 10.1) has two parts. The first reads the analogue test voltage input, displays it and stores it in memory; the second part reads the keypad and displays the key. Several support routines are implemented as include files.

The serial memory access routine, the display driver routine and the BCD conversion routine are allocated reserved GPR ranges in the register label equates. These routines, as well as the analogue port read routine, are included as separate source code files at the end of the main source listing. This allows these routines to be reused in future programs, ideally without modification. Information about the way the routine is used (register requirements, parameter passing and so on) is included in the header to make this as straightforward as possible. The keypad scanning routine was modified to use a mix of Port B and C lines.

```
;;;;;;;;;;;;;;;;;;;;;;;;;;;;;;;;;;;;;;;;;;;;;;;;;;;;;;;;;;
;
;        Project:            Interfacing PICs
;        Source File Name:   BASE2.ASM
;        Devised by:         MPB
;        Date:               28-01-13
;        Status:             Updated for VSM v8
;
;;;;;;;;;;;;;;;;;;;;;;;;;;;;;;;;;;;;;;;;;;;;;;;;;;;;;;;;
;
;        Program to exercise the 16F877 BASE module
;        with 8-bit analogue input, LCD, keypad
;        and serial memory
;
;;;;;;;;;;;;;;;;;;;;;;;;;;;;;;;;;;;;;;;;;;;;;;;;;;;;;;;;;;

            PROCESSOR 16F877A      ; Clock = XT 4MHz
            __CONFIG 0x3731        ; Standard fuse settings
            INCLUDE "P16F877A.INC"  ; standard register labels


; User register label allocation ;;;;;;;;;;;;;;;;;;;;;;;;;;;;;

; GPR 20 - 2A       local variables
; GPR 30 - 32       KEYPAD subroutine
; GPR 40 - 42       ADC input routine
; GPR 60 - 65       SERMEM serial memory driver
; GPR 70 - 75       LCDIS display driver
; GPR 77 - 7A       CONDEC BCD conversion routine

; Analogue input allocation

;    Vin0    PORTA,0
;    Vin1    PORTA,1
;    Vin2    PORTA,2
;    Vref    PORTA,3
;    Vin4    PORTA,5
;    Vin5    PORTE,5
;    Vin6    PORTE,6
;    Vin7    PORTE,7

; Serial memory pin allocation

;    SCL   PORTC,3
;    SDA   PORTC,4

; Display pin allocation

;    RS    PORTD,1
;    E     PORTD,2
;    D4    PORTD,4
;    D5    PORTD,5
;    D6    PORTD,6
;    D7    PORTD,7

; Keypad I/O pin labels

      #DEFINE     Col1    PORTB,0
      #DEFINE     Col2    PORTB,1
      #DEFINE     Col3    PORTB,2

      #DEFINE     RowA    PORTC,0
      #DEFINE     RowB    PORTC,1
      #DEFINE     RowC    PORTC,2
      #DEFINE     RowD    PORTC,5

; Misc outputs

      #DEFINE     Sound   PORTB,4
      #DEFINE     LED     PORTB,5
      #DEFINE     Spare1  PORTD,0
      #DEFINE     Spare2  PORTD,3

; Register labels ......................................

LCDport   EQU     08      ; assign LCD to Port D
LCDdirc   EQU     88      ; data direction register
Temp      EQU     20      ; temp store
Tabin     EQU     21      ; Table pointer
```

Program 10.1
BASE module test program.

```
; MAIN PROGRAM
;--------------------------------------------------------

        CODE    0               ; Default start address
        NOP                     ; required for ICPD

; Port & display setup ---------------------------------

        BANKSEL TRISA           ; Select bank 1
        MOVLW   B'11001111'     ; Port B code for
        MOVWF   TRISB           ; keypad inputs & sounder
        MOVLW   B'10011000'     ; Port C code for
        MOVWF   TRISC           ; keypad, memory & USART
        CLRF    TRISD           ; Display port is output

        BANKSEL PORTA           ; Select bank 0
        CLRF    PORTD           ; Clear display outputs
        CLRF    HiReg           ; select memory page 0
        CLRF    LoReg           ; select first location
        CALL    inimem          ; initialise memory
        CALL    inid            ; Initialise the display

;--------------------------------------------------------
; MAIN LOOP
;--------------------------------------------------------

start   BSF     LED             ; switch on run LED
        MOVLW   0x07            ; Select AN7 input

        CALL    adin            ; read analogue input
        CALL    condec          ; convert to decimal
        CALL    putdec          ; display input
        CALL    store           ; store input in memory

        BCF     LED             ; switch on run LED
        CALL    putkey          ; Fixed message
        CALL    keyin           ; scan keypad
        CALL    send            ; display key
        GOTO    start           ; and again


;--------------------------------------------------------
; SUBROUTINES
;--------------------------------------------------------
; Display input test voltage on top line of LCD
;--------------------------------------------------------

putdec  BCF     Select,RS       ; display command mode
        MOVLW   080             ; code to home cursor
        CALL    send            ; output it to display
        BSF     Select,RS       ; and restore data mode

; Convert digits to ASCII ------------------------------

        MOVLW   030             ; load ASCII offset
        ADDWF   Huns            ; convert hundreds to ASCII
        ADDWF   Tens            ; convert tens to ASCII
        ADDWF   Ones            ; convert ones to ASCII

; Display voltage on line 1 ----------------------------

        CALL    volmes          ; Display on line 1

        MOVF    Huns,W          ; load hundreds code
        CALL    send            ; and send to display
        MOVLW   '.'             ; load point code
        CALL    send            ; and output
        MOVF    Tens,W          ; load tens code
        CALL    send            ; and output
        MOVF    Ones,W          ; load ones code
        CALL    send            ; and output
        MOVLW   ' '             ; load space code
        CALL    send            ; and output
        MOVLW   'V'             ; load volts code
        CALL    send            ; and output

        RETURN                  ; done

; Store voltage in serial memory -----------------------

store   BSF     SSPCON,SSPEN    ; Enable memory port
        MOVF    ADRESH,W        ; Get voltage code
        MOVWF   SenReg          ; Load it to write
        CALL    writmem         ; Write it to memory
        INCF    LoReg           ; Next location
        BCF     SSPCON,SSPEN    ; Disable memory port
        RETURN                  ; done
```

Program 10.1
(Continued)

```
;-------------------------------------------------------------
; Display key input on bottom line of LCD
;-------------------------------------------------------------

putkey    BCF      Select,RS          ; display command mode
          MOVLW    0C0                ; code to home cursor
          CALL     send               ; output it to display
          BSF      Select,RS          ; and restore data mode
          CALL     keymes
          RETURN                      ; done


;-------------------------------------------------------------
; Display fixed messages
;-------------------------------------------------------------

volmes    CLRF     Tabin              ; Zero table pointer
next1     MOVF     Tabin,W            ; Load table pointer
          CALL     mess1              ; Get next character
          MOVWF    Temp               ; Test data...
          MOVF     Temp,F             ; ..for zero
          BTFSC    STATUS,Z           ; Last letter done?
          RETURN                      ; yes - next block
          CALL     send               ; no - display it
          INCF     Tabin              ; Point to next letter
          GOTO     next1              ; and get it


keymes    CLRF     Tabin              ; Zero table pointer
next2     MOVF     Tabin,W            ; Load table pointer
          CALL     mess2              ; Get next character
          MOVWF    Temp               ; Test data...
          MOVF     Temp,F             ; ..for zero
          BTFSC    STATUS,Z           ; Last letter done?
          RETURN                      ; yes - next block
          CALL     send               ; no - display it
          INCF     Tabin              ; Point to next letter
          GOTO     next2              ; and get it

;-------------------------------------------------------------
; Text strings for fixed messages
;-------------------------------------------------------------

mess1     ADDWF    PCL                ; Set table pointer
          DT       "Volts = ",0       ; Text for display

mess2     ADDWF    PCL                ; Set table pointer
          DT       "Key = ",0         ; Text for display


;-------------------------------------------------------------
; INCLUDED ROUTINES
;-------------------------------------------------------------
; KEYPAD DRIVER        Scans 3x4 keypad once
;        CALL keyin    Returns with ASCII in W
;                      0 = no key pressed
;
          INCLUDE "KEYPAD2.INC"
;-------------------------------------------------------------
; LCD DRIVER           Operates 2x16 alphanum. LCD
;        CALL init     Initialises display
;        CALL send     Sends a character in W to display
;
          INCLUDE  "LCD2.INC"
;-------------------------------------------------------------
; NUMBER CONVERTER     Converts byte to 3 digit decimal
;        CALL condec   Receives 8-bit binary in W
;        Returns       BCD digits in 'huns','tens','ones'
;
          INCLUDE  "CONDEC2.INC"
;-------------------------------------------------------------
; ADC DRIVER           Read selected analogue input
;        CALL adin     Receives channel number in W
;        Returns       8-bit input in W
;
          INCLUDE  "ADIN2.INC"
;-------------------------------------------------------------
; SERIAL MEMORY DRIVER    Writes and reads 24AA128 EEPROM
;        CALL inimem     To initialise
;        CALL writemem   To write location*
;        CALL readmem    To read location#
;
;      Before writing or reading:
;      *# Write high address into 'HiReg' 00-3F
;      *# Write low address into 'LoReg' 00-FF
;      *  Load data send into 'SenReg'
;      #  Read data received from 'RecReg'
;
          INCLUDE "SERMEM2.INC"
;-------------------------------------------------------------
          END                        ; of source code
;-------------------------------------------------------------
```

Program 10.1
(Continued)

The directive DT has been used here to create the data table of ASCII codes required for the display of fixed messages. It generates a sequence of RETLW instructions for each code, which is accessed in the usual way by modifying the program counter with ADDWF PCL. The table is terminated with a zero, which is detected by the output routine to terminate the message.

The include files were created by modifying the demonstration program for each interface into the form of a subroutine. This entails deleting the initialisation which is common with the main program, and using a suitable label at the start (same as the include file name), and finishing with a RETURN. This is a simple way to start building a library of utilities for the base hardware that can be reused in different applications.

As can be seen, the software design philosophy is to make the main program as concise as possible, so that ultimately it consists only of a sequence of subroutine calls. This makes the program easier to understand and debug. The subroutines in the main program are mainly concerned with operating the display, while the included routines are specific to particular interfaces. These are not printed here, but are similar to the stand-alone demo programs, and can be inspected in the actual source code files provided on the support website (www.picmicros.org.uk).

If the program source code is written in Proteus VSM v8, a project (base2.pdsprj) must be created in a suitable folder, specifying the MCU to be used. A project fileset is then automatically created (see Chapter 3), plus a folder named after the selected MCU, in this case 'PIC 16F877A'. A source code file named 'base2.asm' must then be added to the project in this folder and opened in the edit window. The include files that provide driver routines for the peripherals should also be stored in the same MCU folder. When the source code is assembled, a 'Debug' folder is created and a 'debug.cof' file is saved in it. This is the file that is attached to the virtual MCU for circuit simulation. An object code file 'base2.o' file is also saved in the same folder, containing the downloading code.

When the simulation version of the any application is correct, a PCB may be prototyped (using Proteus ARES or otherwise) and the program downloaded to hardware, as previously described in Chapter 3. MPLAB utilities must be used for the final stages of downloading and in-circuit debugging. The ICPD unit should be connected to the target hardware, and the source code reassembled in MPLAB (the assembler version should be the same). The firmware is downloaded and tested using the same debugging techniques as for the simulation testing: run, pause, step and break points (only one in the 16F877A).

10.5 Weather Station

To illustrate system integration, a low-cost weather station measuring temperature, light, pressure and humidity will be designed, based on the general purpose BASE module described above. The application specification is summarised in Table 10.3.

Table 10.3: Weather Station Specification.

Input	Range/Description	Display (Max. 8 Characters)
Temperature	−25°C to +75°C	XX degC (7)
Light	0−9999 lux	XXXX lux (7)
Pressure	850−1100 millibar	XXXXmbar (8)
Humidity	0−100% Relative Humidity	XXX %RH (7)

Feature	Specification	Details
Precision	8bits	<1%@mid range
Storage	10 days @ 12 samples/hour	11520 bytes
User	Run, Recall data	Essential
Interface	Display/set date and time	Desirable
Host Interface	Data upload to database and spreadsheet	Desirable

The input variables will be sampled at intervals of 5 min (12 per hour) and data stored for a period of up to 10 days. In run mode, each variable value will be displayed on the LCD. If sampled at 8-bit resolution, one sample for each sensor = 1 byte of data. Over 10 days, the system will store $10 \times 24 \times 12 \times 4 = 11,520$ bytes of data. The user should be able to reset, run and read back data manually. A block diagram helps to define the features of the system and determine the interfacing requirements (Figure 10.6).

A sensor was selected for each measurement based on the range required, ease of interfacing and cost. An analogue interface was then designed to provide the gain and offset required for each. The LM324, a quad op-amp package, is used as the basis for these interface designs. This can be upgraded to a more recent device at the implementation stage, without significantly affecting the design parameters.

To keep the system power requirements as simple as possible, the designs assume a single supply of 5V, giving a limited swing of approximately 0−3.5V at the output of the LM324. The interfaces are designed to produce a maximum output of 2.55V, which is suitable for 8-bit conversion at 10mV per bit. The simulation schematics for the weather station interfaces are filed under VSM project WEATHER2.

10.5.1 Temperature Sensor Interface

The default choice for the temperature sensor is the LM35 type. Its performance is adequate for this application, and it can be connected directly to the PIC ADC input. The LM35C allows negative temperatures to be measured. The sensor negative supply is connected to ground via a diode to lift the 0°C output to around 0.7V. Negative temperatures are then represented by positive voltages (Figure 10.7).

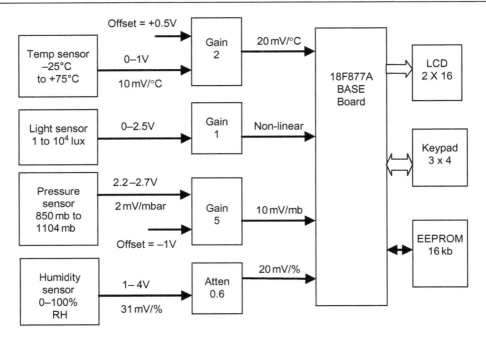

Figure 10.6
Block diagram of weather station.

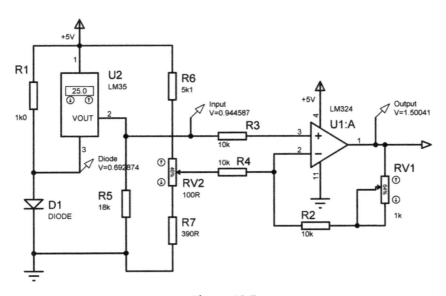

Figure 10.7
Temperature sensor interface.

The non-inverting amplifier has an offset input to compensate for this, resulting in an output of 1.00V at 0°C, rising to 2.50V at 75°C. This corresponds to an overall sensitivity of 1500/75 = 20mV/°C. The simulation allows the circuit components to be adjusted on test to achieve the correct gain and offset, which interact as previously described.

The sensor is represented by a simulated component that allows the set temperature to be manually controlled. The preset feedback resistance will be adjusted for a gain of 2.00, giving an overall sensitivity of 20mV/°C. The circuit depends on the reference diode temperature being constant; as it will drift by 2mV/°C, it is fed with extra current via R1 to raise the self-heating effect and reduce the influence of any external temperature change. The temperature sensitivity reduces with current in any case.

The gain and offset in the simulation were adjusted in turn to give outputs of 0.500V at −25°C and 2.500V at +75°C (within 1mV). When converted with a 2.56V reference, the temperature range will be represented by binary numbers equivalent to 50−250, with 100 representing 0°C.

10.5.2 Light Sensor Interface

An LDR interface is shown in Figure 10.8(a). The sensor is connected to a voltage divider and an inverting amplifier with a gain of −2, which keeps the maximum output voltage to less than 2.5V for 8-bit conversion, and provides a positive-going output as the light level increases. The light level is varied in simulation using + and − buttons that move the virtual light source in relation to the LDR. The properties of this input can be adjusted so that each step corresponds to a particular LDR resistance. These values were set to 1M (dark), 300k, 100k, 30k, 10k, 3k, 1k0, 300R, 100R and 30R.

The output obtained in simulation varied from 0.34 (dark) to 2.56V (maximum light level), recorded in Figure 10.8(b). A set of 256 ADC input values can be obtained from the simulation by substituting an HG (high granularity) pot for the LDR. The resistance at each point represents a light level that can be calculated from the relationship $L = 10^{4/3(5-\log R)}$ lux using a spreadsheet.

The displayed values will range from 0 to 9999 lux, so the table value will have to store a 4-digit BCD code, requiring two locations. The table offset must then be incremented in steps of two, and the basic table routine modified to pick up two bytes per table read.

10.5.3 Humidity Sensor Interface

The Honeywell HIH-3610 humidity sensor has integrated signal conditioning so that an output between 0.8 and 3.9V is produced, representing a change in relative humidity of

(a)

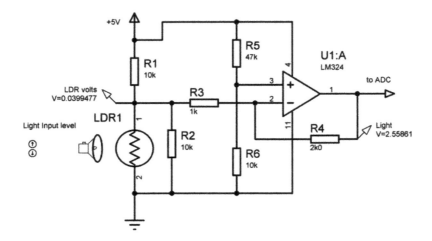

(b)

Step	LDR (RΩ)	Light Level (L lux)	ADC input (Vi)	Daylight level
0	1M0	0.0	0.34	Dark
1	300k	0.2	0.35	
2	100k	1.0	0.36	Dusk
3	30k	5.0	0.40	
4	10k	21.5	0.52	Dull
5	3k0	107	0.84	
6	1k0	464	1.38	Cloud
7	300R	2311	2.03	
8	100R	10,000	2.39	Sun
9	30R	49,793	2.56	

$$L = 10^{\{4/3\ (5-\log R)\}}\ \text{lux}$$

Figure 10.8
Light sensor interface: (a) schematic and (b) output values.

0−100%. This can be fed direct to the ADC, but a non-inverting stage will be used to shift the signal range to a more convenient value. Zero volts output from the LM324 amplifier cannot be obtained, so the output is shifted down to the range of 0.5−2.50V, giving 20mV/% humidity. The offset must be removed in software, by subtracting 50_{10} from the 8-bit binary input.

The sensor is only specified to be about 4% accurate, so the hardware gain and offset adjustment can be used to trim for individual sensors. The sensor can be supplied with individual calibration data if an accurate output is needed.

An interface for the humidity sensor is shown in Figure 10.9. The maximum and minimum output of the sensor is represented by switched test voltages. These were calculated from the sensor specification:

> *Zero offset = 958mV = minimum value*
> *Sensitivity = 30.7mV/%RH*
> *Maximum value at 100% = 958 + (30.7 × 100) = 4028mV*

The gain and offset of the non-inverting stage were adjusted to give an output of 0.50V at 0% humidity and 2.50V at 100% (20mV/%) for 8-bit conversion.

10.5.4 Pressure Sensor Interface

To measure absolute pressure around 1000 mbar, low-cost pressure sensors are available that incorporate a piezoelectric strain gauge bridge with integrated signal conditioning in a compact, robust package. A pressure range of 850–1106 mbar is a convenient measurement range of 256 mbar, allowing an 8-bit conversion at 1 bit/mbar. Standard atmospheric pressure will then occur at a reading of 163.

The SensorTechnics HCX002A6V sensor measures pressures up to 2000 mbar. It produces a fully conditioned output of 0.5V at zero pressure and 4.5V at 2000 mbar with a 5V supply. The output is therefore 2.5V at 1000 mbar. The sensitivity is (4.5−0.5)/ 2000 = 2mV/mbar, so over the required range of 256 mbar, the output voltage range will be 512mV. To utilise the full conversion range, this needs to be increased to 2.56V with an amplifier stage with a gain of 2.56/0.512 = 5.

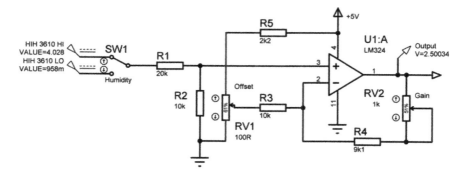

Figure 10.9
Humidity sensor interface.

A non-inverting amplifier with input offset, similar to the humidity sensor interface, but with a gain of 5, is needed (Figure 10.10). Three test input levels will be used. The required range starts at 850 mbar, or a sensor output of $850 \times 2 = 1.700$V. An offset voltage of 0.5V must be added, so the lowest input value is 2.200V. To avoid op-amp outputs less than 0.1V, the low value will be increased by 20mV to 2.220V. The high value will be $2.2 + 0.512 = 2.712$V. A mid-range check value will be 1000 mbar = 2.500V = 1.500V output.

The total sensor possible error is quoted at 0.35% of full scale, or better than 0.2% within the output range. This corresponds to ± 2 mbar, neglecting any interfacing errors, so it is not a high-precision measurement, but this can probably be improved upon by careful calibration.

(a)

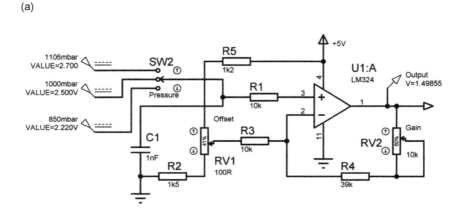

(b)

Pressure P (mbar)	Sensor output (Vₛ)	ADC input (Vᵢ)
0	0.500	-
850	2.200	(0)
860	2.220	0.10
1000	2.500	1.50
1100	2.700	2.50
1106	2.712	2.56
2000	4.500	-
$V_s = 0.002.P + 0.5$ V		
$V_i = 5 (V_S - 2.200)$ V		

Figure 10.10
Pressure sensor interface: (a) schematic and (b) test inputs.

10.6 Design Support

The Microchip Analog and Interface Product Selector Guide lists numerous other sensing and interfacing devices such as serial output temperature sensors, motor controllers, wireless communication transceivers and power management chips. The reader is encouraged to review all available devices and revise the basic design suggestions made in this book to take advantage of continuing improvements in performance and cost of microcontroller system components. In particular, the integration of standard communication interfaces into peripheral devices makes the design of useful microcontroller hardware that much easier as the technology progresses. The Microchip website has a wide range of resources which should be explored fully before undertaking new design work. This book is just a starting point. I hope it has been useful.

Questions 10

1.	State five criteria for selecting an MCU for a given application.	(5)
2.	State three characteristics that a mains d.c. power supply must have.	(3)
3.	State the meaning of the acronym XLP and its significance in mobile applications.	(3)
4.	How are parallel, serial and analogue signals shown in a block diagram?	(3)
5.	Suggest three problems associated with a high-speed clock.	(3)
6.	Outline how an I^2C sensor could be added to the BASE board while retaining the serial memory.	(3)
7.	Calculate the output of an LM35C sensor at 25°C if the output is 1.000V at 0°C.	(3)
8.	Explain briefly why the LDR is more difficult to interface than the LM35 temperature sensor.	(3)
9.	Calculate the light level if the resistance of an NORP12 LDR is 2kΩ.	(4)

Total (30)

Assignments 10

10.1 Interface Analysis
Analyse the temperature, light, humidity and pressure interfaces by obtaining an expression for the output of each amplifier in terms of its inputs, and confirm that they are correct by comparison with the simulated values shown in the schematics.

10.2 Refrigeration Controller
Complete the implementation of the refrigeration controller specified in this chapter. Produce a schematic and demonstrate the simulation of the control program implemented in stages:

1. Temperature control at default value
2. Temperature display of default value
3. Set temperature and display
4. Sensor averaging and fault detection

Select the most appropriate PIC MCU for the final design and produce a costed parts list.

10.3 Rain Gauge Design

Investigate and design a system for measuring rainfall. The cumulative rain for each day should be displayed continuously. At midnight the total should be logged and the gauge should be reset. Do not design the controller itself, but specify its requirements to operate the gauge. Compare alternative sensors for the gauge and identify the advantages and disadvantages of each option.

10.4 Multiprocessor Systems

Investigate the parallel serial port in the PIC 16F877A, and show how it could be used for passing data between two PIC MCUs in a dual processor system. Outline a program and estimate the maximum speed of data transfer achievable in bytes per second. Compare this with SPI and I^{2}C as multiprocessor communications systems, in terms of speed, flexibility, and ease of hardware and software design.

Answers to Questions

Answers 1

1. Processor, memory and I/O (3)
2. A microcontroller has processor, memory and I/O on one chip, while the (3)
   microprocessor needs separate memory and I/O chips to form a working
   system.
3. Output address from program counter on the address bus, select memory (3)
   location containing instruction code and copy it back to the instruction register
   via the data bus.
4. Flash ROM can be electrically re-written many times but is non-volatile. (3)
5. The data has to be converted to serial form in a shift register and transmitted (3)
   one bit at a time on a single line, while parallel data is transferred 8
   (or more) bits at a time.
6. Ports A and E default to analogue input on power up or reset. (3)
7. The Special Function Register is dedicated to a specific function, while the (3)
   General Purpose Register can store any user data.
8. From data sheet Table 13.2:
   Instruction code = 00 0000 1000 1100
   Op-code = 0000001
   Register operand = 000 1100 (3)
9. The port data direction registers are only accessible in bank 1, so the bank (3)
   select bit RP0 must be set.
10. The hardware timer runs simultaneously with the program execution, so the (3)
    MCU can proceed with the program while the delay count is made.
11. The first column contains the address of the memory location where the (5)
    instruction will be stored, usually starting at zero. The second is the
    hexadecimal machine code that will be downloaded to the chip. The third is
    the line number in the source code text file. The fourth is the source code
    mnemonic (instruction and operand). The last may be used for comments.
12. Plug the programmer unit into the host PC (USB) and the target board (5)
    (6-pin connector). Write the program in MPLAB and assemble it. Select the
    software simulator and debug the program. Select the programmer and connect
    to the target. Download the program and run in the hardware (set MCLR).

Total (40)

Answers 2

1.
   (a) Load the working register with the number 0FFh (sets all bits) (2)
   (b) Jump to a subroutine starting with the label 'delay' and return afterwards (2)
2.
   (a) CODE: indicates start of program code (2)
   (b) EQU: declares a constant label (2)
   (c) #INCLUDE: insert source code text file (2)
3. GOTO, SLEEP; program will run through blank locations and repeat (3)
4. Address, register (2)
5. Power-up timer enable, watchdog timer enable, clock oscillator type select (3)
6. A bit test is used to determine whether the next instruction is skipped or not (5)
   (BTFSS, BTFSC). This is usually followed by a GOTO or CALL to change the
   program sequence. If this instruction is skipped, program execution continues
   on the original path. Often the zero flag is tested to control a branch. The zero
   flag test is combined with a decrement or increment in DECFSZ and INCFSZ
   to provide counting loops and similar sequences.
7. Program jumps to subroutine code, executes and returns; macro code is (4)
   inserted each time by the assembler. A subroutine the program uses less
   memory, but a macro using the same code is faster.
8. Start/end, process/sequence, input/output, branch/selection, subroutine/ (5)
   function.
9. A subroutine is a programmed jump (CALL) and return; the return address is (5)
   stored automatically on the stack, so that when the routine has been
   completed, a RETURN instruction causes the return address to be replaced
   in the program counter, taking the execution point back to the instruction
   following the call. The interrupt is an asynchronous external event which forces
   a jump to program address 004, from where an interrupt service routine is
   executed. This is terminated with RETFIE, return form interrupt, to take the
   execution point back to the original position. The stack is used in the same way
   as in the subroutine to store the return address.
10. C can be used for any type of MCU and is easier to understand and more (3)
   powerful than assembler.

Total (40)

Answers 3

1. They are the standard mathematical models for electronic components. (3)
2. The clock settings for simulation are set in the MCU component properties (3)
   dialogue.
3. It is the reset input which must be high to allow the chip to run. It is also the (3)
   programming voltage input, so must be isolated from the target supply.
4. $10 \times 10^{-9} \times 25 \times 10^3 = 250\ \mu s - 4\ kHz\ clock - 1\ kHz\ output$ (3)

5. MPLAB simulates only the MCU and produces only numerical results. (3)
   VSM simulates the whole circuit and displays an animated schematic.
6. Step into steps into subroutines and step over runs through them at full speed. (3)
7. The breakpoint stops the program at a particular line so the MCU status can be checked or a timing measurement made. (3)
8. The oscilloscope displays a limited number (4) of analogue signals of any amplitude. The logic analyser displays numerous digital (logic) only. (3)
9. The netlist specifies all the components in a circuit and their connections. It is passed to the PCB layout system to create a circuit board. (3)
10. In-circuit debugging allows the program to be tested in the real hardware to complete the final fault finding stage. (3)

Total (30)

Answers 4

1. If the switch is connected between the input and 0 V, the pull-up resistor, ensures that the input is high when the switch is open. (3)
2. Capacitor, software delay, timer delay. (3)
3. Hardware timers allow timing operations to proceed simultaneously with other program processes, giving a more efficient use of the processor. (3)
4. The timer pre-scaler is a digital frequency divider which reduces the frequency of the input clock by a factor of 2, 4, 8, etc., which increases the timer range by the same factor. (3)
5. The segments must be illuminated in the correct combination to display digits 0 to 9. The data table provides the required binary output code for each digit. (3)
6. The BCD display has an internal hardware decoder so that it displays the digit corresponding to the input binary code (0–9). (3)
7. The rows are connected to MCU outputs and set high. The columns are connected to inputs and pulled high. Each output is taken low in turn. If a key is pressed, a low input is detected on that column, identifying the key. (3)
8. The LCD can operate with 4-bit input, receiving 8-bit control and data codes in 2 nibbles. An enable input strobes the data in, and a register select input indicates if the input code is a command or display data. (3)
9. ASCII is a standard seven bit code representing characters found on a keyboard. They are used in serial communication of text data. (2)
10. (4)

Total (30)

Answers 5

1. $10010011 = = 128 + 16 + 2 + 1 = 147_d$ (3)
2. $1234_{10} = = 10011010010$ (using division by 2 and remainders)   (3)
3. $3FB0_{16} = = 0011\ 1111\ 1011\ 0000 = = 16304_{10}$ (using binary weighting)   (3)
4. $1001 \times 0101\ (=9 \times 5) = 0101 + 0101000 = 101101 = 1 + 4 + 8 + 32 = 45$ (3)
5. $99 \div 2 = 49r1$, $49 \div 2 = 24r1$, $24 \div 2 = 12r0$, $12 \div 2 = 6r0$, $6 \div 2 = 3r0$, $3 \div 2 = 1r1$, $1 \div 2 = 0r1$
    $99 = = 01100011 = = 64 + 32 + 2 + 1$. For -99, sub 1: $01100011\text{-}1 = 01100010$ (6)
    and complement: 10011101 Answer $-99 = = 9D_{16}$
6. 16-bit FP number $= 0\ 00100\ 1001000000$
    $MSB = 0 = +ve$ number
    Exponent $= 01100 = 12$; Offset 15: $12 - 15 = -3$; Exponent
    multiplier $= 2^{-3} = 0.125$
    Significant fraction $= 0.5 + 0.0625 = 0.5625$; Significand $= 1.5625$
    Result $= 1.5625 \times 0.125 = 0.195$ (6)
7. 56_{10} BCD $= = (5 \times 10) + 6$
    $= = (0101 \times 1010) + 0110 = 101000 + 1010 + 0110 = 111000$
    Check: $0110011 = 32 + 16 + 8 = 56_{10}$ (3)
8. 41 h, 7Ah, 23 h   (3)

                                                                    Total   (30)

Answers 6

1. 12-bit ADC gives $2^{12} = 4096$ steps. $100/4096 = 0.024\%$ per step   (3)
2. The full-scale input is divided into $2^8 = 256$ steps for conversion to binary.
    With a 2.56 V reference, this converts into exactly $2.56\ V/256 = 10$ mV per step   (3)
3. Three bits are used to select 1 of 8 input channels AN0 − AN7.   (3)
4. If the 10-bit result is left justified, the high 8 bits of the ADC result are   (3)
    placed in the ADRESH register, with the low 2 in the high bits of ADRESL.
    If right justified, the low 8 bits are placed in ADRESL, and the high bits in
    the low 2 bits of ADRESH.
5. Gain and input resistance are infinite, output resistance is zero.   (3)
6. A common single supply (5 V) can be used for MCU and interface.
    It restricts output swing and may not reach zero.   (3)
7. $G = 19/1 + 1 = 20$ (3)
8. (a) $V_s = 2\ (1.0 + 0.5) = 3.0$ V (b) $V_d = 2\ (1.0 - 0.5) = 1.0$ V   (3)
9. The capacitor slows down the output transient response and reduces the   (3)
    cut off frequency.
10. Current in offset resistor $= (1.0 - 0.5)/10\ k = 0.05$ mA
    Voltage across feedback resistor $= 0.05\ mA \times 22\ k = 1.1$ V
    Output voltage $= 1.0 + 1.1 = 2.1$ V   (5)

11. Gain = 22/10 = 2.2
    GBWP = 2.2 × BW = 1 MHz
    BW = 45 kHz   (3)
12. When output = 0 V, V_L = 15/25 × 5 = 3 V, V_H = 25/35 × 5 = 25/7 = 3.57 V   (5)

                                                                Total   (40)

Answers 7

1. Two of: simple design, high off resistance, good isolation.   (2)
    One of: slow, high power consumption, bulky.   (1)
2. Base current = 4.6 − 0.6/1000 = 4 mA
    Collector current = 4 × 50 = 200 mA
    Volt drop = 0.2 × 10 = 2 V   (4)
3. It is a voltage controlled current source, with high input impedance.
    Zero and +5 V applied at the gate will switch it off and on.   (3)
4. The DC motor needs a commutator to reverse the armature current on each (3)
    half revolution, so that the torque is developed in one direction only. Discharge
    and wear cause unreliability.
5. The thyristor is equivalent to a latching pair of transistors, which passes direct (3)
    current when the gate is pulsed. The triac is formed of back to back thyristors
    and switches alternating current.
6. The software option can be implemented by the MCU toggling an output with (3)
    a delay. Alternatively, a separate hardware oscillator based on the 555 timer
    chip can be switched on an off by the MCU.
7. Pulse Width Modulation uses a pulse waveform to control a current switch (3)
    connected to the load. If the ON time (duty cycle) increases as a percentage of
    the overall period, the average current in the load, and hence the power
    dissipated increases.
8. See Figure 7.12(c). The switches in the bridge (FETs) are turned on in pairs to (3)
    allow the current to flow in either direction in the motor.
9. 360/15 = 24 steps/rev.
    Speed = 96 steps/sec → 96/24 = 4 revs/s   (3)
10. It can be made physically small with a high power to weight ratio and reliability (3)
    due to the absence of commutator and brushes.
11. 200 slots/100 ms → 2000 slots/s → 2000/50 = 40 revs/s → 40 × 60 = 2400 rpm. (4)
12. The DC motor drive is simpler in construction, more efficient, and higher (5)
    speeds and torque are possible, but it needs a feedback system for
    position control and a gearbox for low speeds. The stepper can positioned
    without feedback and holds its position but is less inefficient and is complex
    to drive.

                                                                Total   (40)

Answers 8 (3 Marks Each)

1. No clock is sent with the data signal.
2. To increase the signal to noise ratio, and the distance sent, by increasing the signal amplitude.
3. Line attenuation and noise limits the distance in proportion to the sending amplitude. SPI signals are sent at TTL levels (5 V) only, while RS232 uses amplitude up to 50 V p-p.
4. ASCII (9) = 39 h = 0111001 Parity bit = 0 (MSB last) Sequence LSB first = 10011100
5. Slave select is a hardware input to an SPI device which enables slave transmission, generated by the master controller. I^2C uses software addressing, where the required device and location are selected by an address sent on the serial data line.
6. In I^2C, a control code and address must be sent before the data, making up to 5 bytes in all, plus control bits. In SPI, only data bits are sent as the slave device is selected in hardware (slave select). Also, SPI can run at 5 MHz clock, I^2C only 1 MHz
7. CAN was developed first, with special control units and was more expensive. LIN was designed to do the same job more cheaply, using the USART port on any MCU with a line transceiver.
8. Data = 10110010
   0 xor 0 = 0, 0 xor 1 = 1, 1 xor 1 = 0
   Result = 01 10 01 01 10 10 01 10
9. The differential signal gives reversible current in a pair of wires that can be more reliably detected at the receiver than a single ended voltage. The twisted pair ensure any interference on the line cancels out.
10. Collisions occur unpredictably in ethernet CSMA operation, so signals transmission time is unreliable. Industrial networks use alternative methods such as token passing which give predictable response times.

Total   (30)

Answers 9

1. More reliable, faster, remote sensing
2. An opto-sensor counts the steps on the strip to position the print head.
3. The rate of change of the output divided by the rate of change of the input, corresponding to the gradient of the characteristic.
4. Accuracy is the extent to which a measurement is consistent with the agreed standard, precision is the smallest output change measureable.
5. Any 3 of: temperature sensing resistor (metal film/wire), semiconductor junction (p-type and n-type silicon), thermocouple (dissimilar metals), thermistor (solid semiconductor), resistance (platinum wire).
6. 4.

7. 100 kΩ
8. Strain gauges are connected as a bridge circuit to provide a differential output that eliminates the large offset voltage when used with a single supply, to maximise the output amplitude and to provide inherent temperature compensation
9. The instrumentation amplifier is a differential configuration, which eliminates offset in the source voltage, has a high input impedance suitable for the high source impedance of the strain gauge bridge and has a high gain suited to the low sensitivity of the strain gauge bridge.
10. 2.35 V = 10 cm, 0.41 V = 80 cm
    Sensitivity = $(2.35 - 0.41)/(10 - 80) = -27.7$ mV/cm
    Voltage change from max = $2.35 - 1.50 = 0.85$ V
    Distance change = $850/27.7 = 30.69$ cm
    Distance = $10 + 30.7 = 40.7$ cm

Answers 10

1. Five of: number of I/O pins, program memory size, peripherals available, data memory, instruction set, developer expertise, cost.
2. Accurate stable voltage, sufficient current, low noise and ripple.
3. XLP = Extra Low Power (MCU)
   Minimises power consumption in battery powered applications.
4. Parallel — block arrow, serial — single arrow
   Analogue — single arrow with labelling and representation of waveform
5. Three of: high frequency crosstalk, high power dissipation, unreliable transmission down long connections, limited component speed.
6. Connect to the same I^2C bus as the memory and assign the sensor address 1 via its hardware address pins.
7. LM35 output changes by 10 mV/°C
   Output = $1.000 + (25 \times 0.01) = 1.25$ V
8. LM35 is linear and has a calibrated output of 10 mV/°C.
   LDR is non-linear, covering a wide range of light levels and resistance values.
9. $L = 10^{4/3(5 - \log R)} = 10^{4/3(5 - 3.3)} = 184$ Ω

PIC 16F877A (16LF877A) Selected Electrical Characteristics

			Notes
Power supply	Nominal V_{SS}	0V	
	Nominal V_{DD} with respect to V_{SS}	5V	
	RAM data retention minimum voltage	1.5V	
	16F877A supply voltage (0−20MHz clock)	4.0V−5.5V	1
	16LF877A supply voltage (0−4MHz clock)	2.0V−5.5V	2
	16LF877A supply voltage (4−10MHz clock)	3.0V−5.5V	
	16F877A maximum supply current (20MHz clock)	15mA	3
	16LF877A maximum supply current (4MHz clock)	2mA	
	16LF877A maximum supply current (32kHz clock)	35μA	
	16F877A typical sleep supply current	1.5μA	4
	16LF877A typical sleep supply current	1μA	
	Absolute maximum V_{DD}	+7.5V	
	Absolute minimum V_{DD}	−0.3V	
	Absolute maximum power dissipation	1W	
	Absolute maximum supply current	250mW	5
I/O pins	Logic low TTL input max (V_{DD} = 5V)	$0.15.V_{DD}$ (0.8V)	
	Logic high TTL input min (V_{DD} = 5V)	$0.8V + 0.25 V_{DD}$ (2.0V)	
	Logic low Schmitt input max (V_{DD} = 5V)	$0.2.V_{DD}$ (1.0V)	
	Logic high Schmitt input minimum (V_{DD} = 5V)	$0.8.V_{DD}$ (4.0V)	
	Output voltage low maximum	0.6V	
	Output voltage high minimum (V_{DD} = 5V)	V_{DD} −0.7V (4.3V)	
	Output rise/fall time typical	10ns	
	Maximum source current	25mA	
	Maximum sink current	25mA	
	Input leakage current	±1μA	

Clock	Frequency ranges		
	RC	0−1MHz	
	XT	100kHz−1MHz	
	HS	4−20MHz	
	LP	5−200kHz	
	With external clock		
	RC, XT	DC−1MHz	
	HS	DC−20MHz	
	LP	DC−32kHz	
Timers	Oscillator start-up timer period	1024 clock cycles	
	Power-up timer period (typ)	72ms	6
	Watchdog timeout period (typ)	18ms	
	Suggested minimum clock input pulse duration	100ns	
	Timer output maximum rise/fall time	50ns	
EEPROM	Erase/write cycles typical	1M	
	Erase/write cycle time typical	4ms	

Notes:
1. PIC 16F877A is listed as a mature product and is not included in the main catalogue of current devices.
2. PIC 16LF877A is a low-power option (2003 data sheet) but limited to 10MHz maximum clock.
3. PIC 16F887 recommended for new designs, has similar supply/clock characteristics.
4. Watchdog timer disabled.
5. Ten outputs sourcing 25mA each.
6. Add together for total start-up time.

Selected Direct Current Characteristics

The electrical characteristics listed above are extracted from the PIC 16(L)F877A data sheet (DS39582B 2003), to which reference must be made for detailed information. The interaction between selected operating parameters is illustrated in Figure A.1.

The default operating conditions for designs in this book are 5V supply, 25°C temperature and 4 MHz (XT) clock, using the standard 16F877A. A low-power version is also available, the 16LF877A, which has significantly different characteristics. One disadvantage of this chip is that its maximum clock rate is 10MHz, while one benefit is its minimum supply voltage of 2.0V.

In general, the threshold voltage at the port inputs varies in proportion to the supply voltage ($0.15 \times V_{DD}$ for TTL inputs) and the temperature. Similarly, the output current and voltage

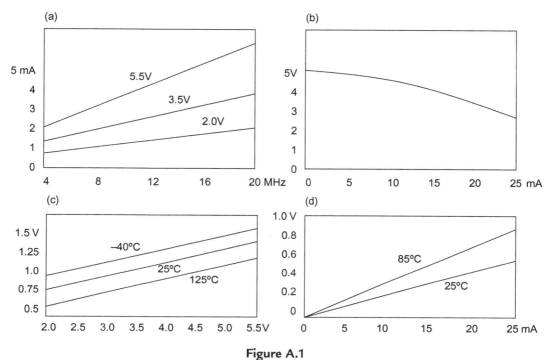

Figure A.1
(a) Supply current versus clock frequency at various supply voltages (HS mode). (b) Output voltage versus output source current at 5V supply (typical performance). (c) Input threshold voltage versus supply volts over the operating temperature range. (d) Output voltage versus output sink current at 5V supply at high and low temperatures.

are dependent on temperature. Many characteristics are fairly linear, making de-rating calculations straightforward. Note that the operating temperature is likely to be much higher than room temperature in any large IC, so the temperature dependence of any characteristic is particularly important.

Digital Signal Transmission

This appendix reviews some basic concepts in information systems for those who have not studied microprocessor systems in the context of a broader electronics course or have limited knowledge of communication principles.

Frequency Components

Digital signals have the signal characteristics of a square wave, containing higher frequency components. These are described by the Fourier series for instantaneous signal voltage of the square wave with equal high and low (mark and space) periods:

$$V_i = V_f \sin(wt) + V_f/3 \sin(3wt) + V_f/5 \sin(5wt) + V_f/7\sin(7wt) + \ldots$$

where V_f is the fundamental amplitude and ω (omega) the angular frequency ($\omega = 2\pi f$).

Therefore, a microprocessor clock signal at a frequency of, say, 1MHz with a high value of 5V and with a low value of 0V has a fundamental sinusoidal component at 1MHz, amplitude 2.5V, and odd harmonics at 3MHz (amplitude 2.5/3V), 5MHz (2.5/5V), 7MHz (2.5/7V) and so on. It also has a d.c. (zero frequency) component at +2.5V which can be added to the series sum above. This can be demonstrated mathematically using Fourier analysis and practically using a spectrum analyser.

These high-frequency components may be significant when designing digital circuits using microcontrollers. The most commonly required precaution is decoupling (fitting capacitors across) of the power supply rails so that these harmonics do not affect correct functioning. In discrete logic or multichip systems, it is advisable to decouple with a low-value ceramic capacitor (say 100pF) adjacent to each chip. This is in addition to the large reservoir capacitor normally fitted across the output of the power supply.

Decoupling also protects against disruption caused by current pulses drawn from the power supply when the transistors within the circuit ICs turn on and off. This is a potential problem because all the gates tend to switch simultaneously with the clock in synchronous circuits, and the transistors draw nearly all their current in pulses, when switching state. Short current pulses have even more high-frequency energy than square waves.

Transmission Line

Since a perfect square wave needs an infinite bandwidth to be faithfully transmitted, it will always be distorted to some extent in a real system. The main problem is that physical conductors always have a certain amount of resistance, capacitance and inductance, due to the imperfect conducting medium, stray capacitance between the signal path and ground (or adjacent signal conductors), and the magnetic field generated by an alternating signal current.

The most important result is that there is a limit to the clock frequency that can be used, since the signals take time to rise and fall. As the period of the signal pulse reduces, it will tend to become triangular in shape before falling off in amplitude, ultimately below the threshold level of the input, at which point the data is lost.

If the rise time is fixed, the smaller the voltage change in the signal, the higher the frequency of operation that can be used. This is one reason why the lower operating voltages of current devices is beneficial, although the main reason is to reduce power consumption. Unfortunately, this reduces the signal to noise ratio, assuming internal noise and external interference are fixed values.

The PCB track or wired connection can be represented by lumped R, C and L components, as seen in Figure B.1. If a pulse waveform is applied, the shunt capacitance (the dominant effect) in conjunction with the series resistance causes a finite rise time at the receiving end. Input capacitance at the input of the receiving device will add to the effect. The inductance and capacitance form a resonant circuit, producing a damped oscillation on the corners of the square wave. The overvoltage spike can induce transients in adjacent tracks and cause malfunction by causing valid logic levels to be exceeded. The effect is exaggerated in this test circuit.

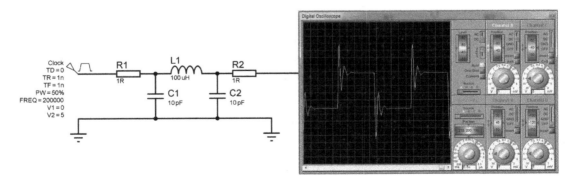

Figure B.1
RLC track model.

PCB Tracks

The tracks on a printed circuit board (PCB) are flat conductors formed by depositing a copper conductor or leaving it behind after etching away the rest of a copper layer. On a single-sided board, it is normal to leave as much copper as possible on the upper layer, connected to ground, and make the supply tracks as wide as possible. This provides a low resistance path for the current pulses in the supply caused by synchronous operation. A ground plane also provides a low resistance path to discharge signal transients and screens the components on the other side of the board.

A thinner track will reduce the stray capacitance, but will also reduce the current capacity, as it will have a higher resistance and lower power dissipation. Lower currents are more susceptible to interference and stray clock, but use less power in battery-powered applications. Using the minimum track size consistent with the current requirements (low for most signal connections) saves copper and increases data bandwidth.

The same considerations apply to the internal dimensions of the circuit ICs. These are also formed in layers, so the smaller the components, the higher the clock speed can be, assuming the extra power dissipated does not raise the temperature above the maximum tolerated by the IC transistors. The power dissipated is broadly proportional to the clock speed since most of the power is consumed when the gates switch over.

A lower operating supply voltage also reduces power consumed. Since power is given by V^2/R, where R remains the same for any given chip, power dissipation is proportional to voltage squared. Therefore, if the supply is reduced from 5 to 3.3V, the power is reduced to $3.3^2/5^2 \times 100\% = 44\%$ of the original value, or less than half. This is the reason that low-power chips have been developed that work at reduced voltages, such as those fabricated using Microchips XLP technology.

Longer Connections

Connections between boards are either via ribbon cables, backplane or cables. The ribbon cable is usually less than 1m, and can be regarded as similar in characteristics to on-board bus connections with parallel signal paths that may be subject to crosstalk, i.e. a signal being picked up in attenuated form on an adjacent conductor. This is not usually a major problem at low frequencies, but higher frequency signals will normally need a coaxial or other form of screened cable where the signal conductor is surrounded by a grounded metallic sheath. As noted earlier, its high-frequency harmonics in any digital signal are potentially troublesome. A backplane typically also has a mass of parallel connections with similar characteristics to the on-board busses.

The most common type of cable is twisted pair, with optional screening. This has a useful combination of low cost with good noise immunity and is used in a wide range of systems, from RS232 to Ethernet. Pairs of small solid wires are twisted together and screened separately or in groups. Standard Ethernet 100 base T cables (bit rate = 100 M, baseband transmission, T = twisted pair) have four pairs. There is a variety of terminations, but the most common are the cheap RJ-45-type 8-way plastic connector used in standard networks, and the more expensive D-type connector with a robust metal housing and retaining screws, which is more likely to be used in industrial systems.

The main advantage of the twisted pair is that any external interference impinges on both conductors simultaneously, due to their physical proximity, producing an induced noise signal in equal and opposite directions in the cable. If differential signalling is used, the interference seen at the receiver inputs is added equally to both signals and has no effect on the net differential signal amplitude. Screening attenuates the interference by discharging any external induced voltage to ground.

Coaxial cable usually has a single central signal conductor with a cylindrical insulation layer and screen, which carries the return signal, grounded at one end (grounding at both ends will create an earth loop, which will cause its own problems). It was previously used in networks, but was found to be expensive and unreliable, and was replaced by a twisted pair with cheap RJ-45 connectors.

Coaxial cable has always been used to connect instruments such as oscilloscopes and signal generators, as well as radio frequency equipment. In networks, it may still be used in high-performance industrial networks. However, single-ended systems with a single signal conductor will always require more careful system design than the differential signalling system of twisted pair cabling, which is physically and electrically robust at relatively low cost.

Characteristic Impedance

In high-frequency and RF systems, the transmission line model of any cable is even more significant. By analysis of the lumped transmission line model, it can be shown that at high frequencies the impedance of an infinite lossless line is $(L/C)^{1/2}$, which has the dimension of resistance, ohms, known as the characteristic impedance (Z_0 or R_0). Coaxial cable is usually designed with a standard characteristic impedance of 50Ω (RF and aerial leads) or 75Ω (instrumentation).

If such a cable is left open circuit or short circuit at the far end from a signal source, the signal will be reflected back and disrupt the transmission, because the signal energy has nowhere else to go. As a result, such cables need to be terminated with a resistance of the same R_0 so that the signal is absorbed at the receiver; this is known as impedance matching. In other words, the next stage in the transmission system must have an input resistance of this standard value. If the receiving amplifier has a high input resistance, a terminating resistor may be fitted across the input. Matching is of particular importance in designing aerials, so wireless data links need to be designed with this in mind.

Network cabling must have standard properties so that the signal transmission is reliable. Ethernet operating up to 1GHz uses Category 5 cable. This has 4 twisted pairs, with a characteristic impedance of 100Ω, loop resistance of $0.188\Omega/m$, capacitance of 52pF/m and inductance of 525nH/m. This results in a propagation delay of about 5ns/m. Crosstalk is reduced by making the number of twists per meter (50−70) different in each pair. The wire diameter is about 0.5mm, typically with PVC insulation. The maximum recommended run length between repeaters is 100m.

Decibel Measurement

When designing communication systems, the frequency response or signal to noise ratio in a transmission system needs to be predicted or measured. The decibel is a logarithmic unit that compares the output (V_o) and input (V_i) of an amplifier, attenuator or filter according to the formula:

$$\text{Gain/attenuation} = 20 \cdot \log_{10}(V_o/V_i) \text{ decibels}$$

Say an amplifier has a gain of 10, then V_o is 10 times V_i, and

$$\text{Gain} = 20 \cdot \log_{10}(10) = 20\text{dB}$$

The gain or attenuation therefore increases or decreases by 20db per decade (factor of 10), as shown in Table B.1. Note that the scale compresses the ratios into smaller numbers

Table B.1: Decibel Ratios.

Gain/Attenuation	Ratio	dB
100,000	10^5	100
10,000	10^4	80
1000	10^3	60
100	10^2	40
10	10^1	20
1	10^0	0
0.1	10^{-1}	−20
0.01	10^{-2}	−40
0.001	10^{-3}	−60
0.0001	10^{-4}	−80
0.00001	10^{-5}	−100

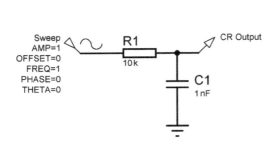
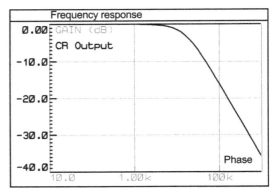

Figure B.2
Frequency response of CR network.

which are easier to handle. Also, the overall gain and attenuation of a system can be found by adding the figure for each stage, rather than multiplying the gain/loss figures, to obtain the overall gain or attenuation.

Frequency Response

The frequency response of a network is usually plotted with the frequency in decades on the horizontal axis, and gain or attenuation in decibels on the vertical. Figure B.2 shows the frequency response of a simple CR network, a simple low-pass filter. It represents the general response caused by stray capacitance on any signal conductor. The gain or attenuation is constant up to a certain frequency and then falls away at 20dB per decade. The cut-off frequency is defined where the output has fallen by 3dB. The straight line approximation of this first-order response breaks at this point. In this way, the overall response of a more complex system can be described in straight line segments, which rise or fall at 20dB per decade, or multiples of this figure.

Signal to Noise Ratio

There are two main sources of noise in transmission systems: external and internal. The external noise derives from high-frequency electromagnetic radiation, which may be from broadcast signals or other equipment that is nearby. Motors which draw a large current and create discharge at their brushes are notoriously troublesome. Mains ripple is a potential low-frequency noise source. Internal noise from conductor materials is usually only a problem in very sensitive measurement systems.

The quality of a data signal can be specified in terms of its signal to noise ratio. It is self-evident that if the noise and interference on a signal are greater than the signal itself, it will be difficult to recover the information. Signal to noise ratio (*S/N*) is measured in decibels. Since most noise is high frequency, a low-pass filter may improve *S/N* ratio in a baseband system. Fortunately, stray capacitance acts as a low-pass filter, so it is not always undesirable. Simple low-pass filtering using decoupling capacitors also has the effect shown in Figure B.2. This will however also reduce the 'sharpness' of baseband digital pulses by filtering out the odd harmonics and reduce data bandwidth.

Speed and Distance

In baseband digital transmission systems, there is a trade-off between distance and speed. Generally, the longer a cable, the greater is the attenuation of the signal due to the combined resistance and capacitance of the line. It may act as a first approximation to the actual characteristics, as the low-pass CR filter. Since high frequencies are attenuated by the shunt capacitance in the line, there will always be an upper limit to the data rate.

The response to a step (pulse edge) input is an output with a finite rise time, so there is a time delay before the signal transition reaches the receiver threshold. This can also be interpreted as the filtering out of the high-frequency components of the pulse wave. The rise time is fixed for a particular data link, so if the frequency of the data is increased, the pulse will not reach the receiver threshold before the drive is reversed and the data is lost, so there is a limit to the bit rate that is possible. The slew rate of the driver is also limited, which adds to this problem.

The longer the line, the greater the potential interference level, and the worse the signal to noise ratio, as it will be cumulative along the line. In order to improve the signal to noise ratio, the signal amplitude can be increased. However, with a fixed rise time, the received signal will take longer to reach the threshold voltage or current. Therefore, in general, the longer the line, the lower the bit rate that is possible. Line attenuation can be countered by regenerating the signal at regular intervals with additional line drivers, as in a network hub.

Encoding

Binary data can be transmitted locally in its original TTL form (5V = 1, 0V = 0), under the control of the system clock (or a signal derived from it). When transmitted between system devices in serial form, the receiver needs to be able to read the data coherently, so some timing system is needed. Asynchronous data (e.g. RS232) re-synchronises after each byte. Synchronous data incorporates the clock into the data stream.

10 Mbit/s Ethernet uses Manchester encoding whereby the data and clock are XORed together before transmission, and again on reception to recover the data (see Chapter 8). 100 Mbits/s Ethernet uses a slightly more complex encoding system to squeeze the extra bandwidth out of the same physical medium. Each group of 4 bits is translated into a 5-bit code which ensures there is at least transition in each group, to maintain synchronisation (4B5B encoding). In addition, a three-level baseband signal is used, in the sequence +1, 0, −1, 0, +1, and repeat, where a 1 is represented by a step to the next level and 0 by no change (MLT-3 encoding). The reduction in the slew rate required by this scheme allows the higher data rate to be achieved. The bit rate is then four times the fundamental frequency of the signal. 1 Gbit/s Ethernet extends this principle to a five-level signal.

Modulation

Modulation allows wireless transmission of data. The simplest form is amplitude modulation (AM), where a data pulse is converted into a burst of higher frequency signal (frequency shift keying). The pulse interval could be silence, but this cannot be distinguished from a fault condition, so a lower amplitude signal at the same frequency is more practical. AM opens the way to multichannel communication, since different carrier frequencies can carry separate data channels simultaneously. Also, multi-level signals can be used to represent bit groups. Using a radio frequency carrier provides a wireless link.

Frequency modulation uses a change in the frequency of the carrier to represent two states, and could be heard in early audio frequency modems and cassette tape storage. However, phase encoding is more useful, where the relative phase of two carriers at the same frequency is used to represent digital information.

Quadrature amplitude modulation (QAM) uses two sinusoidal frequencies whose amplitude and phase relative to one another can be varied. Each combination of phase and amplitude can represent a bit group. Thus, a 2-bit group with four possible codes can be represented in 4-QAM. 16, 64 and 256 (representing a complete byte in one symbol) and higher combinations are used, at the expense of lower noise immunity.

Wireless Links

Many current wireless applications are based on the IEEE Standard 802.11, which specifies the operation of wireless local area networks (WLANs) using carrier frequencies between 2.4 and 5GHz. Subgroup 802.11g is used for Wi-Fi links, in the ASDL (asymmetric subscriber digital line) stage of the domestic broadband system and also 4G networks.

It uses 14 channels in the 2.4GHz band, 5MHz apart, transmitting simultaneously to achieve a maximum data rate of 54 Mbits/s with adequate noise and crosstalk immunity. It uses orthogonal frequency division multiplexing (OFDM) to encode the data, where each sub-carrier provides a separate QAM channel. Spreading the data over a wide range of frequencies improves reliability and security.

Optical Links

The carrier frequency in a communication system is usually considerably higher than the modulating frequency. Visible light has a frequency of between 400 and 800 THz (terahertz). A signal laser diode can transmit at up to 100 Gbits/s and, using frequency division multiplexing, can generate about 100 simultaneous channels of pulse coded data on an optic fibre data link. The optic medium is immune to RF interference and has low signal attenuation, allowing a fibre length of several kilometres. Optic fibre is now widely used in all communication systems and the internet backbone for these reasons. The main drawback is that it is more difficult to connect fibres together, compared with copper wires.

Index